CW00543565

Thrifty Science

Thrifty Science

Making the Most of Materials in the History of Experiment

SIMON WERRETT

The University of Chicago Press   Chicago and London

The University of Chicago Press, Chicago 60637
The University of Chicago Press, Ltd., London

Published 2019
Printed in the United States of America

28 27 26 25 24 23 22 21 20 19 1 2 3 4 5

ISBN-13: 978-0-226-61025-2 (cloth)
ISBN-13: 978-0-226-61039-9 (e-book)
DOI: https://doi.org/10.7208/chicago/9780226610399.001.0001

Library of Congress Cataloging-in-Publication Data

Names: Werrett, Simon, author.
Title: Thrifty science : making the most of materials in the history of experiment / Simon Werrett.
Description: Chicago ; London : The University of Chicago Press, 2019. | Includes bibliographical references and index.
Identifiers: LCCN 2018026816 | ISBN 9780226610252 (cloth : alk. paper) | ISBN 9780226610399 (e-book)
Subjects: LCSH: Science—Experiments—History—18th century. | Science—Experiments—History—17th century. | Science—Experiments—Economic aspects. | Science—History. | Home economics—History.
Classification: LCC Q182.3 .W444 2019 | DDC 507.8—dc23
LC record available at https://lccn.loc.gov/2018026816

♾ This paper meets the requirements of ANSI/NISO Z39.48-1992 (Permanence of Paper).

For Edward and Oliver

Contents

Illustrations

Introduction

For much of the twentieth century a prominent vision of the future was one of endless scientific progress and ever-grander engineering. First there would be rockets and moon bases, then space stations and interstellar travel. But this has not materialized, and the reality of the twenty-first century is more mundane. An aesthetic that values the salvaged and the secondhand has become salient in an era that used to be imagined as one of silver space suits and perfect, modern living. In London, where much of this book was written, old warehouses and railway arches house a growing number of hackspaces and makerspaces, where communities convene to swap tips on repairing consumer goods, recycling, and programming. Artists like Hilary Powell construct art from the materials found on old building sites. Celia Pym travels around the city making artworks by darning holes in socks, mending sweaters, and patching teddy bears. The crumbling tower blocks and edifices of Britain's once grand imperial capital are photographed for a blog on "Derelict London."[1]

Popular films reflect this shift. In the opening scene of George Lucas's *Star Wars* (1977), C-3PO appears hurrying down the corridor of a rebel spacecraft under attack from a Star Destroyer. What many people may not notice is that the golden-bodied C-3PO has an ill-fitting silver replacement leg. *Star Wars* makes surprisingly much of scrap, salvage, and recycling, with grubby corridors and garbage compactors joining the usually glitzy spaceship interiors typical of science fiction movies at the time. In the 2015 sequel *The Force Awakens*, the two heroes are a sanitary worker and a scavenger. George Lucas claimed this grubbiness helped to ground his world and make it seem real, but the scavenging of *Star Wars* could stand for a broader trend in recent times.[2]

This is evident in the way many people now relate to materials. Environmental literature encourages the "maintenance and repair of existing products" instead of buying into the planned obsolescence of consumer culture.[3] In homes as well as public institutions, people learn the need to "reduce, reuse, and recycle," dropping paper, tin cans, and plastic into specially designated bins. The sciences are not exempt from this development. Similar concerns have been manifested in calls to make scientific research more sustainable. A variety of initiatives begun in the United States have identified scientific research laboratories as some of the least sustainable sites in the modern world, using more energy even than equivalent industrial spaces. In the era of Big Science, when scientific instruments might be built on a vast scale for many millions of dollars, sustainability is a problem.[4] A US government paper recently claimed that scientific research laboratories are some of the most "prodigious consumers of natural resources . . . laboratories typically consume 5 to 10 times more energy per square foot than do office buildings. And some specialty laboratories . . . consume as much as 100 times the energy of a similarly sized institutional or commercial structure."[5] Worried scientists have begun "green labs" initiatives to counter such problems, encouraging energy efficiency, sharing, and recycling to reduce waste.[6]

These public and scientific initiatives have commonly been identified as having very recent origins. As Ruth Oldenziel and Helmuth Trischler have noted, recycling is often seen as having roots in the counterculture of the 1960s, associated with figures such as the San Francisco-based environmentalist Stewart Brand and his *Whole Earth Catalog* or E. F. Schumacher's book *Small Is Beautiful* (1973), or in the "do-it-yourself" attitude of the 1970s, determined to make and remake things in a more sustainable way.[7] But as Oldenziel and Trischler argue, such practices have deeper historical roots, and this book is a contribution to revealing them.[8] Rather than something new, it could be argued that modern environmental practices are more of a re-recognition of thrifty practices that have been going on in one form or another for centuries. This book is an attempt to recover some of these practices in the early modern period, in the hope that they offer ways to think about how we might be thrifty in the future.

Exploring the role of thrifty practices in the sciences also helps to take forward recent historiographical developments in the history of science. For some time now, historians of science have sought to move away from studies focused on innovation in the sciences to consider instead questions of the circulation of knowledge.[9] Such studies have shown, importantly, that the circulation of "facts" has been rooted in the circulation of things. An idea cannot travel without being embodied in some way, and a material infrastructure is required to move things like books, instruments, and specimens from one place to another. This interest in the material in the sciences reflects a broader desire by historians to better understand the role of material things in history. Against histories centered on treating the world as a text or with a focus on social relations, historians increasingly appreciate the agency and evidential value of artifacts, objects, and materials in understanding the past.[10]

To explore "circulation" is to establish a cycle for the movements of things, taking in their collection, their mobilization, travel, arrival, and analysis or use for some scientific purpose.[11] This book began as a question about what takes place next. Once scientific things are used up, what happens to them? What happens when instruments are broken, when specimens become worn and unsightly, when books begin to crumble into dust? Could we talk about the destruction of knowledge rather than its production? Inspiration for this question came from reading David Edgerton's book *The Shock of the Old*, in which Edgerton argues that a focus on innovation in the history of technology has led historians to overlook a great deal of what matters in technological practice.[12] "Old" technologies continue to be used long past the time of their invention and initial appearance and remain important. Practices of maintenance and repair enable these technologies to endure, but tend to be overlooked when we focus on innovation. Similar questions might be asked of scientific things, understanding scientific instruments, apparatus, and laboratories as a particular form of technology. How are telescopes or cloud chambers or volumes of *Nature* kept in good states of repair? What happens to them when they are no longer useful?

Following an initial foray into these questions, my point of entry in this book will be the history of experimental sciences in early modern England.[13] In recent years there has been an interesting shift in the historiography of experiment in this setting. A traditional story equated the rise of experiment with a series of heroic male innovators associated with the Royal Society for the Improvement of Natural Knowledge founded in London in 1660. Figures such as Francis Bacon, Robert Boyle, Robert Hooke, and Isaac Newton forged an experimental method that gave rise to new, reliable, and useful knowledge, paving the way for a British Enlightenment in which diverse phenomena were reevaluated using reason and science. Historians once told this story as one of successive improvements in theoretical knowledge, while more recently it has become normal to see these changes as happening in dialog with their historical context. Moreover in the process of shifting from innovation to circulation, historians have shown how this "experimental philosophy" was not a disinterested pursuit of truth but emerged entangled with a host of social, commercial, imperial, and political interests in the early modern period, serving the expansion of trade and colonial power.[14] Neither was "experimental science" necessarily the province of a few university-educated men. In recent years the history of science has greatly expanded its franchise to show how diverse communities dealt in natural knowledge in the seventeenth and eighteenth centuries, among them navigators, apothecaries, artisans, family members and servants, chymists, colonial subjects, translators, and go-betweens.[15] These might be men or women, European or non-European, university-educated, apprenticed, or self-taught. As such, the history of science has become more global, incorporating a wider range of actors, locations, and broader definitions of experiment.[16]

One place where such innovative perspectives have been particularly revealing is in the history of the home as a site for making and using natural knowledge. Numerous historians have revealed the importance of the home for early modern and modern science, demonstrating that "the domestic sphere is not external to knowledge making, but rather a condition for and a consequence of research."[17] Deborah Harkness explored the tensions in the Elizabethan astrologer John Dee and his family's home resulting from the need to manage a household and conduct philosophical inquiries in the same space.[18] Steven Shapin showed how the seventeenth-century chymist Robert Boyle negotiated the gentlemanly conventions of access to the home when experimenting there.[19] For the modern period Deborah Coen has traced the powerful influence of domestic life on the Exner family in Vienna, showing how the family's personal relationships and communal gatherings shaped and promoted scientific inquiry and embodied the family's liberal perspectives.[20] Much of this work has focused on the social order of the household and the way domestic social tensions or conventions shaped scientific inquiry. But historians have also drawn attention to the importance of materials within the domestic setting. Historians of recipe exchange such as Alisha Rankin, Elaine Leong, Sara Pennell, and Michelle DiMeo have explored how men and women experimented with a great variety of recipes and substances to prepare food and medicines, exchanging these in networks extending within and between families and across generations.[21]

Rankin has revealed how the German noblewoman Anna of Saxony operated a substantial medical practice in her sixteenth-century palatial residence of Annaburg, experimenting in agriculture, distilling, and the preparation of medicaments, making use of her storerooms, domestic utensils, and servants to do so.[22] Such studies demonstrate the varieties of materials and material and experimental expertise that populated the home, and the complex social arrangements through which these were collected, transformed, and maintained.

Thrifty Science builds on the insights of these historians to revisit the more familiar community of experimental philosophers in early modern English science. Given the rich picture we now have of a variety of families experimenting with and managing medical, culinary, and chymical knowledge in the household, these insights might usefully be applied to rethink figures such as Francis Bacon, Robert Boyle, or Isaac Newton. *Thrifty Science* argues that the activities of such figures also relied fundamentally on the households in which they lived. This might be a *social* dependence, a reliance on the support and collaboration of other members of the family and networks of exchange, and where possible I have indicated these interactions. But it was also a *material* dependence, since the physical house, its contents, and the techniques used to manage them enabled and were mobilized to create an enormous variety of apparatus and instruments for investigating nature. This material perspective provides the focus of the book. My goal is to make the household visible in early modern science not so much through marginalized people as through forgotten thrifty practices and things.

Thrifty Science recontextualizes early modern science in the material culture of the home. However, there is not much work done here to situate events in a broader historical perspective. This is not because I think this is unnecessary or unimportant but because I am interested in drawing out approaches and practices relating to materials that require a smaller-scale focus to reveal them. Parts of this book are therefore quite ahistorical, insofar as they do not focus on historical change or the impact of changing political, religious, or social events, but remain concentrated on identifying and explaining different artifacts, sites, and techniques. This book is centered on identifying and making sense of a series of objects and practices that might well then be subjected to a more nuanced contextual reading, but whose long-term trajectories would be less apparent if such contextualization were attempted here.

Turning to the nature of these material practices in the home, the book identifies an approach to material culture in early modern England that I call *thrifty science*. Thrifty science (my own and not a contemporary term) involved an understanding of material things as open-ended and capable of serving a variety of purposes; an often ingenious making use of readily available things to perform experiments; a desire to make sure material objects endured through practices of careful storage, use, and maintenance; and an engagement in circulations of used and secondhand goods. Thrifty science was not restricted to early modern England but seems to have been a prominent approach to materials there.

To better understand this approach to material culture and practice *Thrifty Science* makes use of a substantial literature that existed in the seventeenth and eighteenth centuries on "oeconomy" or the management of the household. "Oeconomy," whose definition will be dealt with in detail in chapter 1, amounted to prescriptions for the good ordering of the household, including both the material things and the people who composed it. Oeconomy offers a window onto the ways people in the seventeenth and eighteenth centuries cared for and used materials and by extension a way to understand experimentation in the home.

An important feature of oeconomy was "thrift." Today thrift means saving money and avoiding expenditure, but in the oeconomic literature of the seventeenth and eighteenth centuries it could also refer to the pursuit of a sensible balance between buying new and using old things, between spending and saving and making the most of the resources you already had. It was not so much a financial prescription as a general notion that encouraged the use of things to enable *thriving*, a closely related term. Thrift, this book argues, helps to make sense of experimental practices and approaches to material culture in England in the seventeenth and eighteenth centuries. Thrifty science entailed both the creation or purchase of new, dedicated instruments such as air pumps and telescopes, and the making use of existing materials, often found in the home, for the purpose of doing experiments. So while some of the material culture made use of in the sciences was purpose-built and specialized (e.g., air pumps, electrical machines), this was by no means all of it. Many apparatus were ingeniously adapted household items, and together with more dedicated instruments, experimenters looked after them through practices they applied to any household possessions.

Thrifty science thus entailed a quite distinct sense of material objects. Histories of scientific instruments have traditionally focused on novel dedicated and specialized instruments. As Deborah Jean Warner put it in her essay "What Is a Scientific Instrument, When Did It Become One, and Why?,"

Although instruments and apparatus have long been used for investigating the natural world, their importance increased markedly in the early years of the seventeenth century with the invention of the telescope and the microscope, the barometer and the air pump, and the pendulum clock, and the rise of experimental philosophy.[23]

The historiography of instruments has shifted from tracing the origins and development of such instruments to a more social and cultural history of their meanings and use. But historians have continued to focus on instruments designed and built to serve some specific scientific purpose. Steven Shapin and Simon Schaffer's influential *Leviathan and the Air-Pump* (1985) charted the emergence of English experimental science through the development of Robert Boyle's air pump, an instrument designed to withdraw the air from a glass receiver, within which various phenomena could be explored.[24] H. Otto Sibum reconstructed the apparatus designed by Victorian physicist James Joule to measure the mechanical equivalent of heat to reveal aspects of "gestural" or embodied knowledge involved in its use.[25] Recently the marine chronometers of eighteenth-century clockmaker John Harrison have raised much historical interest, and others have studied the microscope, barometer, telescope, and magic lantern.[26] Similarly, histories of scientific spaces have tended to concentrate on dedicated sites such as observatories, laboratories, institutions, and museums constructed with the intention that they would be used for scientific inquiry.[27] But this book argues that much experimental activity involved making use of existing things and spaces to study nature, especially the home itself and the furniture and goods that it contained. Kitchens, bedchambers, pokers, gun barrels, wineglasses, and stockings were "made use of" or "put into the service of" experiment, what might today be termed a process of adaptation or repurposing. This was not accidental or done simply for a lack of better resources, but followed a certain logic that it is the goal here to understand. Existing histories of scientific instruments have tended to treat them as more or less static objects, fixed in a certain form with a certain function. But this book argues that thrifty householders viewed objects as capable of undergoing regular changes in their use. Studies of maintenance and repair in modern households have noted the way householders treat objects as unfinished. N. Gregson, A. Metcalfe, and L. Crewe argue that "consumer objects are continually becoming in the course of their lives in the domestic. They are . . . neither finished nor inviolable forms at the points of production and acquisition, but rather are better regarded as continually evolving."[28] Things deteriorate and break down, and remain in a constant state of flux, what the sociologist Karin Knorr-Cetina refers to as the "incompleteness" of objects.[29] Such incompleteness is important to recognize because it prompts a great variety of actions of maintenance, repair, reworking, and recirculating that a view of objects as static will overlook. Incompleteness also helps to explain the value placed on the adaptability of objects in seventeenth- and eighteenth-century households and science. Things were open-ended, their use not predetermined or seen as essential, so that they were capable of being turned to a variety of ends, which thrift encouraged as a means to get the most out of them. In turn, things should be durable, ideally lasting over several generations, so that they could be put to use as much as possible. This incompleteness helped make household items into experimental apparatus, putting them into "service" through various techniques (explored in chapter 2).

So *Thrifty Science* explores an approach to materials that historians have not fully appreciated, in which things were seen as "incomplete," adaptable, and open-ended. The book also examines how approaches to material culture in science changed in the eighteenth and nineteenth centuries. While much of the book is an attempt to survey thrifty practices in science, I also explore the development of an alternative view to thrifty science, which I call "economic science." With roots in ideas of the seventeenth century, this approach to materials rose to prominence and challenged thrifty practices with a more utilitarian approach. With certain qualifications, I distinguish an era of *oeconomy* (the seventeenth and eighteenth centuries) and an era of *economy* (the nineteenth and twentieth centuries), based on these contrasting views.

In their thrifty approach to materials and their care and repurposing of possessions, early modern experimenters might be said to have followed practices today considered as contributing to environmental sustainability. While it is not self-evident that such practices were sustainable or directed toward environmental concerns, they certainly reflect current sensibilities that humans should try to make the best of limited resources. Viewed from the present, experimenters' actions appear quite "green" and may hold lessons for sustainability. Thus, this book is a contribution to an environmental history of science. By charting attitudes and practices that encouraged the stewardship of materials, the book will it is hoped be suggestive for thinking about the future of science in addition to its past. Would it be possible, or desirable, for science to return to a more thrifty practice? Is this, in fact, already happening?

This is of course not the first time the nature of seventeenth- and eighteenth-century science has been linked to the fate of the environment. Carolyn Merchant famously pronounced modern science as the cause of "the death of nature."[30] The Scientific Revolution ushered in a contempt for nature, gendered as a subservient female tortured by masculine experimental science. Legitimating extractive and exploitative mining and deforestation, modern Baconian science sought the conquest of nature in contrast to a more balanced and reciprocal relationship with nature promoted by the sciences of the Renaissance. Merchant was a pioneer in making these connections and in showing that science cannot be left outside the realm of environmental critique. Too often, science is seen as the source of environmental knowledge and wisdom on how to solve environmental crises without being understood as an agent in those crises. Merchant left no doubt that science was implicated in the fate of the environment. However, this book departs from Merchant's thesis by focusing more on practice and material culture. Merchant's argument proceeded largely by considering the rhetoric of the new science of the seventeenth century. But it could be argued that the "new science," at least to begin with, was not as powerful as its protagonists liked to make out. This is not so much because seventeenth- and eighteenth-century science did not fulfill promises of utility and comfort as quickly as was expected, but because in practice the new science was not as new as it might appear. In a nice illustration of his modern credentials, the mathematician and natural philosopher René Descartes was presented in a seventeenth-century engraving seated at a table with his foot resting on a tome of Aristotle (fig. 0.1).

The implication was that Descartes was discarding the ancient as rubbish, as something to be thrown on the floor and swept away with the dust that accumulated there throughout the day.[31] Yet the pen with which Descartes wrote was a repurposed bird's feather. His ink might have been made with stale beer and old nails. The paper of his book (and those on the shelves) would have been made of old rags, and his cloak would have been repaired again and again until it was ragged, at which point it would be reused in paper making. The room is typical of a seventeenth-century house in that it is empty save for a few pieces of simple wooden furniture, and the bricks and stones making the walls and floor were most likely reused from an earlier building and would be used again in future ones. If the rhetoric of the image is one of radical innovation, the material culture of it is about the old, the reused and recycled. The picture of science in Merchant's story, and in the story of the Scientific Revolution more generally, is rather like this image, focusing on the message rather than the material particulars. But as this book will show, from a material perspective the new science was rather significantly and interestingly *old*.

Figure 0.1 Cornelius A. Hellemans, *Portrait of René Descartes*, c. 1687–91. Rijksmuseum, Netherlands.

This is not to say that seventeenth- and eighteenth-century science did not herald the "death of nature," but it is to point out that a more materials-focused reading of the period provides a different picture of events. Historians are making this possible by recovering the astonishing array of old and used goods that marked seventeenth- and eighteenth-century life. Donald Woodward traced recycling practices in seventeenth- and eighteenth-century England, while Tim Cooper has considered them in the nineteenth century.[32] Beverly Lemire, Jon Stobart, Ilja Van Damme, and Laurence Fontaine have revealed the extensive circulation of secondhand goods in seventeenth- and eighteenth-century Europe, while Ariane Fenetaux, Sophie Vasset, and Amélie Junqua have shown how recycling was widespread across Europe in the long eighteenth century.[33] In her influential work *Waste and Want*, Susan Strasser has insisted that preindustrial people made much of the "stewardship of objects," investing time, labor, and ingenuity in things to make diverse uses of them and to make them endure.[34] *Thrifty Science* contributes to this literature by making apparent how important such goods, and the attitudes that made use of them, were for the seventeenth- and eighteenth-century scientific enterprise.

The scope of this book is necessarily limited. To indicate this through a list of disciplines covered would be to miss the point that the definition of what counted as experiment or science was not settled in the time period covered by the book. Nevertheless, since recent work has done so much to elucidate female networks of exchange in medical, chymical, and culinary recipes, the discussion of these activities is limited here. Similarly, historians of natural history have said much about its links to "oeconomy" in the eighteenth century, and I will not rehearse those in detail, although they will be situated as one part of a broader history of experiment and oeconomy in chapter 6. I also recognize that different scientific disciplines may have developed different approaches to thrift, and while this is hinted at, more research will need to be done to trace these specific histories. The geographical coverage of *Thrifty Science* is also rather narrow, centered mostly on the urban centers of England and North America. These locations have specific histories, but the lessons of the book will I hope be made use of to consider other places. England in the seventeenth and eighteenth centuries was the lively setting for significant transformations in domestic culture, architecture, and science. Books of household management flourished in this period, following the rise of an English culture of secular publishing in the second half of the sixteenth century. The beginning of the seventeenth century witnessed a significant expansion in oeconomic literature and various efforts to define a new experimental form of scientific inquiry. Across the seventeenth and eighteenth centuries, English people also traveled to the New World and other colonies to settle and seek their fortunes. Colonial settings have often been presented as locations for thrift, substitution, and "making do," offset against the imperial metropolis where new raw materials and new products were more available.[35] *Thrifty Science* contends that such a contrast is a (mostly Victorian) fiction, which served to distinguish the imperial center but failed to capture the ubiquity of thrift in the seventeenth and eighteenth centuries. For this reason, the book considers thrifty practices in several North American towns in the seventeenth and eighteenth centuries, to provide comparison with England.

Thrifty Science begins by proposing that "oeconomic" literature on household management provides a useful starting point for making sense of the material practices of experimenters in the seventeenth and eighteenth centuries. Oeconomy encouraged thrift in managing a household, which meant a balance of expenditure on new things and making the most of what one already had. Focusing on the seventeenth century, chapter 1 argues that householders experimenting in their homes brought thrifty attitudes to new forms of natural knowledge-making, and while historians have focused on the new and dedicated materials of this period, they have overlooked how experimenters exploited what they already had. This thrifty science involved, in particular, "making use of things"—using everyday items as well as possible, a practice that met with derision from some quarters, but that various experimenters claimed was a critical feature of the new science. The chapter also highlights different opinions on household practices as experimental knowledge. For some, domestic experiment was sufficient in itself as a new form of science, but others argued that it needed to be extracted from the home for testing and accreditation elsewhere. Many things might be "experiments," but only some could be "natural philosophy."

Chapter 2 examines in more detail the fact that most experimental science of the seventeenth and eighteenth centuries took place in the home. While scholars have explored how the social order of the home influenced science, this chapter takes a more materials-focused perspective, demonstrating how the house itself was evolving during the early years of experimental philosophy and how developing rooms, their furniture and utensils, provided social and material resources that experimenters cleverly exploited to make new kinds of natural knowledge.

"Making use of things" was a key value of thrifty science and domestic oeconomy, but what exactly did this mean? Chapter 3 considers a variety of making practices that served good oeconomy by turning existing household items to experimental ends. Reuse, repurposing, "shifts" or improvisations, substitution, and alteration were applied to possessions such as clothes and to the production of experimental apparatus and instrumentation. This meant that all sorts of domestic and mundane materials ranging from coffeepots and china to kites and colanders were put into experimental service, assembled together into ingenious bricolages to generate apparatus. These techniques were not unproblematic and led to ambiguities and puzzles for philosophers, but they constituted a critical feature of scientific material culture overlooked by an exclusive focus on the specialized instruments of the period.

Making apparatus was one thing, but keeping it secure and functional required further labors. Chapter 4 examines maintenance in seventeenth- and eighteenth-century English experimentation. Householders mobilized an array of maintenance techniques to achieve this, and chapter 4 surveys methods of storing, protecting, cleaning, and transporting experimental goods with the end of allowing them to endure unharmed. Maintenance entailed both material and social strategies, and the chapter considers how efforts to order the family and servants were closely integrated with labors to maintain domestic and experimental goods.

Even if maintenance failed and things were broken, they could still be put to thrifty use. Early modern homes engaged in a great deal of repair work, as skilled householders or artisans mended an assortment of domestic possessions, including those used for experimental investigations. Chapter 5 explores these labors and shows how experimental work was situated in a rich culture of repair work in the early modern home. The chapter also examines how broken and unwanted domestic materials were circulated into new uses by various scavengers and traders in what today would be called "recycling." Experimenters made use of recycled materials, contributed to the circulation of old materials, or used them to create apparatus. The chapter again shows how these activities entangled the social and material, since choices about repair and recycling invoked hierarchy, distinction, and gendered divisions of labor.

Chapter 6 explores "secondhand science" or other household circulations of scientific goods and instruments in the seventeenth and eighteenth centuries. The eighteenth century witnessed the rise of manufactures and commercial circulations, and while historians have characterized this era as one of rapidly increasing domestic consumption, the chapter explores a diversity of other forms of exchange centered on the home that flourished in the period. The chapter shows how experimenters engaged in a variety of thrifty, often noncommercial recirculations, such as gift- giving, lending and borrowing, inheritance and bequests, and a bustling secondhand market. These material exchanges helped to build networks of support, established trust and credit, and cultivated relationships of mutual advantage.

A final element of thrifty science to be considered is the auction. Chapter 7 examines the auction as a site where thrifty and commercial practices and values overlapped and existed in tension. The chapter proposes that auctions, hitherto largely ignored by historians of science, were key sites of public science in the eighteenth century, serving as locations not only for the exchange of goods but also for the production of knowledge. Auctions emerged contemporaneously with experimental philosophy and faced similar criticisms for being shifty and ignoble. Auctioneers, like experimenters, used appeals to gentility and erudition to justify their activities. Throughout the eighteenth century, auctions and experimenting were closely connected, and experimental goods were increasingly sold at auction. Once again, social and material practices were integrated, as auctions of aristocratic collections "dismantled" science, giving ordinary scholars access to instruments and specimens hitherto unavailable.

Chapter 8 follows the fate of thrifty science into the nineteenth century. The chapter argues that changes in oeconomy and thrift traced in previous chapters culminated in the nineteenth century in a new approach to materials that I call "economic science." An accelerating trend in the early nineteenth century saw the establishment of a growing number of laboratories separated from the home. "Men of science" sought to distinguish a new, autonomous experimental practice from the domestic thrifty experiment of the previous centuries. They did so in part through the creation of specialized instruments for science that followed an "economic" approach to materials. This valued objects serving a single, dedicated function, in contrast with the open-endedness of earlier "thrifty" instruments. The chapter explores the logic of this new "economic science," the way it emerged in contradistinction to thrifty science, and the way it marginalized, though by no means eclipsed, thrifty science.

The conclusion offers some assessment of how thrifty practices are faring in the sciences today, and how they are being reevaluated among hacker communities taking a less formal, more makeshift approach to science and technology in recent years. Some consideration is given to how sustainable these practices are and how we measure sustainability. The book concludes that "oeconomy" and a more thrifty science might be an important tool for assessing and achieving sustainability in the future.

ONE

Thrifty Science: Oeconomy and Experiment

An online search for "equip a laboratory" soon reveals the statement that "Biotechnology labs need specialized equipment for different types of processes and activities."[1] Several lists follow of the different instruments required. None are familiar, at least to a nonspecialist. Being "specialized" means being "narrow or specific in focus."[2] This is a common requirement for many kinds of instruments, laboratories, and procedures in modern science. According to the philosopher Robert John Ackermann, "Modern science is inconceivable . . . without the highly specialized instruments that scientists use to investigate nature."[3]

To what degree was early modern science like this? Did the practitioners of "experimental natural philosophy" believe they needed specialized instruments and laboratories to do science? The argument of this book is that the seventeenth- and eighteenth-century approach was quite different from that of today. The antonyms of specialized might be multipurpose, flexible, variable, or adaptable, and the material culture of early modern science was much more like that than we have tended to appreciate. In fact only a handful of English printed books used the term "specialized" before the eighteenth century, either as a synonym for "specified" (e.g., "an argument as specialized in the New Testament") or to refer to a form of division in mathematics.[4] Only in the nineteenth century did it become common to speak of specialized instruments. Clearly, early modern experimenters did devise and use what we would call specialized equipment with a single purpose such as the air pump or microscope.[5] Until the nineteenth century, experimenters divided instruments into optical, mathematical, and philosophical categories.[6] Optical instruments such as the telescope employed mirrors and lenses while mathematical instruments were used for making measurements. "Philosophical apparatus" was a term introduced by the German émigré Samuel Hartlib in 1649 and included air pumps and electrical machines, instruments that generated new natural effects such as the vacuum or electric fire. But instrumentation included a great deal more. Much of the material culture experimenters dealt with was valuable because it was flexible. Experimenters were keen to explore a variety of uses of mundane material objects that were ready to hand in order to learn about nature. This is what I call *thrifty science.*

This chapter explores some of the broader approaches to materials in seventeenth- and eighteenth-century England and suggests that if experimenters had a distinctive approach to materials in the period, then this reflected a wider attitude of the time. Of course people had no single "approach to materials," and no doubt the ways in which they understood them were as diverse as materials themselves. But it is possible to identify shared attitudes, practices, and ideas that were different from present ones. One such set of ideas and practices surrounded "oeconomy," an approach to the management of the household that had a significant impact on people's decisions relating to material goods. Householders engaged in the oeconomic management of their possessions, putting things to good use, and a variety of literatures gave them advice on how to do this. The claim here will be that the morals and techniques of household oeconomy provide a window onto early experimental practice. Oeconomic literature provides a way to identify and make sense of thrifty science.

This chapter outlines some of the features of oeconomy and the related value of thrift and considers how some seventeenth- and eighteenth-century householders contributed to oeconomic and experimental thinking. The particular focus will be the way householders gave a value to making the most of materials, by using them for different ends, creating things that could serve many purposes, or by reworking and adapting things for some novel function. Early writings calling for an "experimental philosophy" shared in this approach to materials by invoking oeconomic ideals of making use of things, though they differed on whether household practices might count as a source of experimental knowledge. This chapter is less about specific instances of how thrifty science was practiced than about generating a broad picture of the values and approaches that lay behind such adaptive actions. The various practices of thrifty science and the objects they gave rise to then provide the focus of subsequent chapters.

Oeconomy

In early modern England, householders interested in learning about the good management of the home engaged with a literature of "oeconomy." English ideas about oeconomy were rooted in ancient works such as Aristotle's *Politics* and particularly the fourth-century BCE *Oeconomicus* of Xenophon, describing an art of governance or management that applied to the family and household or estate.[7] Translated into Latin by Cicero, the text became available in English as *Xenophons Treatise of House Holde* published in London in 1532 and going through several editions in the sixteenth century.[8] More works of domestic oeconomy appeared in the seventeenth and eighteenth centuries, at first addressed to the landed gentry and then increasingly to less well-off householders. These provided advice on "husbandry" and "housewifery" and included diverse topics ranging from gardening, cookery, and cleaning to medicine and the management of the family, which included both relatives and domestic servants.[9] To take an example, *England's Happiness Improved* was published in London in 1697 with a second edition following in 1699.[10] It consisted of various recipes and advice for "better profiting a Family, and saving much Charges, &c."[11] Householders could learn to prolong the life of foods or possessions. Bread could be made more substantial and long-lasting by various means, and there were recipes to tenderize poultry, to preserve fowl and lard, to keep flowers fresh, to regloss faded paintings, to whiten linen, and to kill insects. A library washed with strong vinegar and scented with Storax and Brimstone would prevent worms from "eating and spoiling Books."[12] One meaning of oeconomy, then, was a certain kind of stewardship or care for material things, which would ultimately "*encrease plenty, and promote pleasure*," as the subtitle to *England's Happiness Improved* put it.

Oeconomy, then, related to what the historian Karen Harvey has called "the material detail of everyday domestic life."[13] Oeconomy simultaneously entailed moral prescriptions about the good order of the household and identified domestic virtues as a foundation of society. Authors supposed a well-ordered household would generate a well-ordered state, but differed on what "order" should mean.[14] Most books discussed the right management of the family and servants, the place and duties of husbands, housewives, and children in the home. This patriarchal order, in which the man was the master of the house, typically proposed a division of labor between the genders, with women attending to "housewifery" inside the home and men to "husbandry" outside, though in practice such divisions did not necessarily apply.[15] Housewifery included "physick" or the preparation of medicines, cookery, brewing, cleaning, housekeeping, and managing accounts, while husbandry entailed looking after the estate or grounds, gardening, farming, and animal rearing. Oeconomic writing reflected this gendered division of labor, with women writing on cookery, medicine, and advice for maids, mostly in private manuscripts but also in some printed books. Men wrote, and were able to publish, works on husbandry, medicine, estate management, and the management of servants. Crossing these gender divisions in print might lead to criticism, but both men and women read works ostensibly aimed at only one gender.[16]

Throughout the literature, the material and the moral were closely intertwined. Husbandry, for example, was often connected to the toil of Adam in the Garden of Eden and given a spiritual, Christian purpose.[17] Husbandry prevented hunger, generated wholesome labor, and made use of idle land.[18] Thus the household was not just a material site but a moral space for the cultivation of patriarchal and Christian virtue: a respect for order and hierarchy, religious devotion, and civic-mindedness. As audiences for books on oeconomy shifted from landed gentlemen to working households in the eighteenth century, these values changed to ones serving a more commercial society such as good discipline, good credit, and personal morality.[19]

The meaning and scope of "oeconomy" did not remain static in the seventeenth and eighteenth centuries. Over time, writers identified oeconomy less with individual households than with larger systems of material circulation and proposed that a theoretical understanding of these would help them to flourish. Instead of referring to good practice in the home, scholarly commentators insisted that oeconomy should have a more general meaning as a system of management or order applied to some entity as a whole, which it was then their role to comprehend and master. As Richard Drayton has shown, husbandry remained a salient location of such discussions, as did the related field of natural history, because natural resources could be presented as critical sources of wealth to be managed.[20] Already in the 1620s the English writer Gervase Markham argued that theory was a better guide than experience in oeconomy. For Markham, the best writers on husbandry were "Schollers," whose nation flourished "more by their *Theorie*, then by the practise of all the pessants of that Kingdome."[21]

In the eighteenth century, scholars continued to offer schemes and proposals for enhancing wealth through improved husbandry, a role that natural historians also came to occupy. The Swedish naturalist Linnaeus did much to expand notions of oeconomy. His 1749 essay "Oeconomy of Nature" explained how God managed a perpetual cycle of birth, life, death, and decay where nothing was wasted.[22] Promoting state intervention in such an oeconomy, Linnaeus proposed elaborate schemes to identify, transfer, and acclimatize plants around the world to enhance Swedish wealth.[23] Although a domestic botany might be part of these schemes, Linnaeus was taking oeconomy far beyond the home or even state. Natural historical accounts of the exemplary divine or ideal human management of nature followed across Europe and helped inspire numerous experiments in Britain in botany and scientific agriculture. Spurred by anxiety over war, debt, and revolution, a community of late eighteenth-century figures such as Joseph Banks, Arthur Young, Thomas Malthus, and Jeremy Bentham promoted what C. A. Bayly and Drayton have called "agrarian patriotism."[24] This urged the improvement of national finance through state-supported experiments in scientific agriculture and a paternalist management of labor on the basis of enlightened principles. In these schemes households, landed estates, and farms might participate in national improvement through consumption and patriotic labor, while import substitution and self-sufficient resource management established security.[25] This was supposed to happen on a scale far beyond the individual home and would be managed not by householders but by the enlightened paternalists who guided them. By now the notion of oeconomy was no longer equivalent to the thrift of the household. Young was explicit in contrasting homely frugality with these larger pursuits in his *Rural Oeconomy* (1770),

> The reader will allow me to use the words *Rural Oeconomy* in their enlarged sense, and not merely confined to the practice of *frugality*, which is the common acceptation of *oeconomy*. *Frugality* conveys but a narrow idea; a man may undoubtedly be very frugal, and yet a vile husbandman: we must therefore understand by *oeconomy*, the system of GENERAL MANAGEMENT.[26]

A variety of private and state-supported "oeconomic societies" emerged in the second half of the eighteenth century to encourage improvement along these lines, through the rationalization of practical expertise, mechanical invention, prize competitions, and experiments in commodity substitution.[27] All of these efforts served to broaden the definition of oeconomy.[28] As Lissa Roberts has argued, "oeconomy" in the second half of the eighteenth century came to signal an arena in which

> the investigation of nature merged seamlessly with concerns for material and moral well-being, in which the interdependence of urban and rural productivity was appreciated and stewarded, in which 'improvement' was simultaneously directed toward increasing the yields of agriculture, manufacturing and social responsibility.[29]

It was in this context that new notions of "political economy" arose, which also shifted the notion of good management away from the household to the state's guidance of finance, industry, and commerce (and by 1800 dropped the "o" from "oeconomy").[30] James Steuart, the first writer to title his book as one of "political oeconomy," promoted the idea that the state as a whole might be imagined as a great household whose oeconomy needed to be kept in good order.[31] But thereafter, political economy was increasingly divorced from the idea of the household as the model of virtue, though the household might be seen to contribute to prosperity through consumption.[32]

Thrift: A Just Medium of Concern

Ideas of oeconomy thus linked together the moral order of the household and the well-being of the state, but shifted from patriarchal and Christian prescriptions for ordering the home as a model of virtue for the nation to enlightened and "economic" views of the household as a consuming unit contributing to a larger circulation of commodities and labor. Similar shifts occurred in relation to a key element of oeconomy, the idea of thrift. Books on oeconomy understood as household management often emphasized the value of moderation, frugality, and thrift. A broadsheet of 1607 told of the "Good Hows-holder" whose home contained "Not daintie Fare and Furniture of Gold, But handsom-holsom (as with Health dooth stand)."[33] *England's Happiness Improved* gave "Instructions for a decent and frugal House-keeping."[34] Richard Bradley's eighteenth-century *The Country Housewife, and Lady's Director* advised readers on "the Frugal Management of the House, and in the Delights and Profits of the farm" and was published "for the Good of the Public."[35] The term *thrift* was in use in English from at least the thirteenth century.[36] Again, meanings of thrift have changed over time. In the seventeenth and eighteenth centuries, thrift did not simply mean saving money, but was a positive value related to the verb "to thrive." It meant both thriving and saving. "Unthriftiness" entailed waste and social impropriety.[37] As Edmund of Langley proclaimed in *Richard II*, "And he shall spend mine honour with his shame / As thriftless sons their scraping fathers' gold."[38] The Puritan divine William Ames insisted thrift was not about saving money to increase wealth. Thrift should "not proceed from the love of riches, but out of conscience towards God, whose benefits we ought not to abuse . . . For honest thrift doth not hinder, but rather promote liberality and other virtues."[39] The ethical value of thrift was that it made the most of precious resources, understood to be God-given. Wasting was the opposite of thriving, a careless expenditure or use of resources that would lead only to injury. Walter de Henley's medieval *Book of Thrifte*, reprinted in the sixteenth and seventeenth centuries, warned readers not to be like those who "spend & wast more then their lands are worth by the yeare, and when they haue wasted their goods: then haue nothing but sorrow in their mouthes, & doe liue in anguish, neither can they make any shift for their profit."[40]

Joshua Yates and James Davison Hunter have noted that modern scholarship has scarcely touched on the subject of thrift from a sociological, historical, or even an economic perspective.[41] Economic accounts of thrift fail to capture the social, moral, and material elements in its meanings. The consumption theorist David Evans has contrasted *thrift*, a desire to save money in order to consume more later, with *frugality*, a desire to consume less.[42] Although this distinction is helpful, it does not match the historical uses of the terms, which were more interchangeable. *The Art of Thriving* (1674) explained, "whoever would Thrive, must continually have an eye to Frugality, a Vertue that is the Root of all Liberality."[43] In the eighteenth century Samuel Johnson defined 'frugality" as equivalent in meaning to "thrift; parsimony; good husbandry," all key elements of oeconomy.[44]

There were many motivations for being thrifty. Historians have often associated thrifty practice with poverty and a lack of resources.[45] Ayesha Mukherjee has identified a significant anxiety over dearth and famine in early modern England that prompted the learned, particularly Sir Hugh Platt, a contemporary of Francis Bacon, to promote schemes for better stewarding the material resources of the realm.[46] But thrift was not necessarily a reaction to shortages and scarcity, motivated by a lack: it could be a positive value too, undertaken to enable thriving. Householders sought to build the wealth of the family. By spending less than the household made, savings might accumulate to ensure security for the future.[47] Thrift was as much moral as material. James Calvin Davis and

Charles Mathewes trace early modern English thrift to an Augustinian tradition, passed down through the Reformation into Puritanism, that sought salvation in the sacramentality of the everyday. Acts of thrift served to express gratitude to God for His gifts, ranging from material possessions to nature and even time itself. In this sense, thrift was not an economic motive at all but one of respecting and making good use of the world provided by God.[48]

Of course, not all people were thrifty or thought that thrift was a valuable goal. As Lendol Calder has noted, debt was a pervasive problem in the early modern period.[49] The king, court, and aristocracy were regularly lambasted for being too extravagant, or celebrated for it, since splendor was a prerogative of kingship. Neither could the very poor be thrifty, since they were forced into limited expenditure or had to spend everything they had.[50]

Thrift, then, applied to people in the middle, an "aspirational" people, as Annie Gray has put it, who believed that by a frugal oeconomy they could better their lives and improve their households.[51] Success or failure in this regard, or other changes in circumstances, might transform the value of thrift for any household, so it was never a fixed feature of any particular community or class's way of living. Sometimes conspicuous consumption or a little extravagance was required: the chymist and landowner Robert Boyle was noted for being thrifty but wore expensive clothes when he met his tenants.[52] At the same time, it is striking how thrift cut across the classes. Historians of science have done much to show the class differences among experimental practitioners in the early modern period, contrasting wealthy gentlemen, middling consumers, and more humble technicians, for example.[53] Yet thrift provides an example of a value and set of practices that could traverse class. Clearly there were differences in the capacity for thrift of people of different wealth. An artisan might fall into poverty and then need to avoid spending out of necessity more easily than a wealthy gentleman. An aristocrat might maintain possessions on a scale that less well-off individuals could never afford. Wealthy ladies were expected to show "liberality," achieved through thrift, by providing medicines for the poor, which a more humble individual could not do.[54] But thrift was not equivalent to the capacity to save or spend money, and it was frequently not a reaction to scarcity: it was about how you managed what you had. Hence a wealthy gentleman like Boyle and a relatively lowly experimentalist like Robert Hooke both appear to have valued thrift, and both pursued the practices that served it in similar ways. Thrift was, after all, a Christian virtue, a duty to make use of divine gifts, that different classes were meant to share. As Richard Baxter noted in his *Christian Directory*, "it is a duty which the richest man . . . is not exempted from, to be frugal."[55]

The meaning of thrift changed over time, across different communities, confessions, and classes, and in different locations. Certainly thrift was prominent in English works on domestic management from the sixteenth to the eighteenth century. In *Observations and Advices Oeconomical*, Cambridgeshire MP Dudley North insisted that "Thrift is a matter of no small consideration in *Oeconomy*."[56] Sixteenth-century works such as the *Glasse for Housholders* (1542) proposed that since God provided for all men, they should make the best use of His gifts, "yf God sende you riches, neuer sette your mynde or affection on theim. But by & by, with the same liberalite that God sendeth theim, begyn . . . to dispose and to vse theim well."[57] For much of the seventeenth and eighteenth centuries the definition of thrift hinged on maintaining a balance between excess and deficiency, recalling Aristotle's definition of virtue in *Nicomachean Ethics*.[58] Thrift demanded a balance between using too many things, which constituted excess, and not using things enough, which led to avarice and waste. As a broadsheet called *The Art of Thriving* put it in the seventeenth century, "Spare not, nor spend too much, be this thy care."[59] Thrift was equivalent not to saving but to making the best use of things. It lay midway between miserliness and extravagance. Literature on oeconomy invoked the need for a balance between the purchase of new goods and the preservation and use of old ones. Dudley North proposed that clothing should not be changed too frequently just to keep up with fashion, but nevertheless should be changed every ten years so that the wearer would not be completely unfashionable.[60] North's son Roger, himself a keen experimenter, shared this view, scoffing at those who were too miserly, "A Cutler, one of 200,000 £, used to wear sto[c]king's of flesh colour, yt holes might Not be seen, & dyd in borrowd hospitall Sheets, and avaritiousness promts to many Such triviall Shifts."[61] The author of an essay called "Oeconomie and Extravagance" that appeared in the *Gentleman's Magazine* in 1731 summed up the general approach, explaining that oeconomy meant "Wisdom applied to the Practice of private Life; it is situated betwixt Profuseness and Avarice, and consists in a just Medium of Concern, as to exterior Goods, between being over Careful and having no Care at all."[62]

Against this view of balance was another definition of thrift that equated it with saving money and opposed it to the contentious desire for luxury.[63] Emerging in the seventeenth century, this idea of thrift grew in significance in the eighteenth century and would make thrift synonymous with saving in the nineteenth.[64] In 1691 the lawyer Sir George

Mackenzie set up "frugality" as the opposite of "luxury" and "avarice," the vices of a growing commercial culture. Denying that frugality could be equated with moderation, he insisted that it could only mean a strict form of living within one's means, so that "Frugality is the true Mathematick of Christian Morality."[65] Others denied thrift was a national benefit, even if it served the individual. Nicholas Barbon, writing in the 1690s, and later Bernard Mandeville castigated thrift and celebrated prodigality as a "noble sin," since "the Prodigal is a Blessing to the whole Society, and injures no body but himself."[66] For Mandeville, profuseness and extravagance ultimately generated wealth, an idea that he set explicitly against a notion of frugality centered on a balance of saving and expenditure. In *The Fable of the Bees* (1732) he explained his understanding of what people normally meant by frugality: "what is generally understood by it . . . consists in a Medium between Profuseness and Avarice, rather leaning to the latter."[67] Mandeville, however, equated frugality with saving, and hence a meanness that led to a lack of spending and of stimulation of wealth-creating labors. "Abundance of moderate men I know that are Enemies to Extremes will tell me, that Frugality might supply the place of [Vice] . . . by equally avoiding both Extremes, [men] might render themselves more happy."[68] But "Frugality is like Honesty, a mean starving Virtue, that is only fit for small Societies of good peaceable Men, who are contented to be poor so they may be easy; but in a large stirring Nation you may soon have enough of it."[69] Mandeville cast frugality as equivalent to saving and thence stagnation, so that the home could not be a model for a prosperous nation. Later, in his 1772 *Wealth of Nations*, Adam Smith insisted on the opposite: frugality was critical to national prosperity, because savings could be spent on capital, which generated new wealth. "Parsimony, and not industry, is the immediate cause of the increase of capital."[70] Although this contrasted with Mandeville's views, Smith nevertheless agreed with him in defining frugality as equivalent to saving, as "the principle which prompts to save."[71] In a growing commercial culture, the Aristotelian notion of thrift as a balance between excess and deficiency was thus giving way to a more utilitarian notion of thrift as equivalent to saving money.

Despite these changes, which would really take effect in the nineteenth century (as chapter 8 will explore), an approach to thrift as a balance between expenditure on new goods and preservation of the old endured and was widespread in the seventeenth and eighteenth centuries. The proliferation of new goods in seventeenth- and eighteenth-century England has been well documented, as have debates about luxury, but the history of thrifty stewardship much less so.[72] This history entailed a great variety of household practices around material possessions that sought to expand their usefulness and make them endure. The particular techniques might include adapting things to new ends, reworking and reusing old items, and maintaining, cleaning, carefully storing, and repairing goods. All of these techniques will be examined in detail in subsequent chapters. But they hinged on a thrifty principle of extending the use of things or what early moderns called "making use." The remainder of this chapter examines the idea of making use more closely, and then shows how it permeated a variety of ideas about experiment in the seventeenth century.

Before we turn to making use, however, it should be noted that even for those who desired it, thrift was typically an ideal and not necessarily a reality. North reckoned the whole family needed to apply thrift to possessions to secure a well-managed household. This entailed the careful stewardship of provisions, stores, building works, furnishings, and "Household-stuff."[73] But when he discussed a way to be thrifty with clothing, he lamented how rarely such practices were to be met with. He described the tradition

> of having of a Wardrobe in considerable Families, wherein was kept such Houshold stuff as seldom came into use, and there was also preserved all the old Vestments of the Master and Mistris, which had been any thing costly. This was a Storehouse, out of which might be taken at any time materials towards the making of new house Furniture, and Saddles, or such things, but now it is grown a shame with us, to preserve any such frippery, as they call it.[74]

Some imagined thrift could lead to ruin if its demand for making use was misunderstood. In the eighteenth century a Philadelphia journal carried "an account of a buyer of bargains" by a man who complained that while his wife was expert in oeconomy and always searched for bargains, this led her to accumulate things that she did not need.[75]

> Whatever she thinks cheap, she holds it the duty of an oeconomist to purchase; in consequence of this maxim, we are encumbered on every side with useless lumber. The servants can scarcely creep to their beds, through the chests and boxes that surround them. The carpenter is always employed in building closets, fixing cupboards, and fastening shelves; and my house has the appearance of a ship stored for a voyage across the Atlantic.[76]

From the Aristotelian perspective of maintaining a balance, the man's wife had failed in the basic requirement of thrift, that while oeconomy demanded careful purchasing and care and preservation of material goods, it also expected that they be used and used well. As the husband wrote, "The common vice of those who are still grasping for more, is to neglect that which they already possess."[77] If this were the case, then even the best oeconomy would only hinder good living, since a care for materials unused would lead only to more accumulation.

It is the great care of her life that . . . the carpets should be taken out of the chests once a month, and brushed, and the rolls of linen opened now and then, before the fire. She is daily enquiring after the best traps for mice; and keeps the rooms always scented by fumigations, to destroy the moths.[78]

And yet, "She is imagining some distant time in which she shall use whatever she accumulates."[79] Saving for the future was catastrophic if that future was always being deferred.

The Many Uses of Things

If things should be used, it was a good idea to use them well. But what constituted good use? In *The Order of Things* Michel Foucault wrote of the "prose of the world" in Renaissance Europe, the seemingly limitless network of hidden meanings and connections buried in things, which the discerning eye might reveal through "signatures," marks implanted by God into things to show their hidden correlations.[80] Such a view had a material and oeconomic analogue in the value placed on the diversification and multiplication of the uses of things. All things contained within themselves potential uses that it was the task of the good householder to discover. A feature of oeconomic advice was to highlight the various ways in which materials and objects could be made serviceable in the home, on estates, and on farms. The sixteenth-century farmer and poet Thomas Tusser wrote works on husbandry and housewifery in short, memorable verses, reprinted through the seventeenth and eighteenth centuries.

Vse all thing with skill
Good vsage with skill, being sober with all,
make huswiues to shine as the sunne on the wall.[81]

Husbands and housewives alike could be ingenious in the diverse uses to which they put things. Works on cookery prompted housewives to make the most of food by making use of everything available.[82] In *The London and Country Cook* (1749), Charles Carter wrote that "my recipes are all suitable to . . . a frugal table: And . . . if they be carefully observed, they will prevent . . . the waste of many good materials."[83] If a thing could be put to many uses, then this was a valuable property. Many materials in the home were not restricted to specialized functions. String, paper, and fabric were used in cookery, distilling, and the preparation of medicines. After they were introduced in the mid-seventeenth century, teacups and saucers were as good for baking as for drinking tea (hence "cupcakes").[84] Books on husbandry also celebrated the diverse uses to which different materials could be put. The 1725 *Dictionaire Oeconomique* noted of the beech tree, "It's useful for many things" such as for making "Dishes, Trays, Rimbs for Buckets, Trenchers . . . Chairs, Stools, Shovels and Spade-Grafts."[85] There was value in the capacity of things to be turned to a wide, seemingly limitless, variety of uses. It was the opposite of "specialization."

Today it is common to speak of "adapting" something to a new end. The term "adapt" originated in English in the sixteenth century to mean the process of taking something having one use and using it for another. It was more common, however, to speak of "making use of" a thing in a new capacity, a phrase originating in the 1570s. "The most usuall and best way for tythinge of hey is to make use of reade-weeds for wikes."[86] We still say "to make use of" today, but it seems to have become synonymous with the simpler verb "to use." People in the seventeenth and eighteenth centuries were perhaps less likely to see one term as a substitute for the other. They gave more emphasis to the *making* in "making use," and so the term "*make* use" carried a more active sense of choosing to give a new use to something that it might not normally have had. Things that could be *made use of* in different ways were considered valuable, as "good for," "commodious," "profitable," and "serviceable to many uses." To make use of things was a godly act. As the Puritan Richard Baxter wrote,

> We must see that nothing of any use be lost through satiety, negligence, or contempt; for the smallest part is of God's gifts and talents, given us, not to cast away, but to use as he would have us; and there is nothing that is good so small, but some one hath need of it, or some good use or other may be made of it.[87]

In line with this logic that everything had its use was a reluctance, in early modern language, to speak of "waste." Generally, there was little sense that once an object no longer fulfilled one function it was nothing more than waste. Not much was truly useless. To use a modern phrase, things could always be "repurposed." This diversity of use was reflected in the ways that the term "waste" was employed in the seventeenth and eighteenth centuries. It is notable how rarely the term "waste" refers to a substance, to useless material to be thrown away. The term did exist, appearing in English from the Old French and Latin *vastus* in the thirteenth century, but it referred most commonly to "waste" land, meaning land that was desolate and not in use (as opposed to being use-*less*). By the seventeenth century, the most common use of the term besides this was probably as a verb in medicine. Spirits in the joints, blood, flesh, pus, moisture, bones, and other materials were liable to become thin, reduce, or "waste" away.

Most things, however, tended not to be identified as waste. Terms we would associate with waste today instead identified the service to which some discarded substance might be put. Hence "rubbish" referred to the debris of ruined buildings, which was used to fill ditches or foundations. Samuel Johnson noted how the French manufactured niter "from the rubbish of old mortar and plaister of buildings."[88] The word was possibly connected to "rubble." "Trash" meant twigs, splinters of wood, hedge cuttings, rags, or straw that could be used in compost or to make a fire. "Garbage" was offal, used for animal food. "Dust" was "Earth or other matter reduced to small particles."[89] "Litter" was "the straw laid under animals, or on plants" or "things thrown sluttishly about."[90] "Junk" was old ropes and cables, usually finding some nautical use. Only "refuse," a fifteenth-century coinage from the Old French verb *refuser*, carried the exclusive sense of worthlessness, of being refused. Johnson defined it as "Unworthy of reception; left when the rest is taken."[91] Most terms indicated something still in circulation as a practicable material, not a form of waste. Some of these terms, like "rubbish," were certainly used to mean something "vile and worthless" as Johnson put it.[92] But this was a moral evaluation as much as, and perhaps instead of, one of utility. Rubbishy things were more "worthless" than "useless."

People did, of course, throw things away. Even today a cursory sift on the banks of the Thames at places like Greenwich reveals an astonishing profusion of animal bones, oyster shells, and bits of broken pottery, glass, and clay pipes that Londoners have thrown into the river over the centuries (fig. 1.1).

But many of the leftovers of domestic life could be transformed into materials for cooking, building, gardening, agriculture, and other ends. As such, the household occupied a place within diverse circulations of

Figure 1.1 Assorted clay pipes and pottery fragments collected from the foreshore of the Thames, part of the waste thrown into the river over the centuries. Photograph by the author.

old materials, managed by a range of traders and laborers who collected, processed, bought, sold, and reused the matter that households discarded. As an encyclopedia of 1764 explained, these people were

SCAVENGERS, two officers annually chosen in every parish in London and its suburbs by church wardens, constables, and other inhabitants, to hire persons called rakers, with carts, to clean the streets, and carry away the dirt and filth, with the ashes and dust from every home.[93]

In sum, oeconomy identified the management of material things as an intrinsic part of the good order of the household. Oeconomy was both a social and a material order, putting into service family members and servants as much as the things they possessed or needed to care for. This management demanded a prudent balance between purchasing new goods and making the most of things one already owned. As Tusser put it,

Though Ladyes may rend and by new ery day,
good huswifes must mende & by newe as they may.[94]

Careful maintenance, cleaning, and repair kept things serviceable, and when their normal use was no longer possible they might be adapted to serve for something else. Materials had many uses, and it was the task of the oeconomic householder to figure them out.

Oeconomy, Experiment, and Knowledge

It might be supposed that "making use of things" was also an *experimental* form of life. After all, it meant deciding whether things were suitable to different ends through an appreciation of their properties, and then trying out different uses to see if they worked. As the chymist Joseph Priestley defined "experiment" in the eighteenth century, it was about putting "an endless variety of things into an endless variety of situations."[95] Historians have argued that many household activities performed by men and women in the early modern household were experimental, ranging from cookery and distilling to husbandry and the preparation of medicines.[96] While this discussion has often focused on particular practices, "making use" was a broader notion informing domestic material culture generally. So while different seventeenth-century writers had different ideas about the kind of knowledge that household practices and the books and manuscripts recording it constituted, they shared in a thrifty approach when it came to material culture.

Seventeenth-century books on oeconomy often recommended personal experience, proofs, and trials as a source of improved knowledge, so that oeconomic writing might be seen as a genre of experimental inquiry. Gervase Markham's *Cheape and Good Husbandry for the VVell-ordering of all Beasts, and Fowles* (1614) explained how to avoid sheep rot, using a salt called adraces: "With this *Adraces* rubbe the mouthes of all your Sheepe once a weeke, and you shall neuer need to feare the rotting of them, for it hath beene well tried; and, as I imagine, the experiment was found out from this ground."[97] Manuscripts of household recipes compiled by families also discussed the testing and adaption of various instances of practical knowledge. As Elaine Leong has shown, Lady Ann Fanshawe wrote her signature beside recipes she had tested in a collection copied from her mother, crossing out those she deemed to have failed.[98] Householders and writers on oeconomy thus commonly identified their exploratory labors as experiments.

Not everyone, however, agreed that such labors could be designated as a new kind of "natural philosophy." Natural philosophy, as taught in the universities since the Middle Ages, sought to reveal the causes behind natural phenomena, being largely based on the writings of Aristotle and his medieval commentators. In the seventeenth century, some people argued that this form of natural philosophy was redundant, because it was impractical and often erroneous. Instead a "new science" would take the practical knowledge of everyday life, the kinds of things known by artisans, chymists, apothecaries, farmers, and householders, and make it the basis of natural philosophy. How this was to be achieved was not agreed upon. Several authors, whose views will be discussed below, looked to the experimental wisdom of the household as a resource for this new science, though they approached this differently. One position, almost exclusively male, argued that while householders might experiment and have practical knowledge and expertise, this was not "natural philosophy" until the causes and general principles underlying these experiments were properly known. For these men, knowledge from the home (and the workshop, field, and other places, for that matter) needed to be taken outside to a scientific academy to be investigated further, tested, and accredited by its members. Results might be published, invariably by a male author. For others, however, the expertise of the home might be more self-sufficient as a definition of the new science. These men and women cultivated experimental expertise in various domestic skills, and were less interested in distinguishing it by seeking the causes or general principles behind, e.g., recipes for food, medicines, or methods of husbandry. They placed much emphasis on recording, circulating, and discussing such knowledge, in manuscript as much as print, and on an exchange in which both men and women participated. Examples of these two positions will be offered below. Critically, they both shared a thrifty approach to materials even as they differed on whether domestic labors were natural philosophical or not.

A desire to distinguish household experiments from philosophical knowledge is evident first in the work of the early seventeenth-century lord chancellor, Sir Francis Bacon, whose works *The Advancement of Learning* and *Novum Organum* did much to inspire experimental philosophy among seventeenth-century scholars. Seeking an alternative to the scholastic natural philosophy current in the universities, Bacon promoted instead a form of "operative" or experimental knowledge that would interrogate nature to enable improvement and profit of various kinds.[99] For Bacon, as is well known, the new science should be a hybrid of everyday practical knowledge, which was useful but unsystematic, and the scholastic natural philosophy of the universities, which was systematic but impractical. Household expertise was useful, Bacon supposed, but insufficient on its own to count as experimental knowledge. In the utopian

New Atlantis, Bacon thus proposed a state-sponsored institution named Salomon's House for testing and accrediting the natural philosophy behind many kinds of artisanal, magical, and everyday knowledge.[100]

The thrifty oeconomic virtue of making use of things is nevertheless evident in Bacon's work and formed a key element of his proposals for an experimental philosophy. Bacon's own household was actually far from being thrifty. His wife, Alice, had a reputation for excessive spending, and Bacon was of course famously dishonored for embezzling funds. As an early biographer lamented, "By what fatality was it that so extraordinary a man did not add to his other virtues that of a reasonable oeconomy?"[101] Nevertheless, Bacon aspired to good oeconomy and wrote on married life, children, building, gardens, and finances in his *Essays* of 1625. Like others, Bacon promoted moderation: "Ordinary expense ought to be limited by a man's estate; and governed with such regard, as it be within his compass."[102] Oeconomy had both financial and epistemic rewards. He who developed a "habit of frugality . . . gaineth as well upon his mind, as upon his estate."[103]

Bacon made much of the oeconomic value of making use of things in his writings on law and the improvement of learning. In 1596 he explained his preference for writing in aphorisms on the basis that "this delivering of knowledge . . . doth leave the wit of man more free to turne and tosse, and make use of that which is so delivered to more severall purposes and applications."[104] Bacon promoted a similar value when he turned to the development of his experimental method for the sciences. One area that he thought needed investigation if the sciences were to be renewed was what he termed "polychrest instances." In the *Advancement of Learning,* Bacon proposed that an element of the new "operative" knowledge he was proposing as an antidote to scholasticism should be "Kalendarium eorum Experimentorum, quae maxime Polychresta sunt, et ad aliorum inventionem faciunt et ducunt" (a calendar of those experiments which are most Polychrest, and which conduce and direct to the invention of others).[105] Polychrest meant many-uses, *poly-chrestos.* One should seek out processes whose understanding might explain many things and open up many more experiments, which would "ad infinita pertinet" (bear on an infinity of purposes).[106] In the *Novum organum,* Bacon elaborated on this idea. He identified "Polychrest Instances" with thrift, as being "Instances of General Use. They are those which relate to a variety of cases and occur frequently; and therefore save no small amount of labour and fresh demonstration."[107]

The notion of polychrests here recalled the values of good oeconomy. For Bacon, processes, rather than objects, remained central to the definition of polychrests, and these he must have taken in partly from household experience. They included, for example, preservation techniques, "by exclusion of whatever impedes and disturbs," such as sealing a vessel to keep things out; compressing and stretching (condensing and expanding); heating and cooling; or leaving bodies alone over time, for instance in the aging of wine to allow mixtures to separate out. Polychrests also included the regulation of motion, the "consents or aversions" of things, or some combination of all of these processes.[108]

Bacon certainly identified practical household expertise as a starting point for natural inquiry, promoting the investigation of soil and compost, orchards and gardens, kitchens and cookery, among other things. But he denied that household expertise could itself count as proper scientific knowledge, because it was too unsystematic. Bacon's project, then, was to propose a more systematized investigation into phenomena and their causes, which he called "experimental philosophy." Bacon also expected this philosophy to be accredited outside the home. This may help to explain the presence, in his utopian *New Atlantis* (1627), not only of an ideal scientific academy, Salomon's House, but also of an ideal family. His well-known utopia *New Atlantis* was organized around two distinct institutions, the "feast of the family," a ritual led by a patriarch to establish "the good estate of the family," and Salomon's House, an institution dedicated to experiment.[109] *New Atlantis* was then both an experimental and an oeconomic utopia, but kept the two apart. If the thrifty home was open-ended, in both material and social terms of who could do experiment, the academy was more closed. Salomon's House is notable for its gendered division of labor, being all male, probably on the model of the English universities, though relying on women as servants; its scale, being much greater than any individual household; and its variety of endeavors, which included artisanal, domestic, magical, and scholarly enterprises of many different kinds. Bacon clearly saw experimental philosophy as something that would be far removed from any individual's home.

Another protagonist often associated with the rise of experimental science was Samuel Hartlib, a German émigré who made it his business to exchange "intelligence" on a huge variety of experimental and natural historical observations, trials, recipes, and investigations among a circle of friends and acquaintances in the middle decades of the seventeenth century.[110] These persons included the writer on agriculture and metallurgy Gabriel Plattes, the Surrey landowners John and Mary Evelyn, and the Irish gentleman chymist Robert Boyle and Boyle's sister Katherine Jones, Lady Ranelagh. The Royal Society included numerous members of the Hartlib circle when it was founded in 1660. For Hartlib, there was little to distinguish the experiments of various householders from natural philosophy in the way that Bacon proposed. His monthly collections of ideas or *Ephemerides* gathered from various quarters included the views of men and women on everything from husbandry and chymistry to new inventions and solutions to political and religious questions. Hartlib himself compiled from other authors and published works on husbandry, as did his friend Gabriel Plattes.[111] Like Bacon, Hartlib also wished to see the creation of new state-supported institutions separate from the home to aid the growth of a new science, but these should be focused on managing and enabling exchange rather than on transforming everyday knowledge. For Hartlib's circle the everyday knowledge of the home was much more directly equivalent to "science" and "experiment"; its authority did not need to be reconfigured, as Bacon demanded. In Gabriel Plattes's utopia, *A Description of the Famous Kingdom of Macaria* (1641), the ideal scientific institution sounded much more like a regular home than Salomon's House:

> they have an house, or Colledge of experience, where they deliver out yeerly such medicines as they find out by experience; and all such as shall be able to demonstrate any experiment for the health or wealth of men, are honourably rewarded at the publike charge, by which their skill in Husbandry, Physick, and Surgerie, is most excellent.[112]

Despite this difference from Bacon, a thrifty approach to materials, achieved by diversifying the uses of things, is again evident in the Hartlib circle's writing. The introduction to Hartlib's book on husbandry began,

> He that will have profit must use the Means, they must not sit, and give aim, and wish, and repine at other increase: There must be Observation to mark how others thrive, Inclination and Imitation to do the like by endeavour and charge; and if one Experiment fail, try a second, a third, and many.[113]

The book went on to explore a variety of "domestick experiments and secrets" drawn from several authors, advocating various methods of using plows, creating orchards, cultivating crops, and so on. Experiments deployed old materials from the home in agriculture or diversified the uses of crops. Turnips, for example, could be cultivated and eaten by humans, with anything remaining given to cattle, whose dung could be used to manure the fields for the next season of growing.[114] All sorts of things could be made use of as fertilizers. "Dunging and manuring lands" was deficient in England because "they do not imploy all they know to the best use."[115] Hence old rope, wool, and linen rags, soot, urine, or pigeon and hen dung could all be used to make manure. Husbandmen could also use "*Ashes* of any kind," or "*Horse-dung*, [which] the Gardiners of *London* much commend for divers uses."[116]

The "divers uses" of things were also explored by Hartlib's acquaintances. The inventor William Wheeler wrote to physician Benjamin Worsley in 1650 that "I have found out many rare uses of the Tarras [a pumice stone normally imported from Holland] not yet knowne to the world."[117] In his *Ephemerides*, under the heading *Dispensatoria Oeconomica* Hartlib recorded various uses of Tarras such as making vessels for storing wine and beer over long periods.[118] Hartlib's circle also promoted mechanical inventions that served many uses. Henry Rivers described a wind-powered engine that could grind corn, grain, seed, wheat, oats, peas, beans, rapeseed, woad for dyers; bark for tanners; apples, pears, colors for potters, painters, and glasshouses; tile, stone, and earth for plaster; sugarcanes, ginger, spices, and snuff tobacco. "The uses it may be applyed unto are many."[119] Hartlib's correspondent John Evelyn allied such inventions with Bacon's polychrests. He called for the commemoration of any "Authors and Perfectors of . . . Polychrests, Inventions and things of universal or multifarious Use: Such as these (how trivial and mean soever the Instances may seem) would have had their Statues in my Lord Verulam's Solomons's-House."[120]

So while Hartlib and Bacon differed on the status of oeconomic activity as experimental knowledge, they both shared an approach to materials that valued the many uses of things. As in other works on domestic oeconomy, the Hartlib circle identified the household as a location for increasing the prosperity of the whole realm, through agricultural improvements in Christian husbandry and housewifery.[121] As Carl Wennerlind has shown, the Hartlib circle's confidence in the abundant potential of nature to offer "infinite Meanes of Reliefe and Comfort" to mankind led them to innovative ideas about the expansion of credit and growth in the state.[122]

Families writing on oeconomy seem to have approached the improvement of knowledge along similar lines to Hartlib. Many householders, husbands and wives, wrote about oeconomy, either in collections of recipes or manuscripts kept at home or circulated among family and friends or, less frequently, in printed publications. The boundaries of oeconomic writing in the home were fluid and open-ended, as various family members might contribute to collections of recipes and practical knowledge compiled over several generations.[123] Printed works on oeconomy tended to divide more strictly along gender lines, with most published by male authors, though women contributed a number of works on housewifery, medicine, and cooking over the seventeenth and eighteenth centuries. Like Hartlib, these men and women were much interested in the preservation and circulation of useful knowledge, and given that their manuscripts might be circulated and read extensively, print might not differ much from manuscripts in audience and distribution.[124] Although they might not argue explicitly for the circulation of "intelligence," such writers recorded, tested, and circulated natural knowledge among men and women. Their writings were also thrifty, in the sense that they also sought to diversify the uses of things and make use of what was available in a balance with the use of new things.

To take an example, a number of writers championed the use of medicines capable of curing many ills, which in the seventeenth century were referred to as panaceas or *polychrests*, likely the inspiration for Bacon's term.[125] Like Bacon's polychrests such medicaments were valued because they used a single substance to achieve many ends. One manuscript recipe recorded "A Green Oyntment good for the Scabbs Anguish, Swelling of wounds, Tooth-aches Bruises, Over reaches of the Sinews, Vaines cramp stitches either in Man or Beast . . . drops into the eares & stopp'd with black wool helpeth all paine thereof."[126] Polychrests, medical or otherwise, tended to generate long lists of applications. Most recipes, of course, were "specific," being used for one purpose, but they made use of ingredients in diverse ways to create each dish or drug, and a number of the most prized medical recipes were valued for being able to cure many ills with one medication. Such polychrests had been valued since ancient times.[127] In early modern England, writers on physick and oeconomy prescribed diverse such recipes. In 1653, Elizabeth Grey, countess of Kent, published *A Choice Manuall, or Rare and Select Secrets in Physick and Chyrurgery*, which featured the recipe for what became known as the Countess of Kent's Powder.[128] A combination of ingredients made the powder universal. They included "powdered pearls to strengthen the heart; white corral to cure bloody fluxes; hartshorne to resist poison, pestilence and rhymes [*sic*]; and saffron to defend against frenzies and weakness of sight."[129] A list served to reveal the many uses of the powder, Kent indicating it was "good against all malignant and pestilent Diseases, French Pox, Small Pox, Measles, Plague, Pestilence, malignant or scarlet Fevers, good against Melancholy, dejection of Spirits."[130] A century later, the contents of *The Country Housewife's Family Companion* (1750), a book intended to make "Money go the farthest," still centered on another universal medication, the "famous . . . Quicksilver Water, which . . . will cure the Pox, King's Evil, Leprosy, Scald-head, Itch,

Scurvy, and Mange in Beasts; so that all Persons . . . ought to drink of this Water."[131] As Alisha Rankin has noted, some of these "wonder drugs" included rare and exotic ingredients that cost a great deal, prompting criticism and controversy. Expensive or scarce ingredients did not preclude a recipe from being thrifty, however, so long as the ends justified the means.

Robert Boyle and Oeconomy

Thrift encouraged experimenting, by varying the uses of things and finding out what they could do, though authors took different views of the equivalence of such domestic ingenuity with natural philosophical knowledge. Bacon sought to mark experiment as reliant upon but superior to common practice, whereas others made less of a distinction. A thrifty approach to materials cut across these positions, seeking to make the most of materials and find their various and multiple uses. This value placed on the many uses of things would endure in the work of a new institution—the Royal Society for the Improvement of Natural Knowledge, founded in 1660. The Society, inspired partly by Hartlib's work and by Bacon's vision of Salomon's House, sought to establish a location for generating and accrediting natural knowledge outside the home. A community of men would thus meet to verify "matters of fact" demonstrated to them by a curator of experiments, who first tried the relevant experiment at home before "showing" it to the assembled fellows. Hypotheses as to the causes behind verified phenomena might then be considered. Establishing this institution proved difficult. As Shapin and Schaffer demonstrated, the new "experimental philosophy" of the Royal Society relied on the trust and authority of gentlemanly sociability to gain authority and consent.[132] Experiments brought into the Society also needed to be made spectacular, to appeal to an audience, distancing them from their domestic origins.[133] Nevertheless, the Royal Society promoted a mixture of the two positions outlined above. Its fellows engaged in correspondence and exchange of recipes and other experimental ideas, and did so with male and female relations and acquaintances. At the same time, they sought to focus their work in a distinct institution, and they excluded women from its meetings. Indeed, the Royal Society was at pains to represent experiment as a masculine pursuit. As Cowley put it in verses opening Thomas Sprat's *History of the Royal Society* (1667), "Philosophy, I say, and call it, He / For whatso'ere the Painter's Fancy be / It a Male Virtu seems to me."[134]

As in the case of Bacon and Hartlib, fellows of the Royal Society also took a thrifty approach to materials. In *Micrographia* (1665), Robert Hooke, the Society's first curator of experiments, asserted that with the aid of the new science, "the World may be assisted with variety of Inventions, new matter for Sciences may be collected, the old improv'd, and their rust rubb'd away."[135] Moreover, it was the "great prerogative of Mankind" that "we are not only able to behold the works of Nature . . . but also have the power of . . . improving them to various uses."[136] Hooke was delighted that "the mechanism of Nature is usually so excellent, that one and the same substance is adapted to serve for many ends."[137]

Exemplary of this position was the chymist and experimenter Robert Boyle, who did much to shape the Society's experimental methods. Robert Boyle was immersed in the busy knowledge exchanges of a wealthy household. He never married but in London lived with his sister, Lady Ranelagh, an active participant in the Hartlib circle and an accomplished experimenter and compiler of medical and chymical recipes.[138] Michelle DiMeo has shown how Ranelagh influenced Boyle's ethical writings and how both developed a chemical interest at the same time, discussing experiments and exchanging recipes.[139]

Both Ranelagh and Boyle were noted for their thrift, despite being extremely wealthy. Boyle praised Ranelagh for using cheap medicines such as juice of horse dung even when expensive ones were easily within her grasp. She was "a great Lady, who though . . . wont to have the attendance of the skilfullest Physitians, scruples not, upon occasion, to use . . . this homely Remedy, and prefer it to divers rich Cordials."[140] Similarly, Boyle, despite being one of the richest men in England, was praised by John Aubrey as "very temperate, and virtuous, and frugal."[141] In 1671 Boyle penned an essay entitled "Of Men's Great Ignorance of the Uses of Natural Things: or, That there is scarce any one Thing in Nature, wherof the Uses to human Life are yet thoroughly understood."[142] In the essay, part of his exploration of the "usefulness" of natural philosophy, Boyle simultaneously celebrated making use of everyday things to learn about the natural world and, like Bacon, sought to distinguish experimental philosophy from ordinary household labors. The essay included long lists of observations of mundane, everyday things that yielded useful knowledge, from eggs, water, vegetables, jars of oil, vinegar, urine, and plaster to plates of glass. He echoed exactly the oeconomic notion that things were not tied to one use but could, and should, serve many: "we are not always to suppose, that because a natural body has such an use on some occasions, the same body cannot on other occasions be employed to uses, that seem of a quite differing . . . nature."[143] His essay then made the revelation of uses a central task of experimental philosophy and marveled at the seemingly limitless potential of things to be adapted to new ends. "I doubt . . . fire, will have all its uses discovered, before the last great fire shall dissolve the frame of nature."[144]

At the same time, like Bacon before him, Boyle did not want to stress too much equivalence between everyday domestic practices and experiment. If experiment were identical to the activities of householders, it would have no legitimate place of its own. Hence, he proposed that a systematic and theoretical understanding of things was the distinguishing feature of experiment. The "vulgar" or common people failed to notice important aspects of the behavior of bodies as they were too familiar, whereas the experimenter would discover the causes behind them. Experiment was thus not the same as common oeconomic practice, but certainly relied upon it. This epistemic divide was manifested in the physical separation of the Royal Society from the home and perhaps offers one reason why the Society was at pains to represent experimental philosophy as a masculine pursuit. Male exclusivity followed the tradition of the universities and patriarchal culture, but it also echoed Bacon's division of the family and academy in *New Atlantis*. Excluding women, both physically and rhetorically, may have served to secure the epistemic exclusion of oeconomic knowledge from experimental philosophy, even as the latter relied in practice very much on the former.[145] It is notable in this regard that Lady Ranelagh was never admitted to the Royal Society, indicating the limits of Boyle's willingness to collaborate with his sister on experimental investigations.

Proponents of diverse visions of a new science and "experimental philosophy" all evoked a thrifty approach to materials in their work, even if they did not agree on whether expertise in oeconomy and household skills amounted to experimental knowledge or not. Figures ranging from Elizabeth Grey, recording her medical experiments in manuscripts, to Hartlib, seeking institutions to circulate better "intelligence," to Robert Boyle, publishing works which he identified as going beyond household expertise to discover the hidden causes behind natural phenomena, shared a common attitude to materials that sought to make the most of them by diversifying their uses or valuing their capacity to do many things.

Of course not everyone agreed that household practices warranted further investigation, whether they were experimental, scientific, or not. For most people, no doubt, household activities were much too mundane and everyday to be considered as having any link whatsoever to natural philosophy. Skeptics of both male and female experimenters quickly recognized the homely nature of their works and attacked them as trivial labors that had no place among university professors or professional physicians. Such attacks might focus on gender. University-educated physicians castigated women compilers of recipes for interfering in what should be a male province of expertise. To make medicine with mundane materials in the home was a threat to health: "the Experience . . . of phisick is dangerous . . . for the materiall subjecte of phisicke is not tilestones, or common stones, dyrt, woode or leather."[146] Satirists mocked experimenters for making science out of the mundane and everyday. Jonathan Swift's 1726 *Gulliver's Travels* famously lampooned the triviality of experimental knowledge, for generating "sun-beams out of cucumbers." Significantly, Swift located his satire not in a dedicated institution like the Royal Society's Gresham College but in a series of homes given a new use, his "Academy of Lagado" being "not an entire single building but a continuation of several houses on both sides of a street, which growing waste was purchased and applied to that use."[147] Another criticism of making mundane things into science was that it was precisely not thrifty to do so. In 1696, Mary Astell attacked "the character of a Vertuoso" for taking too much interest in trivial, everyday things and giving too much value to things people ordinarily treated as rubbish.

> He preserves carefully those Creatures, which other men industriously destroy, and cultivates sedulously those Plants, which others root up as Weeds. He is the Embalmer of deceas'd Vermin . . . A piece of Ore with a Shell in it is a greater present than if it were fine Gold, and a string of Wampompeag [Native American shell beads] is receiv'd with more joy, than a Rope of Orient pearl, or Diamonds wou'd be.[148]

For Astell, because the virtuosi wasted their lives in the idle pursuit of these trivial things, "all they teach Men is, but a specious expensive method of throwing away both Time and Money."[149] Nevertheless, even though Astell criticized it, she still equated the new science with a making use of things in the home.

Conclusion

Not all experimenters were thrifty in seventeenth-century England. Aubrey recorded of William Harvey that "For twenty years before he died he took no manner of care about his worldly concerns, but his brother Eliab, who was a very wise and prudent manager, ordered all . . . faithfully."[150] Nevertheless, in the course of the century many experimenters brought together thrift and the study of nature. This chapter has sought to reveal aspects of this thrifty attitude. Oeconomic literature offers a useful means to understand how early modern householders made sense of materials, how they valued goods and possessions and considered the question of how to make use of them. Oeconomic literature advised householders to take care of their possessions and to balance the purchase of new things with the good use of the old. Thrift and frugality were important values that promoted this balanced approach, making sure things were used well, but not eschewing the purchase of new things when necessary. To use material things well meant caring for them and getting the most out of them by "making use." A variety of experimenters in households shared this approach to materials, even as they differed on the status of household knowledge as natural philosophy. They valued machines, medicines, and materials that could be put to many uses (the more the better) and valued finding out new uses for things by means of observation and experiment. Such an approach to materials was the basis for an array of practices that characterized "thrifty science." The following chapters explore these practices in more detail, beginning with "making use" of the home itself.

TWO

Making a Home for Experiment

Historians of science have often drawn attention to the value of the home as a scientific site. Steven Shapin noted of seventeenth- and eighteenth-century science that "The overwhelming majority of experimental trials, displays, and discussions that we know about occurred within private residences."[1] Shapin revealed the social bricolage of the new science, how Robert Boyle negotiated the etiquette of spatial access among gentlemen with the needs for privacy in pursuing experimental inquiries. Gadi Algazi has explored how scholars shifted from a monastic model of celibate scholarly labors in the Middle Ages to one of occupying a family home in the Renaissance. Like Shapin, Algazi stresses how scholars needed to create a space of isolation from the family to work, "From now on, creating invisible divisions within shared domestic spaces and abstracting oneself from relationships perceived as distracting would become a central theme of scholarly discourse."[2] Deborah Harkness, in a study of the family home of John Dee in Mortlake, Surrey, also points to the tensions that family life generated between isolated scholarship and domestic sociability.[3] In contrast, Alix Cooper has emphasized the value that families had for scientific labors in the home:

> Only with the full support of the household, and in particular with the participation of family members, could many of the laborious, "Baconian" tasks of early modern science, which tended to require extensive information gathering and many years of labor, be brought to fruition.[4]

Although these studies have identified the household as an important epistemic space, they have typically explored the social roles, conventions, and tensions that characterized the pursuit of natural inquiry in the home: terms of access, privacy, and the gendered division of labor. Work on the history of recipes has highlighted the importance of the family, understood to include relatives and their servants and domestics, in collaborating on the production of knowledge.[5] While these social perspectives need to be kept in mind, more attention needs to be given to the physical house itself as a productive resource for experimental inquiry. Historians have paid attention to rooms such as the study, the library, and kitchen as spaces for experiment, but the thrifty use of these rooms and the deployment of the material culture they contained have received less attention.[6] So this chapter explores how experimenters made use of their dwellings and exploited possessions and furniture to do experiments. Here the study, kitchen, cellar, and bedroom will serve as examples, though the parlor, privy, granary, storerooms, corridors, and other spaces might yield similar stories. It may be imagined that scholars used these spaces because there was nothing better available, and in some cases this may have been true. But, as in the case of Boyle discussed in the previous chapter, it would be wrong to suppose that if experimenters had the money or resources they would construct more purpose-built laboratories and specialized instruments. While such things were being employed in this period, many experimenters set out to make use of more mundane things surrounding them, and they did this mostly at home.

As the previous chapter indicated, a diverse range of men and women undertook experiments in the home, involving a variety of practices that might or might not be deemed "natural philosophical." Access to literature on these experiments varies depending on the kinds of practices being examined, and in what follows I have relied for the most part on the many available sources relating to the experimental works of men associated with the Royal Society. On occasion I refer to the experimental works of women or the collaboration of husbands, wives, servants, and other family members, to indicate that it was not only men who practiced "thrifty science".

Making Use of the Home

Today we readily distinguish "adapted" and "dedicated" scientific spaces, the former being a site converted for something other than the original use and the latter being purpose-built or having a specialized scientific function, as in the case of a particle accelerator or an operating theater. Early moderns certainly created sites dedicated to the sciences and pursued much investigation in these spaces. In England the Royal Greenwich Observatory was constructed to conduct astronomical and navigational work in the 1670s. The Ashmolean in Oxford, built in 1683, included a chymical laboratory in its basement.[7] The Royal Society was not housed in a dedicated building, borrowing rooms in Gresham College, Arundel House, and Crane Court, but it did undertake the specialized function of accrediting knowledge claims and experiments developed elsewhere.[8]

The decision to create these sites for science no doubt depended on many things—the kind of science involved, the scale of the operation, the nature of the patronage or economic relation that demanded it, and the intellectual and epistemic goals of those who would work in it.[9] Astronomical observatories manifested a patron's power over the heavens, and this may be why numerous European monarchs sponsored purpose-built observatories over many centuries.[10] Another reason for creating buildings dedicated to astronomy was scale. Astronomy benefited from typically large, heavy instruments that had to be positioned carefully and remain steady, so a permanent location in a sturdy building could be advantageous. Chymical laboratories like the Ashmolean might also require dedicated space because the sizable furnaces needed were impracticable in a domestic dwelling.[11]

While much scholarly attention has been given to these dedicated sites, less attention has been given to sites that served several uses, and above all scholars' homes. Experimenters might travel to the Royal Society to seek credit for their work from others, but most of this work was done at home. Inside their houses, most early modern experimenters made use of rooms and their furnishings and utensils for scientific ends.

It is important to recognize that the house itself, like the laboratory or observatory, was undergoing change between the sixteenth and eighteenth centuries. Typical medieval houses were a single story in height and contained a large hall around which inhabitants went about their lives, with humans residing at one end and animals at the other. But the seventeenth and eighteenth centuries witnessed a "Great Rebuilding" in which the ceiling of the hall was converted into a second story, opening up space for more rooms.[12] Animals retreated to outbuildings, and humans gained more privacy in separated and increasingly differentiated rooms, with brick chimneys and hearths replacing the central fireplace. Householders contributed to this distinction by purchasing a growing number of new goods, from furnishings to kitchen utensils, becoming available through the growth of manufactures.[13]

Until the end of the eighteenth century, however, the house remained an open-ended, ambiguous space, inside which it was not yet entirely determined how space should be divided and which activities belonged to which room. Since several rooms contained a fireplace, for example, they might all equally serve for the preparation of food. Often townhouses had a workshop or retail shop on the ground floor that could also be a bedroom for servants.[14] As such the house recalls what the sociologist Karin Knorr-Cetina refers to as "incomplete objects," namely, those objects which are open, not yet defined, and which "appear to have the capacity to unfold indefinitely." Knorr-Cetina explains,

> The everyday viewpoint, it would seem, looks at objects from the outside . . . These objects have the character of closed boxes. In contrast, objects of knowledge appear to have the capacity to unfold indefinitely. They are more like open drawers filled with folders extending indefinitely into the depth of a dark closet. Since epistemic objects are always in the process of being materially defined, they continually acquire new properties and change the ones they have.[15]

Knorr-Cetina's "objects of knowledge" refer to entities studied in scientific research, whose identity may emerge gradually through various processes. But it could be argued that this idea is very evocative of how early moderns thought about material things, including their own homes. Rather than view houses and their contents as having some predetermined and singular function, they understood them to be open-ended and capable of revision, reworking, and repair over time.

This incompleteness is crucial to recognize, because it meant that the house was open to new practices such as the pursuit of experimental philosophy. The latter emerged at a point in time when there were a growing number of rooms in houses, providing space for experimental pursuits. These rooms were distinct enough to offer novel possibilities for use in relation to science, which possibilities will be explored below. But they were not sufficiently differentiated to restrict their use to specific purposes such as sleeping, cooking, entertaining, and so on. Consider the case of Robert Hooke's quarters in Gresham College. An inventory made after his death did not identify any room as a bedroom or kitchen, but there were beds in the parlor, a garret, a cellar, and in a room next to the stairs. There were cooking utensils in several rooms. Neither was any room designated a "laboratory," but one was called a

"Cellar or Workshop" while the parlor contained globes, lodestones, and three telescopes.[16] There was no reason in the seventeenth century not to make a room in one's home into a space for experimental pursuits, because the conventions for governing how space ought to be used were still very flexible. Over time, rooms would be identified as kitchens, bedrooms, and laboratories, and it was only later, in the nineteenth and twentieth centuries, that the idea declined of the laboratory as a possible room in the house and experimental research became an inappropriate domestic labor. At this point, as chapter 8 will relate, the laboratory might move into a separate location from the home.

So unlike the bedroom or kitchen, the laboratory ultimately "failed" as a domestic room. Prior to this, however, domestic space and scientific space were usually one and the same, as, indeed, were the various new furnishings and utensils that occupied rooms with increasing frequency across the seventeenth and eighteenth centuries. Only after scientific and domestic space were divided in the nineteenth century might it look as if rooms in the early modern home were "adapted" to science, while other buildings, such as observatories, were "dedicated." In fact, using the home and its furnishings for science was just a part of the history of domestic development.

So what made a good house for an experimental philosopher in the seventeenth century? There was no single answer, of course, as accommodation ranged from university rooms to rented lodgings to private country seats, from London townhouses to rural parish church establishments. But individuals might consider the capacity of spaces for philosophical work when they chose a home. When Robert Boyle planned to reside in London after the Great Plague in 1666, Daniel Coxe sought out ideal quarters for him. One of these, three miles from Westminster, had "a handsome front, and Courtyard, a spacious Hall wainscoted a parlour, Kitchen, & other offices below the Stairs. Above there is a faire dining room wainscoted . . . with 5 or 6 lodging Chambers, & 3 or 4 Garrotts." The annual rent was forty-four pounds a year, and there were "all desirable Conveniences for a Laboratory," though Coxe did not say what these were.[17] In the end, from 1668 Boyle lived with his sister, Lady Ranelagh, on Pall Mall, with a laboratory constructed outside by Robert Hooke.[18]

Experimenters were ingenious at adapting many kinds of space, and whatever their size, domiciles could quickly fill up with philosophical things. John Evelyn described John Wilkins's lodgings in Wadham College, Oxford, as containing a "variety of shadows, dials, perspectives, and many other artificial, mathematical, and magical curiosities, a waywiser, a thermometer, a monstrous magnet, conic and other sections, a ballance."[19] Spaces of work, of family life, and of craft were not yet distinct. In the early eighteenth century, Royal Society demonstrator and instrument maker Francis Hauksbee kept his home together with his wife, Mary, and five children at a workshop in Wine-Office Court and later Crane Court off Fleet Street.[20] From 1738, the instrument maker George Adams lived with his wife and two daughters, Sarah and Ann, in Racquet Court, near Fleet Street.[21] Adams referred to the place where he made orreries, microscopes, and other instruments as "my own House."[22] Many other forms of artisanal production took place in people's homes.

Houses provided room for philosophical conversation, performance, and display. Space might be divided along gender and class lines.[23] Although they were seen as interchangeable in the early seventeenth century, by the eighteenth a separate still room could be the province of the housewife for experiments in distilling and preparing medicines, while a laboratory might be reserved for men, though this was not always the case.[24] In the eighteenth century, besides experimental investigations, public lecture courses and tutoring took place in private residences. The lecturer Erasmus King's house in Duke's Court, on the site of what is now the National Portrait Gallery, included experimental rooms and a "philosophical museum" upstairs, while downstairs there was a lace "warehouse" or "chamber" belonging to his wife.[25] The downstairs rooms at Henry Cavendish's Clapham home included an observatory and transit room, a drawing room put into service as a laboratory, and another room acting as a forge. Servants provided household and experimental support. Cavendish's house was managed by a housekeeper, a maid, a cook, a gardener, and an instrument maker.[26]

Houses offered opportunities for individual and collaborative experiment and exchange. Men and women learned from one another. In 1666 the Hartlib circle member and fellow of the Royal Society John Beale recorded in a letter to Boyle how he learned the properties of heat (one of Bacon's polychrests).

> Last autumne, I had leave to spend a whole day in observeing the Sweete Chymistry of a Lady, Who is most perfect in the Arte of making Marmalads, & conserving fruite in all their approved colors; And she hath taught me to confesse The miracles of Heate; Tis sure the philosophers stone that can doe & undoe all things.[27]

The house was a sociable venue where individuals could come together for discussion. In the eighteenth century Isaac Newton asked Hans Sloane to have Francis Hauksbee bring an air pump to his house on Jermyn Street in London, "then I can get some philosophical persons to see his experiments, who will otherwise be difficultly got together."[28] The concentration of homes in urban locations also enabled communication and the sharing of instruments. The Royal Society often ordered experiments, dissections, and meetings to take place at fellows' homes, and had the operator, Richard Shortgrave, deliver apparatus and specimens to the relevant houses.[29] Preeminent Newtonian public lecturer and curator of experiments at the Royal Society John Theophilus Desaguliers was able to meet fellow freemasons in the Rummer and Grapes Tavern near the London home on Channel Row he shared with his wife, Joanna. Their house included a thirty-foot-long room with a fire. Desaguliers ran a school there, lecturing on experimental philosophy, relying on his wife to take over when he was away.[30]

Not all dwellings were suitable for experiment. Some enterprising chymical experimenters added a furnace to their homes.[31] But chymical works indoors could produce unwanted smoke and smells, so outside space was often more preferable. Robert Hooke built a laboratory at the back of Lady Ranelagh's home in Pall Mall, and Isaac Newton's chymical laboratory was located in the garden of his Trinity College residence.[32] However, much experimental work happened inside the home, and philosophers were adept at fashioning spaces serving their families and science. The natural theologian William Derham occupied the church and rectory of Upminster in Essex from 1689. Derham made astronomical observations from a platform erected on the church spire, and in the garden of his nearby two-story home High House he observed birds, bees, and flies. A thermometer on the wall and a clock and barometer enabled meteorological observations, and a room in the house contained collections of birds and insects.[33] William and Caroline Herschel's five-story terraced townhouse on New King Street in Bath included a workshop on the basement level leading out on one side to the garden and on the other to the kitchen. The workshop housed specialized equipment such as a furnace for making mirrors. The Herschels observed in the back garden.[34] Scientific space in the home was by no means simple or basic. In the seventeenth century the Shirley family of Staunton Harold Hall in Leicestershire kept a large still room with space for several operators that contained a screw press for extracting oils, four stills, several pairs of scales, and various tools.[35] By the end of the eighteenth century, the laboratory in Joseph Priestley's Birmingham home contained over six hundred pounds' worth of purpose-built optical, chymical, electrical, mathematical, and philosophical instrumentation.[36]

The Study

Just as scholars made use of their homes for scientific research, so they had the contents of particular rooms in the house serve various experimental ends. Changing urban and domestic conditions contributed to this. While the "Great Rebuilding" generated more rooms in the early modern home, urban expansion put pressure on space. Early modern London suffered from soaring house prices, high rents, and overcrowding, so there was a need to make good use of rooms and resources.[37]

Perhaps the room that historians have most commonly associated with scholarly purposes was the study, a distinctive space of work and contemplation to be found in the houses of clerics, lawyers, merchants, and scholars.[38] The word "stody" first appeared in English at the beginning of the fourteenth century, being a room for reading and writing often adjoined to a bedchamber or located in attic spaces above it.[39] The term derived from the Old French *estudie* from the Latin *studium*, meaning an enthusiasm to apply oneself to something. Any chamber or room could be "fitted up" as a study, but such spaces were probably restricted to persons who could afford to employ space in such a manner. Scholars used studies for reading and writing, where they could enjoy a relatively high degree of privacy compared with other rooms. The aristocracy used studies for managing their estates, while the clergy used them as places to cultivate piety and compose sermons.[40] A closet, or small room, might also be adjoined to the bedroom, as a place for prayer and a store for medicines or curiosities.[41] Wealthier persons might include a library in their homes, as Samuel Pepys did.[42]

Furniture in these spaces included shelves, tables and chairs, and of course books. Shelves were open and sometimes covered by a curtain to keep off dust. Glass-fronted bookcases appeared first at the end of the seventeenth century. Fabric draped along the top edge of a shelf might serve to brush dust off books as they were removed.[43] Pepys's library contained twelve bookcases, a cane-back chair, steps for reaching high shelves, and a map of Paris hung on the wall with ribbons.[44] Books, maps, and pictures in studies and libraries might be functional, decorative, or inspirational.[45]

The study was often identified as a space of solitude, which as Shapin wrote of the early modern laboratory was associated with the privacy of a monastic cell, free from disturbance and distraction.[46] The Czech scholar Comenius defined the study as "a place where a Student, a part from men, sitteth alone, addicted to his Studies."[47] Closets were likewise places of silent and private prayer. However, as Shapin notes, a genteel expectation that homes should be accessible to visitors often lay in tension with the desire for private contemplations.[48] The Calvinist minister John Dury complained to Samuel Hartlib that his thoughts were much disrupted on having to share his chambers with another traveler during a visit to Bremen:

> wee liue in all friendlines as wee ought together in the brotherhood of the Gospel but only you may intimat that the narrownes of our lodging doth deprive mee of that retirednes and ability to enioy myself to entertaine those secret and deepe thoughts which otherwise I could haue . . . For when hee is with mee in my study (which is as oft and long as hee pleaseth) I cannot bee vnsociable vnto him, and all society of discourses which I can haue with him is a distraction from that which I must thinke vpon to perfect my taskes.[49]

In the late eighteenth century, Henry Cavendish also suffered from social obligations to make space accessible. As a gentleman, Cavendish opened his library in Bloomsbury for others to use, but found visitors conflated access to the library with access to himself. Cavendish felt he had to stress that "what was accessible was the library and not its owner."[50]

The Kitchen

It could be concluded that the study was not in fact an ideal space for experimenting in the early modern home. Certainly it was not preferred over any other room in the house as a location for this purpose. The study was, if anything, exceptional because it was one of the few places in the house that contained relatively few furnishings to serve the development of experiments. The particular use of the study was perhaps then as a place to record experiments, either in family recipe books or as papers to be read and published at another site. The experiments recorded, however, were happening throughout the house.

One of the chief places where this happened was the kitchen. The kitchen only gradually emerged as a distinctive domestic space in the early modern period, and as Sara Pennell has shown, its contents and uses were not limited to food preparation.[51] Not yet a "backstage" site of service, the early modern kitchen was accessible, opening onto the house and the street, and probably enjoying much social traffic. Kitchens were also flexible. Medieval houses relied on a central fireplace and vent for smoke, in which a large cauldron or pot was placed directly on the fire.

But by the eighteenth century coal-fired hearths with brick chimneys had replaced central fireplaces, prompting the use of a variety of new pans and utensils for cooking with the ovens, hobs, and spits that a side hearth made possible. The kitchen thus changed from a generic living space to a room increasingly dedicated to the preparation of food.

Cookery was exemplary of early modern "making use," as men and women turned scraps and leftovers into new meals while preserving and making the best of various foodstuffs.[52] Divines evoked the Gospel of Saint John to recall the importance of making use, since "Even Christ when he had fed thousands by a miracle, yet commanded his Disciples to *gather up the broken bread or fragments, that nothing be lost.*"[53] As a transitional space, according to Pennell, the kitchen "contained within it the possibilities of transformation" marked by the continued presence of books, birdcages, clocks, weapons, and other items.[54] This flexibility suited an oeconomic attitude to materials that valued making and varying the uses of things. Francis Bacon reckoned that the kitchen would be "most . . . serviceable" to polychrest science:

> The preparation of meats, bread, and drinks, if it be well ordered and agreeable to this intention, is of very great importance. And although it be a thing mechanical and savouring of the kitchen and the cellar, yet it is worth more than the fables of gold, precious stones, and the like.[55]

Certainly, the architecture of the kitchen and its furnishing and utensils offered opportunities for ingenious experiment, while the foods and processes it contained provided subject matter for philosophical speculation. Wendy Wall has shown how women like the seventeenth-century cookery writer Susanna Packe experimented with recipes, writing after them *probatum est* (it has been tested/proved).[56] In the 1680s Elizabeth Hirst wrote that she "often & effectually experimented" her recipes.[57] Robert Boyle kept work diaries over many years containing recipes that made no distinction between chymical, experimental, medical, and culinary procedures. His 1655 diary contained receipts for perfume, artificial marble, a herbal purgative, ginger beer, and Sir Kenelm Digby's lemon sherbet.[58]

Experimenters used a variety of pots, pans, and kitchen utensils in their work. Such wares were themselves undergoing a transformation in the seventeenth and eighteenth centuries, as the development of the side hearth led to the proliferation of new cooking paraphernalia instead of a single all-purpose cauldron.[59] Medieval wooden vessels and utensils, which may have been little use for containing chemicals, were replaced by pewter, tin, and silver.[60] By the early eighteenth century, besides china and glassware, experimenters could exploit a range of new metal tea kettles, coffeepots, strainers, colanders, skewers, ladles, and gravy spoons.[61] While utensils proliferated, they were not dedicated to a single function. As the archaeologist Annie Gray has noted of early modern kitchen utensils, "even when an object is clearly intended for the kitchen, a multitude of uses may still be possible."[62] Experimenters exploited the possibilities. Desaguliers tied feathers to an upright skewer to electrify them.[63] George Adams used a tea kettle and tobacco pipe to extract fresh water from salt water.[64] The electrician Tiberius Cavallo electrified chocolate in his kitchen:

> Chocolate . . . as it cools in tin pans . . . becomes strongly electrical: when turned out of the pans, it retains for some time this property, but soon loses it by handling. Melting it again in an iron ladle, and pouring it into the tin pans as at first, will for once, or twice, renew the power; but, when the mass becomes very dry, and powdery in the ladle, the Electricity is revived no more . . . but if then a little olive oil be added . . . it will be found to have completely recovered its electric power.[65]

A brick hearth was the principal feature of any kitchen, and fire served a great array of experiments. The physician Herman Boerhaave, whose *Elements of Chemistry* appeared in English translation in 1735, identified fire as principal among "the instruments of chemistry."[66] The kitchen hearth furnished not only a steady heat, because it was the only fire in the house kept burning all day, but also variable temperatures in different parts for cooking and experimentation. A manuscript recipe book recommended that a mixture for fever should be "infused in for 24 hours on warme Embers."[67] For the Middlesex clergyman Stephen Hales, variable heat enabled a series of experiments on making sea water potable through distillation.

> These Distillations were made with great ease, in small Pint and Half-Pint Retorts; which were placed in common Pipkins filled with sand; and then placed on a Trivet over the Kitchen Fire, for half an Hour, to be gently warmed; and then, the Trivet being taken away, the Pipkin was set on a hot place on the Fire.[68]

Hales also used his kitchen bellows for experiments, while Cavallo recommended using kitchen pokers for an experiment to magnetize iron from the earth.[69] Another practice was to warm the leather cushion of an electrical machine by the fire before using it to generate a charge.[70]

The social order and tacit expertise of the kitchen also contributed to investigations. Experimenters often described putting instruments "into service" or as "serviceable." Instruments could thus provide service, and servants could serve as instruments, with no strong boundary separating the two.[71] Dudley North advised male householders, "In the choyce of particular Servants much care should be used . . . Oeconomy is an Art and every Artist ought to be curious in the choyce of his Instruments."[72] Experimenters described their servants as instruments. In the very first volume of the *Philosophical Transactions*, John Beale described how his cook generated a shiny substance from mackerel in a pot of water. If the cook's hand moved in the bowl "as the Dairy-maid do to gather the Curds for Cheese," then this increased the shininess.[73] Beale reckoned kitchen observations of cheese making, brewing, and bread kneading would lead to profound insights, telling Boyle, "If thus we began softly, & in vulgar liquids & by obvious examples, wee might possibly arrive in due time to the configurations of all stones, mettalls, & other solids."[74] Boyle's servants also contributed to experiments.

> Yesternight when I was about to go to bed, an Amanuensis of mine, accustom'd to make Observations, informed me, that one of the Servants of the house, going . . . into the Larder, was frightened by something of Luminous that she saw . . . where the meat had been hung up before: Whereupon . . . I presently sent for the meat into my Chamber, and caused it to be placed in a corner of the room capable of being made considerably dark, and then I plainly saw . . . that the joint of meat did in divers places shine.[75]

Boyle used a candle, glass, and china cup to manipulate the luminous meat and found he could read letters in the title of a copy of the *Philosophical Transactions* he had to hand by its light. In this case, moving between rooms facilitated experiment, with the kitchen providing the phenomenon and the bedroom a dark place to investigate it. Perhaps Boyle wrote up the episode in his study.

The kitchen has sometimes been characterized as a uniquely feminine space, but this was surely not the case.[76] Women, as Pennell and Leong have shown, certainly developed high levels of expertise and complex practices of recipe recording and exchange in early modern households, but husbands and other householders were not excluded from these practices. Hartlib circle member and Royal Society founding fellow John Evelyn and his wife, Mary, kept a recipe book for many decades that recorded recipes from family members and friends.[77]

The kitchen belonged to the family as it belonged to diverse oeconomic practices, an intimacy expressed in the work of Dutch painter Thomas Wijck. Wijck, who spent much time visiting England between 1663 and his death in 1677, produced many genre paintings of chymists

Figure 2.1 Thomas Wijck, *The Alchemist*, c. 1640s–1670s. Courtesy of Chemical Heritage Foundation Collections.

at work. While genre painters often used chymistry as a symbol of folly, Elisabeth Berry Drago and Michael McKeon have both argued that Wijck's paintings promoted an ideal vision of the family at work.[78] In a painting later entitled *The Alchemist*, Wijck presented a kitchen, inside which chymical and culinary materials were intermixed (fig. 2.1). A husband, in the background, pursues his studies while his wife, a young man, and a child tend to the cooking in the foreground. A taxidermied turtle hangs below a stuffed iguana just above the husband's head to symbolize the comfort and security of home. Wijck's painting was surely an argument for a certain kind of oeconomy, a vision of household order, expressed through an image of material order, where both men and women mixed the pursuit of scholarship, reading, cookery, and distilling, assisted by their children. There is barely any separation between study and kitchen. This space, and the material objects it contains, are central to Wijck's vision. Indeed, Wijck's means of expressing his argument was itself a material object, a painting and not a written text.

Of course this was an ideal, and there were limits to cooperation, at least according to the English satirists. Reports on glow worms and Boyle's claims about rotten veal led the satirist Thomas Shadwell to have his spoof experimenter Sir Nicholas Gimcrack read by the light of a leg of pork in his play *The Virtuoso*.[79] Others lampooned domestic dissections. Kitchen preparation of meat, as Anita Guerrini has shown, perhaps contributed to skills in anatomical investigations.[80] Certainly the physician and Royal Society fellow George Ent carried out dissections at his home during the 1660s.[81] But the playwright Edward Ravenscroft found such an idea absurd and satirized it in his English version of the French play *The Anatomist* of 1697. Beatrice, addressing the doctor about to perform a dissection at home, asks,

> *Bea[trice].* Why do you chuse this back Apartment at the end of the Garden? You us'd to do it [dissect bodies] in the great Hall formerly.
> *Doct[or].* My Wife will have it so, and that's enough.[82]

The Cellar

The place where Ravenscroft's anatomist left his bodies before he dissected them was the "Vault" or cellar. Cellars provided another site for experiments, offering distinctive environmental conditions that experimenters could make use of. Early modern cellars served variously as sites of storage, food preparation, or trade, with external access to the street. Some were converted into taverns or used as separate dwellings for poorer tenants.[83] Robert Hooke's "workshop" was in the cellar, and at his death it contained "A modell of an House, A pair of Smith's Bellows, some odd utensills & Lumber."[84] Cellars might also be cool, still, and free from disturbances, and were often used to store wine and fermenting beverages.

Cellars then offered resources for experimentation. Bacon prompted scholars to explore processes of coagulation, refrigeration, and conservation in deep caves in *New Atlantis*.[85] Nehemiah Grew reckoned "Biting and Hot Plants" such as radishes and onions were made sour by storing them in a corked bottle of water in a cellar for a few days.[86] To make oil of tartary (a concentrated solution of potassium carbonate) the chymical author Alethea Talbot suggested steeping the "crust that sticketh on the sides of White-wine Vessels" in vinegar before draining it in linen bags hung up in "some Cellar or moist place."[87] Robert Boyle undertook experiments on phosphorescence, the freezing of beer, and the preservation of flowers in his cellar.[88] Boyle also made the cellar itself a subject of experimental inquiry. He wrote a short essay in 1665 entitled "A Sceptical Consideration of the Heat of Cellars in Winter, and their Coldness in Summer."[89] Using his own cellar to take readings of a weather glass, Boyle argued that the temperature of a cellar depended on its depth and location, rather than the season. "Ordinary cellars and vaults" being close to the earth's surface tended to increase in temperature in the summer, whereas deeper cellars maintained a steadier temperature. This was confirmed by the reported observation "that in sharp Winters small Beer would freez in Cellars, that were not very deep, but would continue fluid in those that were."[90] Like Beale, Boyle also collaborated with women in generating philosophical knowledge of the cellar. In an essay on effluvia of 1673, he discussed the belief that beer soured during a thunderstorm, which he put down to the acidizing effect of some particles excited by thunder. Evacuating air from a vessel of beer seemed to prevent this, and Boyle noted another means to avoid the problem, "particularly employ'd by a great Lady, that is a great house-keeper, and is very curious and expert in divers Physical Observations." On "talking with her about the remedies of the Sowring of Beer," Boyle learned how the lady undertook her own thrifty experiment, "by putting, at a convenient distance, under the Barrels [of beer in her cellar], Chaffing-dishes of Coals, when she perceiv'd that the Thunder was like to begin."[91] This seemed to avoid the souring. Perhaps, supposed Boyle, the coal smoke or heat generated in this process disturbed the power of effluvia excited by the thunder, stopping it from penetrating the barrels to sour the beer.

In the eighteenth century, the conditions of the cellar remained a subject of inquiry, and the cellar continued to be a useful site for experiment. In 1741, Peter Shaw translated Herman Boerhaave's *New Method of Chemistry*, which included a section on what Boerhaave called "oeconomical chemistry." This would take in

> the several offices of a house; wherein, as in so many different laboratories, 'tis usually practiced: that is, with regard to the brew-house, the cellar, the store-room, kitchen, dairy, laundry, and their respective stores, furniture, and apparatus.[92]

Oeconomical chemistry should include an "oeconomical history of fermentation, or the management of the brew-house and the cellar," covering studies of the best methods for brewing, making wine, defending the cellar from frost, and restoring wines that had frozen.[93]

The Bedchamber

The atmospheric conditions of the cellar made it suitable for use as an experimental space, whereas the kitchen contained a diverse array of utensils and furnishings that suited experimentation. The bedchamber too served early modern householders as both a domestic and an experimental site. While the kitchen might be used during the day, and the cellar over several weeks at a time, the bedchamber was a room associated with the night. The bedchamber was a place for sleep, for sex, for being sick and recovering, for dressing and undressing, for reading and socializing, and ultimately for dying, but it became a private space only gradually, so business and study were common there.[94] John Evelyn described Robert Boyle's chamber as a philosophical site: "Glasses, pots, chemical and mathematical instruments, books and bundles of papers, did so fill and crowd his bed-chamber, that there was but just room for a few chairs; so as his whole equipage was very philosophical without formality."[95]

Bedchambers make evident the early modern concern to make the most of possessions. While seventeenth-century kitchens were beginning to fill with novel appliances, bedrooms were more likely to contain old materials. Beds, made with flock, feather, or straw mattresses, were one of the most expensive items a family could purchase, accounting for some 20–25 percent of household investment between 1550 and 1774.[96] Many people inherited their bed and passed it on to relatives, Shakespeare being only the most famous case (his will said "I give unto my wife my second best bed").[97] Bedchambers contained other old goods. An inventory of household possessions dating to 1632 listed old carpets, old blankets, an old rug, and old hangings.[98] Experimenters made good use of these things. Bedchambers were places for trying medicines. To clear one's head, Alethea Talbot recommended heating an old brick or tile on the hearth, then taking it to the bedchamber and dropping a white-wine vinegar and bean flour mixture on it, generating fumes that should be sniffed up and that also altered the air in the room.[99]

Boyle found that his bed provided a usefully dark space. To investigate the shine of a phosphoric liquor kept in a vial, he awoke before sunrise and having

> enclos'd both the Glass and my Head between the Sheets, the light seem'd to me to be very considerable, and to enlighten the compass of a foot or more in Diameter, and probably would have diffus'd itself further, if it had not been bounded by the sheets, whose whiteness made the reflection of the light from them appear very prettily.[100]

The tall testers or canopies and curtains favored on English beds in the late seventeenth century also served for experiments. In 1667, the son of Christopher Merret used the tester over his bed as a place to stand a bottle of Rhenish wine into which he had placed some steel filings. After a month the wine had soured into vinegar, but after two years, "it had recovered its former goodness in all things; only it was of the colour of claret-wine."[101] Bed making informed instrument making. An electrical machine purchased by Benjamin Franklin from Benjamin Martin in London in 1766 featured a frame of four posts similar to those of a contemporary bed, with bolt holes covered in brass mounts (see fig. 3.5). Furniture makers may have been employed to make the frames for electrical machines, orreries, and other large instruments.[102]

As rooms associated with sleep, bedchambers also contained candles, frequently made use of in early modern experiment. Oeconomic literature advised householders not to use candles too much: "Go to bed early and rise early."[103] In the seventeenth century, John Aubrey recorded how William Oughtred found it hard to do mathematics at home, since

> [his] wife was a penurious woman, and would not allow him to burn candle after supper, by which means many a good notion is lost, and many a problem unsolved; so that Mr Thomas Henshawe, when he was there, brought candle, which was great comfort to the old man.[104]

Windows

While the cellar was useful as a space that might be closed off from external disturbances, the bedchamber's value lay in part in its openness to the outside world. Bedroom windows provided a convenient place for experiments, where the light, temperature, and flows of air could be controlled more or less at will. The poorest houses had nothing more than oiled paper in a hole to serve for a window, but more common were "casement" windows, dating back to the thirteenth century, which opened out vertically at an angle. They consisted of one or two hinged iron frames with leaded glass, which were fitted into the timber window frame.[105] The window, the word deriving from "wind eye," was a place across which wind, rain, air, light, and old materials passed regularly, when waste was thrown out of them or they emitted noxious smells and vapors. Windows were also symbolic and evocative. Broken windows signified poverty, marginal status, or exclusion from the community, and the traffic across windows was social as well as material, raising possibilities for gossip and wooing or anxieties over privacy and security.[106] Scholars from Henry Oldenburg to Benjamin Franklin debated the merits of sleeping with an open or closed window. Franklin proposed that a humid room stifled perspiration, leading to "nightmares and horrors inexpressible; we fall from precipices, are assaulted by wild beasts, murderers and demons, and experience every variety of distress."[107] It was a good idea, then, to keep the window open.

Experimenters made use of the physical properties of windows. The Winchester writer Ann Shackleford gave a recipe for clear fruit cakes in her *Modern Art of Cookery Improved* (1767), a book based on "long experience, . . . oeconomy and prudence."[108] A candied fruit juice should be placed "upon glass plates, or pieces of glass" and dried in a stove or oven, "or by setting them in a window where the sun comes, keeping the window shut."[109] Isaac Newton famously made a small hole in the shutters on his window to cast a beam of light through a prism.[110] Robert Hooke used the light from his window to make observations with the microscope, varying it to achieve effects. When investigating the rainbow colors in peacock feathers, he noted how "the opening or shutting a Casement and the like, very much diversifies the appearance."[111] Hooke's theory of color was centered on an explanation of such reflective colors. The window thus provided a space where conditions could be manipulated while keeping objects safe and free from disturbances, another useful polychrest.

Outdoors

The examples above provide a sketch of the many ways early modern householders made use of their homes in experimenting. The same was true for the area immediately surrounding the house. The outdoor part of a house might include a garden or yard, a subjoining street or alley, and the roof and outbuildings of the house itself. These might be surrounded by other houses, fields, woods, or farms. Nature, of course, provided resources in the form of common land, and experimenters routinely made use of locally available materials such as wood, earth, water, and stone.[112] Much oeconomical literature discussed husbandry, or the management of outdoor estates. Husbandry made use of what would otherwise be wasteland to cultivate food, livestock, medicinal herbs, or spaces of beauty and leisure for householders. Scholars contributed studies of husbandry from Hartlib and Plattes to John Evelyn's work on forests.[113] In the eighteenth century, Cambridge professor of botany Richard Bradley translated and augmented the Lyons curate Noel Chomel's study of husbandry, the *Dictionaire Oeconomique: or, the Family Dictionary*.[114] Bradley included recipes for breeding and feeding livestock, improving land, farming, and preparing food and drink. The book was much augmented with references to the works of Evelyn, Hartlib, and their circle.

The importance of the garden for natural philosophy, natural history, botany and other sciences has been well documented and needs little treatment here.[115] Suffice it to say that gardens provided the whole family with opportunities and resources for experiment. Aristocratic ladies used ingredients gathered in gardens to make medicines, which they distributed to the poor, an act of liberality afforded by good thrift.[116] Gardens also provided places for communication and collaboration. The Elizabethan gentlewoman Margaret Hoby took walks in local fields accompanied by her maids, planted herbs, fruit, and vegetables for distilling and cookery, and indoors had herbals read to her by a Mrs. Brutnell.[117] Men also busied themselves with the garden. Newton spent time in his garden in Trinity College and the chymical laboratory it was home to. As Humphrey Newton recalled, Newton "was very Curious in his Garden, which was never out of Order, in which he would, at some seldom Times, take a short Walk or two, not enduring to see a weed in it; On the left end of the Garden, was his Elaboratory."[118] Here Newton explored a cosmology founded on seeds, organic growth, putrefaction, and "vegetable spirits."[119]

Gardens might be places of solitude. A lodging recommended to Boyle included an orchard with "a walk for solitary meditations."[120] Gardens were also locations for interactions with animals such as mice, rats, frogs, chickens, cats, dogs, and horses.[121] Insects and animals passed in and out of the house, sometimes as companions, sometimes as resources, and sometimes as pests. Numerous works on husbandry discussed bee keeping and honey making, and men and women both kept bees.[122] Goose feathers, bedbugs, lice, and fleas provided subjects for researches with the microscope.[123] Birds and mice were placed in the receiver of the air pump to study respiration. The poet Anna Letitia Barbauld famously versified over a mouse "found in the Trap where he had been confin'd all night" by Joseph Priestley for use in a chymical experiment.[124]

Another much-used part of the house for experiment was the roof. Roofs, usually sloped to keep off rain, varied regionally across England but were typically made with thatch, stone slates, or wooden or ceramic tiles.[125] They were important as sources of water for experimentation and served as foundations for lightning rods and observing platforms. In 1670 Robert Hooke made a hole in the roof of his Gresham quarters to set up a zenith telescope in an unsuccessful attempt to observe annual parallax, which would serve to prove the Copernican theory. The approach was repeated by James Bradley and Samuel Molyneux in the 1720s, again without success.[126] Experiments with lightning rods sent many a philosopher up to the roof. As with investigations in other parts of the house, this might be work shared by the family. Irritated at missing two good thunderstorms after raising a lightning rod on the side of his house in Philadelphia, Benjamin Franklin asked for assistance:

> I had given orders in my family, that if the bells [which rang when the rod was electrified] rang when I was from home, they should catch some of the lightning for me in electrical phials, and they did so, yet it was mostly dissipated before my return.[127]

The outdoor spaces around the home were thus important locations for experiment and observation. The same might be said, finally, of outdoor spaces further away. Natural historians, insofar as they studied nature "in the field," might travel far from home, often to distant overseas locations. Leaving home could prompt as much thrifty practice as staying inside it, since travelers needed to improvise arrangements and adapt and make use of locally available resources to live on the move. The apothecary Robert Dossie proposed an "expedient" form of botanical illustration of making impressions of leaves on paper with candle smoke, a method "valuable, on account of its being practicable at almost all seasons, and in all places, within the time that the leaves will keep fresh and plump."[128] Naturalists also improvised by making a meal of flora and fauna. In the eighteenth century Joseph Banks studied, and then ate, kangaroo, turtle, and stingray on his voyage with Captain Cook.[129] Eventually, in 1855, the Victorian scientist Francis Galton compiled an entire book on how to improvise on scientific voyages, entitled *The Art of Travel, or, Shifts and Contrivances Available in Wild Countries*, filled with thrifty advice on making do when far from home.[130]

Conclusion

Early modern homes were bustling sites of experimentation involving all members of the family. Different rooms and their material contents offered distinctive opportunities for making use of possessions and led to many experimental investigations of a range of phenomena including phosphorescence, electricity, the nature of generation, magnetism, anatomy, heat and cold, fermentation, crystallization, and putrefaction. These might be undertaken in the various rooms and spaces of the home, then written up in a study, closet, or bedroom in manuscript books of recipes or as papers to be taken off for reading at the Royal Society. This arrangement emerged as the architecture of the home was undergoing a transformation. The medieval home had little inside it for experimentation, but early modern houses contained diverse spaces and things that could be put into experimental service. At this time, rooms and their furnishings were "incomplete," their functions not yet constrained enough to hinder their use for science. Different rooms, moving between rooms, or the whole house could be useful in different ways. While the cellar was valuable for its environmental conditions, the kitchen and bedchamber contained serviceable utensils and furnishings. While the kitchen hearth offered steady temperatures and sources of heat, windows onto the bedchamber and other rooms enabled philosophers to vary conditions of light, heat, and wind inside the house. Richard Sorrenson has proposed that the early modern ship was a form of scientific instrument, whose peculiar conditions and contents could be exploited to make natural knowledge.[131] It might be argued that the early modern home, often compared to a ship, was another kind of scientific instrument.[132] Experimental households used the home as an instrument through which they could manage the ingress and egress of people and material objects, foodstuffs, animals, plants, light, heat, water, humidity, and even electricity, using lightning rods. This was no simple matter, and the material and social conditions of such actions could be difficult to control.

The engagement of experimenters with their homes also had an impact on houses themselves. To thrive, men and women constantly experimented to improve and enhance their own homes. Some contributed to substantial transformations in the design of houses. Thomas Sprat wrote in 1667 of the Royal Society's desire to contribute to rebuilding in the wake of the Great Fire of London:

> This . . . is the fittest Season for men to apply their thoughts, to the improving of the materials of building, and to the inventing of better models, for Houses, Roofs, Chimnies, Conduits, Wharfs, and Streets: all which have been already under the consideration of the Royal Society.[133]

Christopher Wren and Robert Hooke then contributed much to the reconstruction of London after the Great Fire. Before the fire, timber houses protruded over haphazard streets, but afterward, symmetrical, classical-styled terraced housing was introduced together with stricter building codes (though these tended to be ignored). Thatched and timbered constructions were superseded gradually by brick, stone, and tile.[134] Hooke also contributed a new style of window. He is credited, along with Thomas Kinward, master joiner in the Office of the King's Works, with inventing, or at least mechanizing, the sash window in the 1670s, which replaced casement windows in new buildings constructed after the fire.[135] These vertically opening windows relied on a counterbalance system of weights and pulleys to rise and fall, and used large panes of crown glass that let in more light than the smaller, lead-camed casements. After Wren used sash windows for all new royal palace designs, they became a fashionable and distinctive feature of Georgian houses, celebrated as symbols of English inventiveness by Samuel Johnson, and satirized by Swift as unnecessarily cumbersome mechanisms.[136]

Historians have tended to view the interiors of these fashionable Georgian terraces as the sites of "polite science," as spaces where science was consumed in the eighteenth century in the form of decorative instruments such as orreries, globes, telescopes, and other "conversation pieces."[137] Undoubtedly, such dedicated instruments did become common in people's houses in the period, extending the list of utensils and furnishings that were increasingly available to bring into the home. But the purchase of such items was only part of the story of how the home was related to scientific inquiry. The house, with all its rooms and furniture, was itself a productive instrument for making natural knowledge.

THREE

Shifty Science: How to Make Use of Things

In 1759 the former financial administrator Robert Symmer argued for a two-fluid theory of electricity based on experiments with his socks. Symmer reported how removing his silk stockings made a "crackling or snapping noise" and in the dark the stockings would emit "sparks of fire."[1] After experimenting with what he perceived as an electrical phenomenon, Symmer found that when a combination of black silk stockings (which he wore because he was mourning his wife's recent death) and white stockings worn together for ten minutes were removed from his leg, they showed no electrical effects while united. But when separated they inflated and were attracted together from some distance. When they touched, their electrified state subsided and they deflated, but separation restored it again, reinflating them. Subsequent investigations made use of Symmer's room, since he found the separated stockings would adhere to paper hangings (wallpaper), the painted boards on the walls, and a looking glass. He concluded that electrical phenomena might entail two "distinct and counteracting powers," one negative and one positive, that normally balanced one another out.[2]

Symmer's findings were published in the *Philosophical Transactions* of 1759, but he admitted that "The Homely Apparatus of Stockings was enough to disgust the Delicacy of more than one Philosopher."[3] Not everyone valued making household objects into experiments, though the criticism was perhaps more to do with using stockings, which might be considered indelicate, than with using household items in general. After all, Symmer's use of a household item to make experiments was typical of his time. The previous chapter showed how different rooms and their contents offered possibilities for experimenters to make good use of things. But what did making use of things actually involve? This chapter explores in more detail a variety of playful, creative, and ingenious practices through which experimenters put objects into scientific service. Householders appropriated materials for experiment using a range of techniques reflecting good oeconomy. Many of these depended on a form of improvisation or "shift," as early moderns had it. This chapter argues that thrifty science was often shifty science.

These practices were rarely articulated: Symmer never said he "adapted" stockings to experiment, he just used them. In part this may have reflected the ubiquity of practices that were taken for granted. But it also followed the notion that material objects were "incomplete," open to revision and changes in use, with little in the way of a predetermined function. This means that care must be taken in the language used to describe early modern practice. The language of adapting, reuse, recycling, and repurposing supposes that something has been produced for a particular purpose, and when it is no longer capable of serving this use, it may be put to another. But an incomplete object had no strongly designated use from the outset, or at least its uses were always comprehended as potentially diverse and open to change. The vocabulary of seventeenth- and eighteenth-century material practices reflected this by referring not to use but to types of action enacted on a material. The making of *shifts* entailed improvisation, using things expediently to achieve some end, while *alteration* entailed taking some form and changing it, ideally to improve its functioning. *Substitution* was a form of shift involving the use of one thing to stand in for something else. There were many other practices that were not given explicit names but are identifiable if we use modern terminology—practices such as reuse and repurposing. What follows, then, is a survey of these techniques, which pays careful attention to language. Throughout it should be kept in mind that while it is analytically useful to discuss practices of, e.g., repurposing in the seventeenth and eighteenth centuries, no such distinction was made by contemporaries, because objects were understood to be changing their purpose all the time.

Sartorial Oeconomy

Symmer used stockings for experiment, and in fact the practices surrounding clothing in early modern homes help to illustrate the diverse ways in which material objects might be put to good use. Working with clothes was one part of the experimental household described in chapter 1, and making and remaking garments was an occasion for much ingenuity and expertise. In early modern England, it was not a common practice to go to a shop and purchase a predesigned item of clothing of a standard size, then wear it until it was no longer wanted before throwing it away. Certainly some ready-made clothes were bought, but men and women (though it was often considered women's work) also purchased fabric from tailors, made clothes themselves or had someone else make them, and then altered them over time.[4] Clothes were worn many times. Washing and relining clothes kept them fresh. Housewives, or the seamstresses they employed, also took clothes apart to be made serviceable again as a different kind of garment (one account called this "new making," but today it might be called repurposing).[5] This was common for men and women. Samuel Pepys, president of the Royal Society, wore a "grey cloth suit and faced white coat, made of one of my wife's petticoats."[6] Clothes could be altered to fit changing body shapes and waistlines or to suit the latest fashion. Men and women also decorated and accessorized clothing with all sorts of buttons, brooches, edgings, mittens, caps, ruffles, and buckles.[7] These could be purchased separately and added to clothes according to the fancy of the maker, so that what today might be called a *bricolage* of different elements made up each garment. Damage to clothes could be repaired by patching, turning, or relining.[8] Clothes were put into service even when they could no longer be worn. Parts of several old garments might be removed and stitched together to make a new one, or rags could be put to uses such as paper making, a practice that today would be called "recycling."[9]

Clothes, then, are good examples of incomplete objects. They were understood to change and be transformed over time and to serve different uses in different circumstances. There was no neat division here of "production" and "consumption," since things were being made and used and remade again all the time, often over long periods and across diverse sites. In practice the various techniques just described might be closely interrelated. Hence these instructions to the merchants Robert Crary & Co. from George Washington in 1760: "Mrs. Washington sends home a green sack [a form of dress] to get cleaned, or fresh dyed of the same colour; made up into a handsome sack again would be her choice but if the cloth wont afford that, then to be thrown into a genteel night gown."[10] Such techniques were pursued not just to save money. Edwina Ehrman notes that Martha Dodson, who kept detailed accounts of her wardrobe in the eighteenth century, went to great trouble to maintain and rework items of her clothing despite their being inexpensive.[11] The ways women and men experimented with their clothes applied to many other common and household objects. Cheap clay tobacco pipes were made use of in various ways.[12] Another "incomplete object" was the household manuscript recipe book, which the whole family might contribute to and which was constantly being revised and transformed, sometimes over generations.[13]

The contention of this chapter is that experimenters approached instruments and apparatus as similarly "incomplete." Concerned to make the most of material things, they constructed and reconstructed instruments out of materials ready to hand (like stockings) and engaged in diverse practices akin to those they used to materially manage garments and other possessions. As has been argued previously, these practices should be seen not as an alternative to but as complementary to the purchase and use of specialized equipment. As Symmer's critics indicate, not everyone thought "homely" improvisation was appropriate, and experimenters did go to instrument makers to buy specialized instruments like air pumps, telescopes, stills, and microscopes. However, like clothes, these instruments were also subject to reworking and revision and were only a part of larger, more "shifty" experimental apparatus.

Using Again and Converting

The term "re-use" dates to 1797, when it was employed in a digest of laws summarizing acts against the "re-using" of stamps passed in 1772.[14] It is notable that the word "re-use" emerged as manufactures were expanding and prompting a stronger sense of objects as commodities produced for some designated purpose. Technically, then, nothing was "re-used" before about 1800, though people did "use again" or "use repeatedly." For something to be used again, it needed to maintain relevant properties across several uses. It might need to be manipulated to make this possible. "The lute, which has served for one distillation, may be used again, after the burned or decomposed parts have been separated."[15] Different thrifty practices might conflict in determining the value of using something again. A concern over cleanliness led householders to throw away clothes. In her *Lady's Complete Guide* of 1788, Mary Cole, cook to the earl of Drogheda, insisted that garments belonging to persons suffering from "the Itch" were "not to be used again, unless they have been fumigated with brimstone, and thoroughly cleaned, otherwise they will communicate the infection anew."[16]

Material limitations could also dissuade people from using something again. The archaeological excavation of the Ashmolean chymical laboratory in Oxford in 1999 found fragments of cheaply made crucibles whose chemical contents had fused with the container. Since the crucibles had lost their property of providing a "neutral" container for further reactions, they lost their use, and in all probability this led users to throw them away into a rubbish pile outside the window of the laboratory.[17] Crucibles were early modern "disposable" goods. Nevertheless, it was common for instruments to be "used again."

Closely related to using again was what might today be called "repurposing." Once again, the term "repurpose" is very recent, originating in the 1980s in discussions of recording media such as videodiscs and CD-ROMs (e.g., could content from one medium be transferred into another or "repurposed" to save costs?).[18] Since then it has taken on a broader meaning of the adaptation of something to a different use. As the previous chapter showed, repurposing was a regular feature of early modern experiment, though the term was not of course used. Occasionally early moderns spoke of "converting" something "common" (meaning in regular use in the home) to a new use, but this was rare. In practice, the

Figure 3.1 A pipkin stands next to a mortar and pestle on the floor of an alchemist's laboratory, in a detail of the anonymous Dutch painting *Trouble Comes to the Alchemist*, oil on canvas, 17th century. Courtesy of the Chemical Heritage Collections.

same item was simply used for many different purposes. Lady Ayscough used a pipkin, a form of glazed earthenware pot, to prepare "a Rare and Excellent Salve to heal all" in her 1692 book of recipes (fig. 3.1).[19] Robert Boyle, helped by his "Domesticks," used a "large new Pipkin" to investigate the refrigeration of water by mixing it with saltpeter.[20] Hannah Woolley gave a recipe in her *Cook's Guide* (1664) to boil a duck by immersing it in gravy in a "pipkin with a few currans."[21]

Making use was just what was done. To achieve it involved the adaptation of the properties of a thing to a new service to which they were all or partly fitted. The instruments that appeared in the first volume of the Royal Society's *Philosophical Transactions*, published in 1665, included familiar items such as the telescope, barometer, and thermometer. But there were repurposed objects too. In the first set of experiments to be reported by the Royal Society, Thomas Henshaw described manipulating dew with linen cloth, a spoon, drinking glasses, earthenware basins, and a "Jarre-Glass" all kept variously in the garden or in casement windows. Henshaw and other fellows expected to learn about the origins and sustenance of organic life from these experiments.[22] Women experimenting with medical recipes also made use of household goods for the purpose. As Jennifer Stine notes, "Much of the equipment required by the recipes would have been close at hand in a large kitchen, including skillets, a chafing dish of coals, an oven, a strainer, a colander, a mortar, a kettle and a knife."[23] In the eighteenth century Joseph Priestley often made use of household utensils as chymical apparatus which served in his discovery of new airs. An earthen trough could be put into service to collect gases by using a container "commonly used for washing linen," which he passed on to a friend when he was done with it.[24] Priestley used gun barrels and broken tobacco pipes to extract air from solids and carried jars of air around immersed in water on "common tea-dishes."[25] Elizabeth Fulhame performed a series of experiments to create gold cloth by chemical treatment in the 1790s. Although her husband was skeptical of the possibility, she persevered and published a book on the subject in 1794.[26] Fulhame used "a few glass vessels for the solution of metals . . . The cheapest and the most simple of those described by Dr. Priestley answered my purpose."[27] Fulhame used her fireplace to dry silk samples, and after placing them in "Dr. Nooth's machines for combining carbonic acid with water" kept them "in a dark closet, to exclude the action of light" for several months.[28]

These examples indicate that the properties of an object needed to be seen as relevant to the task at hand for a person to make use of it; for instance the darkness of the closet made it ideal for storing Dr. Nooth's machine. For experiment many vessels—tobacco pipes, spoons, cloth, basins—were valued because of their capacity to carry fluids from one place to another. The capacity of a tea dish and jar to make an airtight container enabled them to be used to carry airs. Often, however, the capacities of available objects shaped the possibilities for experiment in unanticipated ways. John Beale noticed how a cooking pot of mackerel generated a shiny substance: "the Children took drops in their hands, as broad as a penny, running with them about the house."[29] Thus an existing arrangement generated a new phenomenon, which the experimenter then exploited (as did his children). Such actions served thrift, which demanded a balance of good use and expenditure. Too much repurposing might be considered ill-advised. William Henly informed John Canton of the electrician Benjamin Wilson, "All Mr W's kitchen utensils are converted into Leyden bottles. If so, I should not much care to dine with him."[30]

Through the process of being converted to new purposes, domestic utensils might evolve over time into dedicated philosophical instruments. The example of Newton's prisms is well-known, the prism beginning life as a toy Newton picked up at the Stourbridge fair near Cambridge. While the original prisms contained bubbles and imperfections, Newton developed them into highly refined glassware, prompting much trouble in debates over replication of his optical experiments.[31] Another example would be the porringer, a shallow bowl with one handle for eating gruel or porridge, first appearing in the early seventeenth century, which from that time was occasionally used for bloodletting and related experiments.[32] By the eighteenth century, dedicated pewter "bleeding-bowl" porringers were made exclusively for bloodletting, with rings graduated on the inside to measure the quantity of blood taken.[33] Pap boats, spout cups, and caudle cups also transitioned from domestic to a more specialized medical use as feeders in the eighteenth century.[34]

Another material commonly made use of by early moderns was paper. Homes contained a variety of paper ephemera including trade cards, bills and receipts, notebooks, letters, envelopes, and playing cards, all of which could be turned to other uses. Playing cards and message cards were printed only on one side and were often made use of. John Evelyn drew designs for chymical apparatus on an old playing card (fig. 3.2).

Edward Gibbon organized his library by making one of the earliest card catalogs, writing entries on the backs of playing cards, which continued to be used for various card catalogs into the nineteenth century.[35] Card was also used to make reusable notebooks. The Yorkshire physician and arachnologist Martin Lister kept a notebook with "reusable pages

Figure 3.2 Chymical furnace designs drawn by John Evelyn on a playing card. Thanks to Vera Keller for this reference. © The British Library Board, ADD MS 78345.

of blank tables" written on with pencil, which he suggested rubbing out with "a piece of new bread . . . to take it clean off."[36]

"Waste paper," the term for sheets of paper left over from the process of binding quires of paper for sale by printers, was another type of paper given diverse uses by early moderns.[37] It served to protect newly printed books fresh off the press and could be turned to various philosophical ends.[38] On the first voyage of Captain Cook, between 1769 and 1771, Joseph Banks dried, pressed, and protected some three hundred and fifty botanical specimens between the pages of the *Spectator*'s printing of *Paradise Lost*, waste paper purchased from a London printer (fig. 3.3).[39] Banks reckoned, "Our Plants dry better in Paper Books than in Sand [i.e., being laid out in the sun on a beach]."[40] Old books and the papers of deceased scholars were also repurposed if they were not saved for posterity. As Anna-Marie Roos has noted, Martin Lister's papers were destined to be used for wrapping groceries and pastry cooking before John Fothergill purchased them in 1768.[41] Medieval manuscripts were used to cover books, to wrap gloves, or to bung holes in barrels.[42]

Paper was reused not necessarily because it was expensive—prices varied greatly depending on the size and quality required.[43] Rather, as with other goods, scholars simply made the most of it. Mathematicians were instructed to do calculations on pieces of scrap paper.[44] Over a lifetime, the mathematician John Pell filled both sides of a multitude of used paper envelopes, old letters, and scraps with tiny calculations, presumably to make use of every sheet. The normal conventions of writing disappeared when scraps were used to write on. Early modern paper scraps recovered from beneath the floorboards of the Museum of the History of Science in Oxford contain a mixture of calculations, doodles, scribbles, and notes drawn at all sorts of angles and scales.[45] Pell drew bubble-shaped lines around calculations to separate them in the cramped space of each scrap.[46] Alternatively, mathematicians might work with no paper at all. Aubrey wrote of William Oughtred that "his head was always working. He would draw lines and diagrams on the dust."[47] Navigators were instructed to practice additions until they could be done "in the Mind without setting them on waste paper."[48] Names and calculations in pencil and chalk crisscross the interior of the lid of an octant case made in New York in 1794, evidence of its use as a convenient calculating space (fig. 3.4).[49]

Figure 3.3 "Paradise Lost"—waste paper used for drying and storing plants by Joseph Banks on Captain James Cook's first voyage around the world, c. 1768–71. Barcode BM001121503. Photograph by the author. Courtesy of the Trustees of the Natural History Museum, London.

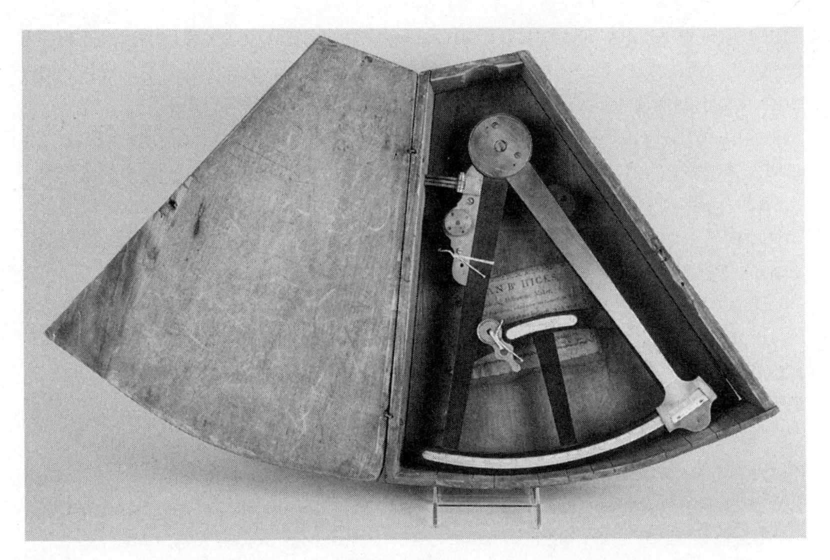

Figure 3.4 Octant case, by Adrian Br. Hicks, New York, c.1794. Harvard University Collection of Historical Scientific Instruments. The lid served as a convenient alternative to paper for calculations.

Shifts

A familiar slogan in Britain during the Second World War asked people to "make do and mend." Householders should reduce purchases of new goods and instead make more of what they already had by darning socks, fixing broken chairs, or making clothes from sheets and pillowcases, for example. The phrase "make do" was unknown in the seventeenth and eighteenth centuries, but the term "make shift" was used to much the same purpose. In early modern English usage, a "shift" entailed improvisation, the opportunistic use of some device or contrivance to effect a purpose, often when a better option was temporarily unavailable or inconvenient. To "make shift" (from which we derive the adjective "makeshift") meant to improvise or manage with what was available. For the poor, as Steve Hindle has shown, "making shift" or managing to get by was a necessity.[50] But for the middling sort, shifts were expedient actions that served a useful purpose. Printed books of household oeconomy reckoned making shift was a sensible, thrifty practice:

Ill husbandrye taketh
& spendith vp all,
good husbandrye maketh
good shift with a small.[51]

The value of shifts was often contested. Across the seventeenth and eighteenth centuries, they were described variously as artful improvisation and as trickery and deceit. In his English dictionary of 1789 Thomas Sheridan identified a "shift" as equivalent to "expedient" as a "means to an end contrived in an exigence" (the term "hack" has come to signal this today).[52] But he also defined "shift" as an "expedient found or used with difficulty, difficult means, mean refuge, last recourse, fraud, artifice; evasion, elusory practice."[53] Shifts, in other words, could be "shifty."

Experimenters praised shifts. John Evelyn recommended the Dutch for their ingenuity in dealing with shortages of fuel. In knowing "by what shifts our Neighbours the Hollanders, do yet repair that defect," the English would

> be invited to exercise their ingenuity: For besides the Dung of Beasts, and the Peat and Turf for their Chimneys . . . they make use of Stoves . . . and truly the more frequent use of those inventions in our great, wasting Cities . . . would be an extraordinary expedient of husbanding our fuel.[54]

Others scoffed that shifts were tricks. Daniel Defoe wrote how "The brickmakers all about London, do mix sea-coal-ashes . . . with their clay . . . and by that shift save eight chaldron of coals out of eleven."[55]

Other terms help to clarify what shifts entailed. Some might involve *improvisation*, either out of necessity, when something was unavailable, or just to see what happened. Others involved the contrivance or *bricolage* of several items together to create something new. To "make a shift" also meant to manage or get by. Meaning "I hope I'll get by," the natural historian John Ray wrote, "I hope I shall make a shift to rub through well enough."[56] Improvisation, in the sense of making do, was necessary in the face of scarcity. Experimenters were definitely "shifty." Henry Oldenburg grumbled to Boyle over the great personal cost to him of editing the *Philosophical Transactions of the Royal Society*, "How strangely therefore I must needs shift for my subsistence, and with what distraction I must performe my tedious work, let any sober man judge."[57] Experimenters offered ingenious means of improvisation. A manuscript recipe book noted that to measure out medical powders for consumption, "take as much of these as will lye upon six pence."[58] Robert Hooke noted of glowworms that they "shine some of them so Vividly, that one may make a shift to read by them."[59] In 1618, the archdeacon of Salisbury William Barlow described a magnetic sailor's compass that could be used to measure the height of the sun. Barlow explained that while this normally required a specialized instrument,

If a man haue no exact instrument to take the height of the sunne, hee may make some reasonable shift, by setting the little bead towards the top of the string, that it may in the forenoone cast a shadow vpon the Fiduciall line of the ruler; and marking that place with any thing (as inke or blacke lead) in the afternoone attend, vntill the shadow of the bead (the bead remaining still where it was) fall vpon that place.[60]

Scarcity of time, labor, or materials was a common motivation for shifts. In *The Art of Distillation* (1653) the chymist and physician John French explained that "in defect of a Furnace, or fit matter to make one, we may use a Kettle or a Pot set upon a Trevet." Like other practitioners of thrifty science, French did not think shifts needed to occur only as a response to scarcity. "The truth is, a good Artist will make any shift, yea and in half a dayes time make a Furnace or something equivalent to it for any operations."[61] The ability to improvise was a useful skill in any situation.

Another kind of shift was "substitution," the use of one material to replace another, to make up for a missing or expensive material or to vary a recipe. The chemical experimenter Elizabeth Fulhame spent much time seeking a chemically produced substitute for gold cloth:

I saw a piece of gold stuff made for the late King of Spain, which was of a purple colour, with gold wire shining through it . . . and it was much admired: I was so struck with its beauty, that I attempted to imitate it on a small bit of white silk; and succeeded, having produced a beautiful purple colour, with gold beaming through it.[62]

As Ayesha Mukherjee has shown, substitution was periodically invoked during times of dearth, offering alternative foods when crops failed, for example. Sir Hugh Platt proposed substitutes for bread in his book on famine of 1596.[63] But even outside periods of dearth or famine, substitution was a common experimental practice, for example in the preparation of medicaments and chymistry. The Elizabethan chymist Grace Mildmay identified earthenware vessels as a cheaper alternative to a glass alembic in chymical operations.[64] The tradition of using substitutes when a medical ingredient was unavailable had ancient roots.[65]

Substitutes could be used as long as relevant properties remained, though determining these might be difficult. In the sixteenth century, astronomers printed paper instruments for pedagogical and calculating purposes. Although these lacked the solidity of wood or brass instruments, they were often sufficiently accurate for many tasks.[66] Substitution was a form of experimentation itself, to see what happened if an alternative material was used in some process, and was also common when certain apparatus was unavailable or inappropriate. In the seventeenth century Newton substituted arsenic for silver in experiments to improve telescopic mirrors because silver's tendency to bubble created distortions in the mirrors. As Anna-Marie Roos has shown, the result was that Newton "discovered the proportions for a good metal alloy that was used [in telescope-making] for the next two hundred years."[67]

Substitution might be undertaken as a short cut, and then shaded into adulteration, though judgments over the boundary were fraught. In the eighteenth century, the apothecary and chymist Robert Dossie gave advice on detecting improper substitutions in medicines.[68] He recorded how mercury was adulterated with tin or lead, detectable by straining it through leather through which the tin or lead would not pass. Adulteration of oil of cloves with oil of olives could be detected by standing the mixture in a cold cellar where the unmolested oils would separate out.[69] It was sometimes not obvious what counted as adulteration and what as experimental substitution. In the early nineteenth century the Royal Society fellow Humphrey Jackson, author of a work on detecting adulteration in beer and wine, was himself berated for adulterating beer, even though Jackson envisaged his work as experimenting with substitutes.[70]

Substitution beyond the Metropolis

A brief excursion is necessary at this point to consider a common assumption about substitution. Historians of scientific instruments have discussed substitution as a distinctive feature of instrument making in Britain's colonies in the seventeenth and eighteenth centuries. For Silvio Bedini, instrument makers in the American colonies were forced to be ingenious "thinkers and tinkers," using substitute materials to make instruments because they did not have access to skills and raw materials available in the metropolis.[71] Similarly, those working in the English provinces have been seen as needing to tinker and substitute more than those in the metropolis.[72]

Undoubtedly, distance from urban centers could be associated with scarcity and the need to make shift with substitutes. In the eighteenth century, discussing the use of rain gauges, the English agriculturalist William Marshall wrote that "gentlemen who reside at a distance from the metropolis may find it difficult to procure a mathematical apparatus" and therefore might make their own rain gauges consisting "of a common Tin Tunnel to collect the Rain-Water; a common glass bottle to receive it; and a small China vessel to measure it in."[73] In New York the physician

T. Gale described an electrical machine made with a "common decanter," probably an onion-shaped glass bottle, as a glass receiver for generating a charge, because specialized glassware was unavailable.[74] The University of Pennsylvania professor of chemistry James Woodhouse noted in advice to would-be chemists in 1802, "Retorts are also procured with difficulty, at this time, even in the great cities of the United States. It is of great consequence then to procure a substitute for them."[75]

Bedini draws attention to the acute need in the American colonies to substitute scarce imports with more accessible materials. Instruments were in short supply, and makers of navigational and mathematical instruments used available materials and skills to copy imported examples. Brass was not available in the colonies until the early nineteenth century. In seventeenth-century Newfoundland, the earliest colonial sundials were made with slate instead of brass.[76] Alternatively, artisans collected old brass and copper utensils and melted them down with a mixture of calamine (zinc carbonate) rather than pure zinc metal, a process known as cementation.[77] The inventory of the workshop of Benjamin King of Salem included nearly eighty pounds of old brass, valued at $15.20, and six dollars' worth of old iron.[78] Iron could also be substituted for brass. The first telescopic instrument made in America, David Rittenhouse's transit instrument of 1769, used iron for what were commonly brass arms, and employed lenses taken from an imported English instrument, since no optical glass was produced in America before the nineteenth century.[79]

The scarcity of other substances prompted ingenious substitutions. Tropical hardwoods used in instrument making such as lignum vitae, rosewood, mahogany, and ebony could be obtained in North America only by importing them from Britain, and this was expensive. While some became available from merchants dealing directly with the West Indies in the eighteenth century, American makers experimented with substitutes, including black walnut, rock maple, pearwood, wild cherry, apple wood, and curled maple.[80] Makers such as Daniel Burnap and Daniel King of Salem also substituted wood for brass, producing wooden surveying compasses, a uniquely American instrument, from the 1740s to the 1820s.[81]

Bedini argued that this scarcity of materials had profound implications for the material culture of science in America, because it "forced the evolution of a new artistic style . . . devoid of minute detail and somewhat austere because of its basic simplicity."[82] Furthermore, "the need to use substitutes for traditional materials led to innovation of design."[83]

When dies and metal type were unavailable, for example, American makers engraved their instruments, "giving an instrument an unusual attractiveness."[84] Scarcity also led to standardization. Unlike European designs, American wooden marine and surveying compasses were constructed with the same diameter, so that the same compass cards could be used for both types of instrument.[85]

Although Bedini is right to point out the problem of scarcity, this should not be taken to imply that substitution was a practice restricted to the English provinces or the American colonies, or that colonial science was inferior to metropolitan for that reason. Substitution was as much a part of metropolitan scientific practice as it was anywhere else. As chapter 1 indicated, the metropolitan culture of thrift was not motivated by scarcity alone. Dearth and scarcity might prompt substitution, but it was common to make use of available materials for other reasons, such as the religious impulse to honor God in the everyday stewardship of materials.

In the course of the eighteenth century substitution extended from the small scale in the production of instruments to much larger oeconomic projects of acclimatization, emulation, and import substitution. Naturalists such as Joseph Banks oversaw efforts to acclimatize plants from New Holland to the English climate. Matthew Jones has recorded the enthusiasm for "emulation," or imitating with a view to improvement, in eighteenth-century manufactures, and Maxine Berg has insisted on the creativity entailed in such activities.[86] British manufacturers engaged in much import substitution, most famously seeking to replace Indian calicoes with homegrown fabrics. The Society of Arts explained in its *Transactions* that it was committed to "encourage in the British Colonies the culture and produce of such commodities as we must otherwise import from Foreign nations."[87] Britons thus engaged in substitution just as much as their American counterparts.

Bricolage

Another form of shift was *bricolage*. The previous chapter identified ways in which rooms and utensils with increasingly dedicated uses were exploited to experimental ends. However, it was often the case that experimenters might put together a whole range of objects from many diverse locations to forge experimental apparatus. This was what today is known as bricolage. As the previous chapter noted, few experimental instruments were wholly new or wholly made from old or domestic things. They were typically a combination of objects from the home, from artisanal workshops, presents from friends, parts of new or old instruments, natural artifacts, etc., all made to function together to some end that depended on the collection, juxtaposition, and interaction of the various properties each item could contribute. In the 1720s, for example, Stephen Hales described how to kill insect infestations on ships by fumigating food with an instrument for producing sulfurous smoke. Hales made this from a copper "porridge pot," which could be filled with burning wood-ashes and brimstone. This was ignited by a hot iron bullet to discharge smoke, which was collected into a funnel-shaped hopper from a cider mill that spread the fumes into the ship.[88] The combination of pot, bullet, and hopper thus created a means to produce and guide the fluid smoke as Hales wished. If existing items did not contribute exactly the properties required, they could be altered. Hales proposed very large organ or blacksmiths' bellows as a "very commodious" means to blow stagnant air out of ships, but since they required heavy labor to compress, he suggested increasing the size of their valves to make this easier.[89]

In bricolage parts of one instrument might be removed and used in another (what is sometimes referred to today as "cannibalization"). In 1797 Greenwich Observatory purchased a ten-foot achromatic telescope from Dollond. In 1816 the object glass was removed by John Pond for Edward Troughton's ten-foot transit instrument, and then in 1851 the same glass was used by George Airy in a reflex zenith tube, where it still remained over a century later in 1975.[90]

One of the more famous bricolages of the eighteenth century was suggested in 1752 by Benjamin Franklin. To show that lightning was the same as common electricity, he proposed flying an altered kite into a thunderstorm. Kites were usually used as children's toys or in fireworks displays.[91] Franklin altered the usual design by spreading a large silk handkerchief across two cedar-wood sticks with a sharp foot-long wire attached at the top. A hemp twine or string hung from the kite, with a door key attached at the lower end, leading to the Leyden jar to collect the electric fire.[92] This mixture of specialized instrument and household items was typical of electrical experiments.

Experienced electricians made use of bricolages of household items and dedicated instruments throughout the eighteenth century. George Adams reported on the Swiss experimenter Horace-Bénédict de Saussure's efforts in the 1780s to test Volta's accounts of electricity in the atmosphere. Saussure mixed kitchenware and electrometers for this purpose. To simulate the way rain and vapors made the air become negatively charged in a storm, Saussure insulated a "small chafing-dish" (used to keep food warm on a table) with silk cords, and placed a coffeepot on it containing a little water. An electrometer was attached to the coffeepot and another to the chafing dish. A pair of bellows was used to stoke the fire under the coffeepot to boil and evaporate the water. The electrometers registered a negative charge after a few minutes, suggesting Volta was correct in his explanation of how rain became charged.[93]

Alterations

A final means of making use to consider here is alteration, when an existing instrument or object might be changed in order to improve it for experimental ends. The term "alteration" derived from French and ultimately from the Latin *alterare*, to change. Eighteenth-century dictionaries simply defined alteration as varying something or making it otherwise.[94] The term was commonly used to refer to the alteration of clothes, either for the purpose of making a garment fit a body shape, or to make it more fashionable. "The Mechanicks wifes and Kitchen maids gowns came trowling in to be new altered, for *out of the fashion, out of the world.*"[95] The term was also used to describe modifications to instruments. A decade after the kite experiment, in 1766, Franklin purchased an electrical machine from the instrument maker Benjamin Martin in London, which he shipped to Harvard College (fig. 3.5). The machine produced electricity from a glass globe rubbed by a silk cushion, but in 1789, after the globe broke, the Rev. John Prince of Salem replaced it with a glass cylinder mounted on glass pillars painted with pitch. He was paid £4 16s. for "mounting the great cylinder and altering the old frame for it."[96] Alterations did not always go well, and those responsible might disagree about the oeconomy of making them. In the early nineteenth century a telescope made by James and Thomas Short purchased originally by Franklin for Harvard University needed repairs and was sent to W. & S. Jones in London. William Jones cleaned and adjusted the telescope, but he also altered it by adding "a new stand upon an improved principle in brass, with mahogany folding legs, graduated circle and arch." Jones charged Harvard £54 12s. for the alteration, which the university did not think very oeconomical. Jones responded to university protests by insisting that the old "stand was unmechanical or ill contrived," whereas the new one was "very material," so much so that it raised the value of the telescope from 50s. to £150, and so justified the expense.[97]

Figure 3.5 An altered electrical machine, c. 1766. The original glass globe was replaced with a glass cylinder on four pillars. Courtesy of the Harvard University Collection of Historical Scientific Instruments.

Conclusion

Historians have asked "what was a scientific instrument" in the seventeenth and eighteenth centuries, but it might be more appropriate to ask what was *not* one.[98] Robert Symmer's use of stockings to investigate electricity was rather typical of his age. Like the clothes that householders altered and reworked in the seventeenth and eighteenth centuries, many material things used in the sciences were common household items. One of Herman Boerhaave's terms for chymical apparatus was, appropriately, the "furniture of an Elaboratory."[99] Scientific instruments might perhaps be better thought of as *processes* rather than as fixed, static objects. We should properly ask "what could *become* a scientific instrument in the seventeenth and eighteenth centuries?" Consider Boerhaave's very open definition of an instrument of chymistry as that "which either has or may be given it a certain motion, which being applied to the Body to be changed will produce such an Alteration in it, as the Art had before determined."[100] This was a very broad definition, and Boerhaave included fire and solvents as instruments in addition to the usual apparatus. Servants would also fit this definition, which invites a sense that instrument might be a verb, to point to this process of becoming. "Instrument" should be thought of as a practice, rather than a thing. It might be said that early moderns "instrumented" things all the time, turning objects to experimental uses.[101] Then a variety of techniques turned household goods to experimental ends. Experiment involved much of what today would be termed reuse, repurposing, adaptation, improvisation, substitution, bricolage, and alterations. The results were consequential, as making use made possible investigations into the nature of everything from electricity and the production of new gases, to the nature of light and colors and the brightness of stars. This was thrifty science, because it made the most of available objects. Indeed, experiment could even furnish a good meal. In November 1792, the London electrician Tiberius Cavallo reported to his friend James Lind in Windsor on another bricolage, made to generate animal electricity with a coin, a fish, and a piece of tinfoil. Place a live flounder between the foil and a silver half-crown and it would produce "strong motions" in the fish. Not only that, Cavallo told Lind, but "living flounders, which are very cheap, may be kept sometime in water, and may be eat after all without the least detriment."[102]

FOUR

The Power of Lasting: Maintenance and Cleaning

According to the oeconomic literature, it was important to take care of things. Richard Allestree in *The Gentleman's Calling* (1660) insisted on the "well husbanding" and "prudent managery" of a person's wealth, which meant "having such a provident care of those goods, and possessions, wherewith God hath blest a Man, as may secure them from that Consumption, to which carelessness and sloth will infallibly betray them."[1] An older collection of verses on husbandry and housewifery admonished,

> Kepe kettles from knocks, & set tubs out of sunne
> for mending is costly, and crackt is soone done.[2]

The problem with putting things to good use is that they were liable to wear out. It was a paradox of thrifty science that making the most of materials could lead them to become damaged. If the "furniture of the laboratory" was simply put away, it would be preserved but useless, yet if it was used too much or carelessly, it might be ruined. Maintenance was therefore necessary to ensure that goods and possessions endured for as long as possible. Goods should be "durable," having "the power of lasting," as Samuel Johnson's dictionary defined durability.[3] This chapter explores the various means by which maintenance was achieved and some of the ways it could fail.

David Edgerton reminds us that historically maintenance has always been necessary but often neglected: "maintenance and repair are matters we would rather not think about. Mundane and infuriating, full of uncertainties, they are among the major annoyances surrounding things. The subject is left in the margins."[4] The same was true in the past as today. In the early nineteenth century an officer on a voyage of exploration complained about the mistaken ideas of Tilesius, the ship's naturalist.

> Tilesius thinks that if he has told a sailor this has to be fastened down, preserved, or put away, that he has done his part and does not need to concern himself with the matter any further. Who is . . . going to look after and be responsible for his specimens in bottles, glasses, etc., filled with spiritus? Since this counselor himself in no way looks after these things, they are noticeably rotting, spoiling, and breaking. Soon, he will wake up from his dream.[5]

Like Tilesius, not everybody paid attention to maintenance, but on the whole, household experimenters were fascinated with finding means to make bodies endure. Maintenance involved the preservation of both natural and artificial bodies and the skills needed to preserve them, and this chapter explores the dialog between people and things that made maintenance possible. It begins by noting how good design was considered requisite to ensuring minimal maintenance. Maintenance was creative and innovative, as people sought out ways to make things endure. An appropriate choice of materials and forms could make instruments easier to keep in good repair. Design was not sufficient, however, to achieve durability, and the chapter considers a series of practices that helped to keep things and people in good order in the seventeenth and eighteenth centuries. In the home, men and women engaged in regular inspections, cleaning, careful storage, and a variety of preservation techniques to do this, and when they needed to make apparatus travel, another set of practices secured them en route. Maintenance was not always desirable. The chapter concludes by showing how experimenters recognized that occasionally accidents could be productive. Making mistakes was an epistemic virtue. At least *some* damage was considered valuable, because it might reveal aspects of nature otherwise concealed. Nevertheless, experimenters normally advocated care in operations and regular maintenance to secure apparatus from oblivion.

Designed to Last

Experiment could be about creating the new, but it was also about making the most of the old. In his programmatic statement for a new experimental philosophy in *Micrographia*, Robert Hooke reckoned that with the aid of the new science, "the World may be assisted with variety of Inventions, new matter for Sciences may be collected, the old improv'd, and their rust rubb'd away."[6] New inventions have received most historical attention, but these were only part of what experiment was about. Much of the apparatus of thrifty science did not last long, because shifts and improvisations entailed making use of things in experiments before returning them to their typical domestic use. However, while a bricolaged apparatus might be transient, its parts could be durable. In both the seventeenth and eighteenth centuries, durability, simplicity, and ease of access were important features in the design of many commodities.[7] Householders might have few possessions, but they put them to good service and so expected them to endure. The *Daily Journal* in 1728 advertised, "Lake's Mathematical Lamps . . . these Lamps are made so durable and useful that they will last for many Years, and not be out of Order or in Danger of breaking."[8] Advertisements for "Porcelane Ware" in 1753 promoted Chinese porcelain as superior to English because, "besides the Beauty of its Colours," it was "as smooth, and as easily cleaned as Glass, and at the same Time bears the hottest Liquors without Danger of breaking."[9]

Similar values pertained to more dedicated experimental apparatus. In his *Elaboratory Laid Open*, the chymist Robert Dossie gave advice on installing furnaces:

> they should be well designed, and judiciously executed; otherwise their defects greatly enhance the expense, and frustrate the intention, of the operations they are to perform; besides their being extremely liable to become, in a very short time, out of repair and uselessly ruinous.[10]

Specific qualities of design were deemed to contribute to durability. Simplicity helped because a simple arrangement was easy to manage and required little skill to keep in good repair. A simple design might involve relatively few parts, connected straightforwardly and in a manner easy to understand. Robert Hooke thus described an instrument for sounding depth at sea in 1691:

> I know it will be objected, that this [design] will make the Apparatus very chargeable and difficult; and (as seeming complicate) to be apt to be out of Order; and few will use the Caution and Circumspection, that such an Instrument will necessarily require: To which I answer, that I can make the whole so easy, and obvious, that the whole Instrument need not cost above a Crown; and that any one, almost, shall be able to make, or to mend it; and any one, that can but write and read, can be able to make

Trial therewith, and keep Account thereof; nor will it easily be so out of Order, but that it may easily be mended, and set to Rights again.[11]

High regard for simplicity in design extended all the way to the court. At St. James's Palace in 1788 the naturalist and artist Mary Delany recorded how she "found the Queen very busy in showing a very elegant machine to the Duchess of Portland, which was a frame for weaving of fringe . . . of such simplicity as to be very useful."[12] A good design made the parts of an instrument both socially and materially accessible, and this was another means to ensure easy maintenance. Difficulties of accessing an instrument threatened damage. Robert Boyle had to be patient when he tried to revive and remove a mouse from the interior of an evacuated glass vessel, "without breaking of the Glass (which I was loath to loose, having then no other of the kind)."[13] But good design could avoid such problems. The Somerset physician and inventor John Allen described a wooden frame for a new type of copper water boiler in 1730, noting it was designed to take account of the "Necessity to take it abroad sometimes to mend the copper within, when defective."[14]

Another way to ensure robust design was to pay attention to the properties of the materials used to make an instrument. In his *Elements of Chemistry*, Herman Boerhaave explained the relative values of different materials for making chymical vessels, in each case giving most value to those that were least liable to crack, melt, or be damaged by the substances they contained. Wooden vessels were therefore best for containing "Salts, saline Bodies, Limes, and calcined Substances" because they kept out damp and allowed substances to "keep."[15] Iron could withstand high temperatures but ultimately melted and was corroded by salts when red hot. Clay vessels could also withstand great heat but tended to absorb saline particles. Glass, however, was "of excellent service" because it did not alter the substances it contained, and allowed heat, light, and magnetism to pass through it.[16] "Whatever is preserved in a Glass vessel, that is perfectly closed, and intire on every part, becomes immutable, and incorruptible. Glass eludes the power of all corrosives, even the Alcahest itself, if there ever was any such thing."[17] Glass was therefore excellently durable and "the principal instrument in the Art of Chemistry."[18]

Inspection and Cleaning

Design took into account a variety of concerns, including cost, practicality, function, ease of cleaning and repair, and durability. But there was only so far that such features could be "built in" to instruments. Once things were in use, they needed to be attended to through practices that stopped them from becoming worn out or damaged. Maintenance thus combined social and material roles, depending on the capacity of people to work with objects in order to achieve their endurance. In the case of glass, for example, volatile reactions needed to be handled carefully to stop even the most durable glass from breaking. As William Smith, provost of the College of Philadelphia, recorded in a chymical notebook in the eighteenth century,

> Effervescence is to be avoided to prevent Dissipation, and waste of Fluid—sometimes the Solvent & solvend are entirely carried over the Vessel, in the form of Frothily great Effervescence. The great Heat generated thereby & the Elastic Air detached will much endanger the Vessel; therefore the solvend is to be added *paulatim* [little by little].[19]

During the service of any instrument, it faced a multitude of external threats to its integrity. Volatile reactions, humidity, extraordinary temperatures, unreliable porters, dust and dirt, rain and damp, pests, vermin, and just plain clumsiness could lead to damage, taking away properties deemed valuable in a thing. Social and material practices combined to avoid this. Users inspected objects for wear, replaced parts that were damaged or missing, and repaired or corrected defects. Regular cleaning and careful storage were essential maintenance techniques, removing unwanted materials from apparatus and preventing them from coming into contact with threats in the first place.

Maintaining philosophical material culture required the collaboration of servants, maids, relatives, and artisans, who no doubt performed much of the requisite labor. New forms of service and housewifery emerged when medieval homes gave way to houses with differentiated rooms and a growing number of goods requiring care. Housewives were now expected to clean and maintain the home, a change that prompted thrifty values in the sixteenth and seventeenth centuries.[20] Servants were encouraged to contribute, through good service, to the maintenance of the household. "I recommend Thrift to you," explained Richard Lucas in *The Duty of Servants* (1710).[21] Female householders no doubt did much of the work of maintenance in the home and in experiments, though as Patricia Fara notes this may be hard to recover.

> Very little information about these domestic situations has survived, so it is impossible to discover how many women collaborated in experimental research and took part in dinner-party debates. This female participation was undoubtedly far greater than old-fashioned

accounts suggest. For one thing, experiments often require more than one participant; astronomers dictate their observations while looking through a telescope; chemists cannot simultaneously pour in reagents and record measurements; physiologists need several hands to hold down specimens and manipulate instruments.[22]

Maintenance no doubt required similarly collaborative labors, and some of these are familiar to historians. Shapin noted the various "invisible technicians"—amanuenses, operators, and assistants—working for Robert Boyle.[23] Robert Hooke was an "apprentice" to Boyle and a servant to the Royal Society, being paid, for instance, to clean the Society's meeting room.[24] Hooke himself relied on female domestic helps for cleaning, cooking and repairs, and sex in his Gresham College quarters.[25] In the eighteenth century Margaret Flamsteed made astronomical observations with her husband John, while later the astronomer William Herschel employed his sister Caroline as an assistant, both relying on numerous servants to manage their telescopes.[26]

What practices did such efforts require? Maintaining household possessions and instruments demanded a skilled eye and expert handling. Oeconomic literature gave advice on looking after household possessions, beginning with the need to make vigilant observations of things to detect wear and damage. Anne Barker's *Complete Servant Maid*, published circa 1770, gave advice to housemaids on keeping a household clean, describing methods for maintaining carpets, locks, furniture, silverware, clothes, bedding, and floors.[27] Barker wrote that to prepare linen for washing, a laundry maid should "First, look the linen carefully over, and then mend every place where you find it torn, otherwise if it is washed, the rents will be much worse than before."[28] Other householders followed this approach. In a 1785 essay on electricity, the instrument maker George Adams offered advice on the maintenance of electrical machines.

> Before the electrical machine is put in motion, examine those parts which are liable to wear from the friction of one surface against another; or to be injured by dirt, that may insinuate itself between the rubbing surfaces . . . If any grating or disagreeable noise is heard, the place from whence it proceeds must be discovered, then wiped clean, and rubbed over with a small quantity of tallow; a little sweet oil or tallow should also be occasionally applied to the axis of the cylinder.[29]

This was *sensuous maintenance*, relying on the senses of the householder or those assisting: sight, to observe wear or damage; hearing, to listen out for unwanted grating and grinding; and touch, to check for loose screws.[30] Electricians should be "attentive" and "take care" with these details, a discipline of the householder as much as an ordering of the instrument. Then small-scale remedies helped put machines in order and ready them for use: tightening screws, lubricating joints, dusting glass.

Inspections might be followed by cleaning. The philosopher of science Thomas Kuhn once claimed that science is mostly a "mopping up operation," and surely, for the early modern period at least, he was right.[31] Early moderns were much concerned with cleaning. In the eighteenth century, the diarist Elizabeth Shackleton recorded her efforts to clean her home in anticipation of visitors. "I wash'd all the China Pots &c in the Store room which was extremely well clean'd out—a very troublesome Job am glad it is over so safe and well. It answers the pains and looks very clean nice and well."[32] Cleaning here was a material and moral duty, in which everyone from masters to housemaids was expected to participate.[33] Catholics and Protestants alike put cleanliness next to godliness. According to the Jesuit Robert Southwell, "God is delighted in cleanness, both bodily and ghostly, and detesteth sluttishness as a thing which he permitteth as a punishment of sin and one of the scourges of hell."[34]

Cleaning was no simple matter and entailed expertise and experimentation to get right. In the eighteenth century Martha Dodson recorded how her laundering of clothes involved "borax, bottles of powder for grease, 'stuff for spots,' stone blew, powder blew, starch and ironing."[35] Householders and servants deployed a variety of tools to perform cleaning tasks, including whisk brooms, brushes, mops, oily rags, flannels, bellows to blow off dust, scouring and emery paper, vinegar, ashes, soap, sand, and dusters. Specialized books on cleaning appeared in England as early as 1583, when Leonard Mascall published *A Profitable Booke, declaring Diuers Approoued Remedies, to take out Spots and Staines in Silkes, Veluets, Linnen and Woollen Clothes*.[36] His first recipe explained how to wash clothes using a mix of "unquenched lyme two ounces, of Oake ashes two ounces of cleare water two pound."[37] Numerous works explaining how to clean glass, silver plate, lace, jewelry, silk and wool, furniture, and marble followed in the seventeenth and eighteenth centuries.[38]

Cleaning experimental apparatus made use of the same materials and techniques deployed to clean other domestic goods. Boyle recorded a recipe for cleaning chymical apparatus in his work diary, recalling the ingredients in Mascall's book. "To wash receivers take about equall parts of wood ashes & quicke lime, boyle them a very little while in water, & immediatly use them scalding hot."[39] Washing powders and solutions could be applied using soft leather. Oeconomic treatises advised using soft leather dipped in ashes or burnt wheat straw to clean silver plate.[40] The instrument maker George Adams used soft leather to clean the glass globules he used for microscope lenses.[41] Rags were an alternative. Alexander Morrice wrote of his Saccharometer, for measuring the specific gravity of worts in brewing, "It is very necessary to keep the instrument clean, which is easily done, by plunging it, after every Experiment, in warm Water, and rubbing it with a Linen Rag, or a Piece of soft Leather."[42] Not all cleaning was done with water. Lady Ann Fanshawe's collection of culinary and medical recipes included directions on distilling flowers, which would presumably lose their essence or be damaged if they were washed. "Wash nothing that you will still; but wipe it with a cleane Cloth."[43] Like wiping, "rubbing" was a common term in the eighteenth century, used to mean what today would be referred to as polishing. Eliza Fowler Heywood's *A New Present for a Servant-Maid* (1771) gave detailed instructions on rubbing floors, silver, and other articles. Locks were to be rubbed with an oily rag, fire stone (sandstone), and white brick powder.[44] Rubbing stairs and skirting boards with an oily flannel would "make them look as if new-painted."[45] Silver plate could be rubbed with a rag dipped in spirit of wine then whiting (powdered chalk). Wooden furniture should be rubbed with linseed oil, then "a dry cloth till they are bright."[46]

The term "polish" seems to have been used more particularly to describe artisanal practice, in which some abrasive material was used to grind a surface to shape or smooth it. It often appeared in reference to the polishing of lenses and mirrors for telescopes. In this case, polishing refined the grinding, which created a smooth and reflective surface.[47] Cements of pitch and ashes mixed with soap or putty diluted in water were used for this purpose. There was some crossover between polishing and domestic "rubbing." William Herschel was induced to try out iron and steel mirrors for his telescope on account of "the very high polish often given to domestic articles" made from these materials.[48] "Polish" was also a social category, since it could refer to the refinement of a person as much as the cleanliness of an object. A manual addressed to mothers in 1784 advocated "Domestic attention. Diligence and activity. Oeconomy" and identified a "fine woman" as having a "true classical and rational polish."[49]

Rubbing was considered particularly important by electricians. It was one of the practices by which electricity was discovered, since objects made of amber and jet were seen to attract dust and straw when they were vigorously "rubbed."[50] In the eighteenth century George Adams reckoned that the quality of effects in electrical apparatus depended on the cleanliness of the apparatus. "Take care that no dust, loose threads, or any filaments adhere to the cylinder, its frame, the conductors, or their insulating pillars," he insisted, "because these will gradually dissipate the electric fluid, and prevent the machine from acting powerfully."[51] Tiberius Cavallo similarly instructed young experimenters in 1777,

> The first thing that the young Electrician should observe, is, the preservation, and care, of his instruments. The electrical machine, the coated jars, and in short every part of the apparatus, should be kept clean, and as free as possible from dust, and moisture . . . Before the machine be used, the cylinder should be first wiped very clean with a soft linen cloth, that is dry, clean, and warm; and afterwards with a clean hot flannel, or an old silk handkerchief.[52]

This was a common prescription, but cleaning could be a puzzle for electrical philosophers. Rubbing would generate an electric charge, but too much or inappropriate cleaning could dissipate it. Edward Nairne's client John Fell made experiments to melt wire in the mid-1790s with a partly coated glass Leyden jar. He insisted that the experimenter should take care

> not to warm the uncoated Part of the Jar, nor to wipe it clean & dry before Using, as this Practice lessens the Intensity of the Charge, for by warming Glass, or wiping it very clean & dry, it becomes electrified, & therefore more readily attracts Dust & Moisture, hence follows a Dissipation of the electric Fluid in Charging.[53]

Cleaning was essential, yet too much wiping could be ruinous. John Fell proposed greasing the uncoated part of the jar to make it receive a larger charge but knew that this "will dispose it to break more readily, at the uncoated Part."[54] Maintenance here required a nice judgment over the balance between cleaning and durability.

Storing and Containing

Another means to make objects endure was storage. Zoë Sofia has proposed that the notion of the womb as a passive receptacle has gendered containers female in the past, so that historians more interested in "masculine" machinery have tended to overlook their significance.[55] But containers were critical to maintenance. Placing things on shelves or in containers established a barrier to external threats such as damp, dirt, and dust. Storage also provided security for what were sometimes valuable items. Storage was again both a social and material technology, since it hinged on managing access to things in order to keep them safe.[56]

The home was the secure container for most people's possessions. This was the original meaning of the term "Mayntenance," derived from the French and introduced into English in the late fourteenth century, to refer to protecting one's house from an external threat. Charles Estienne and Jean Liébault's *Maison rustique; or, The Countrie Farme*, translated by Richard Surflet in 1600, advised, "Your House . . . will be . . . of greatest maintenance, preseruation, and safetie, if you inuiron it round about with water."[57] Subsequently maintenance was extended to mean a broader care for one's possessions, which avoided damage and allowed possessions to endure.

If not by moats, most homes were kept secure by an array of locks and bolts on doors and windows, managed through the attention and discipline of householders, who had to trust one another to keep the house safe. Even the best-protected homes could be vulnerable. In 1781, the astronomy lecturer Adam Walker entrusted his home to a student named Edward Arnold. Walker's maid testified that his "books, telescope, and violin" were locked up in closets, but this did not stop Arnold from stealing them.[58] While Walker was away, a neighbor suspected Arnold was making "an ill use of Mr. Walker's library" and soon found that Arnold and his "man" Samuel Wright were stealing Walker's possessions. These included "a very large telescope, which he said was pawned in Princes-street, for half-a-guinea." The books and violin met the same fate, but the neighbor brought the two men before a justice of the peace and they confessed.[59]

Inside the home, family members kept food, linen, and household goods safe in "receptacles" and "containers," terms dating to the fifteenth and sixteenth centuries respectively.[60] Containers were critical for both storage and experiment and included bottles, barrels, cups, glasses, jars, hogsheads, crates, cabinets, cupboards, and trunks.[61] Archaeological excavation in 2011 of the location of Joseph Black's apparatus store in Edinburgh University revealed domestic vessels used to store chemicals. These included two mid- to late eighteenth-century wine bottles and

> miscellaneous vessels of a . . . domestic nature, such as blue and white transfer-printed pearlware bowls and creamware jugs, that on the basis of residues present had evidently been employed in the preparation of chemical materials.[62]

Containers also kept households safe from hazardous materials. Chymists deposited volatile powders in paper packets, since friction from glass bottle stoppers might ignite them.[63] Some receptacles were improvised on the spot. Boyle recorded how a footboy broke a retort being used to burn lead with spirit of wine to reduce it to a calx. He spilt the resulting calx, "yet we made a shift to save about five grains of it."[64] Household waste could be good for containing things. James Watt made tiny boxes out of deal (i.e., pine) shavings, curled and stapled to form an oval box, or he knocked together boxes out of pieces of scrap wood.[65] Containers also served to keep money secure. Early modern academic funding consisted of a pile of coins that needed to be kept safe. When the Royal Society received monies in the early eighteenth century, it placed them in a "great Iron Chest under three keys."[66]

Containers had to be identifiable to be serviceable. Labels indicated their contents and helped to locate items, though they could get mixed up over time.[67] Watt's garret workshop contained sixty-six white creamware jars with forty-two labels, and a thirty-eight-drawer cabinet of tools, with each drawer labeled. Watt handwrote the labels on rectangles of paper and stuck them on the vessels. He also kept paper packets of minerals tied with string with the contents written on the packet. The curator and historian Jane Insley tried to match the labels with materials contained inside and found many were unidentifiable. Labels were either too vague ("admirabulans") or too specific to Watt to know what he meant ("Eiseuman"). Watt no doubt knew all the contents, but the knowledge was not transferable.[68] In other situations, such as apothecaries' shops, accurate labeling was of course vital. "Many a mistake, and many an error have been committed . . . which have killed the patients instead of relieving them . . . by putting a wrong label round the neck of a bottle."[69]

Containers also kept contents free from dust, damp, and impurities. Francis Bacon took as exemplary of "polychrest" instances those which saved labor by excluding that which hindered or disturbed things. Since exposing things to air and the "rays of the heavenly bodies" could cause decay, "whatever therefore serves to exclude them, may justly be reckoned among things of General Use."[70] He proposed experiments to make vessels and food containers air-tight with cements, oil, wine, wax, honey, and herb juices. He immersed things in mercury to preserve them and considered cellars, granaries, and riverbeds as places conducive to preservation. An act of organic preservation, experiments on the freezing of chickens, famously led to Bacon's demise.[71]

Back indoors, men and women experimented with techniques to keep out dust and disturbances. Margaret Ponsonby recounts the example of Susanna Whatman, wife of Kent papermaker James Whatman, who, writing directions for her servants in 1776, "recorded . . . what time the sun reached particular rooms so that the blinds could be drawn to keep the sun off the furnishings."[72] Books, specimens, and instruments might be placed behind curtains or glass doors in cupboards and cases to keep them dry and dust-free, the transparent property of glass enabling users to see contents at a glance. Glass containers also kept out impurities and allowed reactions to take place undisturbed. Amy Eyton's late seventeenth-century recipe book gave instructions for making cherry water for the passion of the heart. Ingredients placed inside glass bottles could be left to mix and interact over time: "hang in each glase one graine of musk & one graine of amber greace sett it in ye sun 3 weeks or a month."[73] Isaac Newton also experimented with keeping out impurities, when developing his reflecting telescope. The lacquered-paper, wooden, and metal tubes of early telescopes, besides allowing one to change the focus of the instrument, served as a container to protect the lenses from damage.[74] In the case of Newton's telescope, the tube protected a metal mirror or speculum incorporated to avoid the chromatic aberration associated with glass lenses. In a letter of 1672, Newton addressed the criticism that a metal speculum was liable to tarnish and so make the instrument worthless:

> when a metal is once well polished, it will be a long while preserved from tarnishing, if diligence be used to keep it dry and close, shut up from Air: For the principal cause of tarnishing seems to be, the condensing of moisture on its polished surface, which by an Acid spirit, wherewith the Atmosphere is impregnated, corrodes and rusts it; or at least, at its exhaling, leaves it covered over with a thin skin, consisting partly of the dust, which flying to and fro in the Air has setled and adhered it.[75]

A truly airtight container would prevent dustiness and hence tarnishing, and Newton recommended immersing the mirror in spirit of wine (used by servant maids for rubbing silverware) when not in use to avoid this. If a mirror did tarnish, Newton had this advice: "their polish may be recovered by rubbing them with a soft piece of leather or other tender substance, without the assistance of any fretting powders, unless they happen to be rusty; for then they must be new polished."[76]

Containers offered distinctive possibilities for experimental bricolage. A barrel filled with cold water allowed John Evelyn to condense vapors by passing the worm tube of his "refrigeratory" through it.[77] Boyle ingeniously used an empty eggshell buried in sand as a container for preparing Van Helmont's *salia essentialia*, the arrangement allowing heat to be applied evenly from all sides.[78]

While many experimental containers were improvised, specialized containers also protected apparatus from damage and decay. These ranged from simple wooden boxes to the fine cases made by Stephen Demainbray for King George III's instrument collection.[79] Purpose-built cases demanded much art. A wooden case for an electrical machine constructed by George Adams of London in the eighteenth century used green felt pads, corks, swiveling pegs, and wooden fittings to secure the brass and glass of the instrument safely. A hinged lid fitted with hooks provided access, and carrying handles allowed the box to be lifted comfortably.[80] Instrument makers promoted their boxes. When John Fell purchased an electrical machine from the London maker Edward Nairne in 1783, Nairne sent a pamphlet describing the apparatus as being

> contained in a Deal Box 24 inches long, 10 ¾ Inches broad, & 21 ½ Inches deep. Every Article is placed in this Box in the most ingenious, convenient, & secure Manner. Labels should have been fixed near every Part of the Apparatus containing their respective Names, by which means the Difficulty which the Electrician sometimes meets with in replacing each Article would have been entirely obviated.[81]

A well-ordered box supported easy maintenance. Containers were not peripheral to the instruments they contained, and indeed sometimes might be indistinguishable from them. Nairne devised a compound chest microscope consisting of a microscope body suspended on a hinged pillar that was lifted upright from its box and fixed in place for use. The box served as a secure base for the instrument and contained besides six different magnifiers, ivory sliders, and many other accessories (fig. 4.1).[82]

As Bacon had hoped, making sealed containers prompted experimental inquiries into the properties of materials. In 1786, Tiberius Cavallo reported that the King's Clockmaker Benjamin Vulliamy had discovered a "remarkable property of cedar wood; it is that it thickens the oil that is exposed to its effluvia."[83] Vulliamy found that on two occasions a regulator inside a cedarwood case stopped after the oil inside the regulator thickened over several weeks. "He repeated the experiment, but constantly had the same result, which obliged him to have a new case made for the clock, and then the oil remained fluid as usual."[84] Intrigued, Cavallo then made a cedarwood box containing pots of different oils to investigate the effect further, though the outcome is unknown.

Containers could be liquid as well as solid. Immersion in mercury has been mentioned, but all sorts of waxes, varnishes, lacquers, resins, and glazes were used to seal and protect things from fire, damp, and corrosion.[85] Chymical and distilling operations often required seals capable of

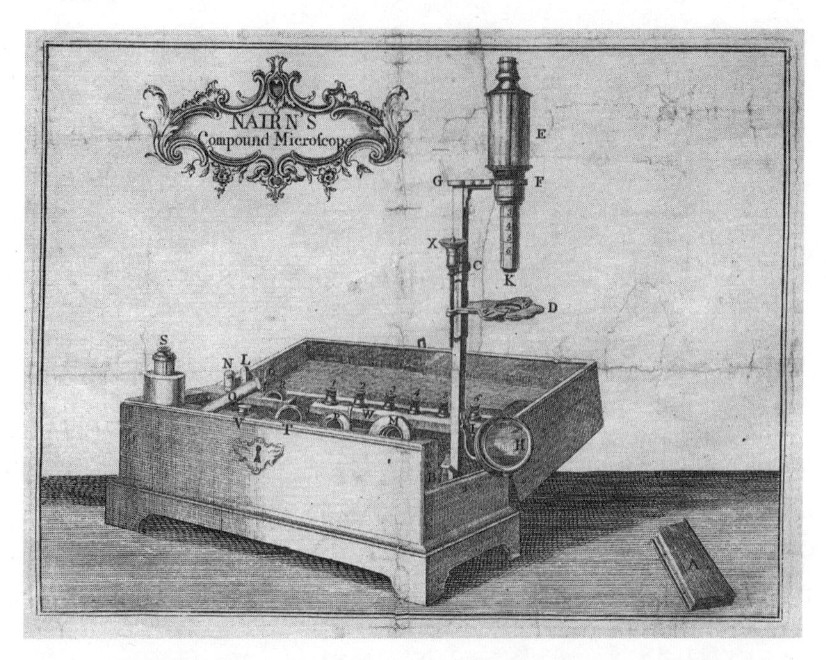

Figure 4.1 Edward Nairne's compound microscope. Detail from Edward Nairne, Directions how to Use the Compound Microscope (n.d.). American Philosophical Society. Thomas Court Scientific Instruments Collection. MSS 509.078 M582.

withstanding high temperatures or extreme conditions. "Lutes," usually a form of clay or cement, were used to seal pieces of chymical apparatus together temporarily or applied to the surface of a vessel to add strength or protection and prevent it from breaking.[86] The Elizabethan noblewoman Grace Mildmay wrote that she prepared medications through distillation, "in a vessel of glass, well luted."[87] In January 1655 Benjamin Worsley gave Boyle a recipe for a lute "that no wind shall breake nor water pierce nor fire dissolve."[88] This was composed of ground tarris stone (the ingredient Hartlib celebrated for its many uses) and "shreds of old jugges or boules that come out of Germany," mixed with quicklime, wood ashes, and buttermilk.[89] The mixture applied over the surface of a vessel would protect it from intense heat. In the early eighteenth century, George Wilson introduced his *Compleat Course of Chymistry* with a lute "to defend a glass in a naked fire," consisting of a stiff paste of sand, clay, tow, and iron smeared over retorts and receivers with a knife to secure them.[90] Lutes were also used in cookery. Hannah Glasse recommended cooking a cod and pork chowder in a kettle, with the lid luted on, over which hot embers were placed.[91] Foodstuffs from the kitchen then served as ingredients for lutes and seals. Alembics could be protected by covering them with slips of paper, bladder, or parchment dipped in mixtures of flour, water, starch, and egg whites. If glass did crack, Wilson explained how it might be "fortified" by wrapping it with linen rags dipped in a syrup made of Suffolk cheese scrapings, *calx vive* (quicklime), and egg white.[92]

To keep ceramic, glass, and earthenware vessels used in chymical furnaces secure from cracks and breaks, operators also heated or cooled them very gradually, a process known as nealing (or annealing). In the seventeenth century Boyle claimed that the term "neal" came from "workmen's language" and spoke of vessels "being first well nealed to prevent cracking" before being used in experiments.[93] William Petty recorded pouring hot distillate into a "Retort being first made hot for feare of breaking."[94] In advice on "the oeconomy of the table" in the 1750s Anne Battam recommended cooling a fish kettle fresh from the fire before adding fish to it, "or else it will crack it."[95] Sir Kenelm Digby theorized the process by which nealing worked on glass. When glass was heated, fire entered into it in place of air, making the glass glow red hot, and when glass cooled fire was released and replaced by air. If fire was replaced by air too fast, it would shake the material until it broke. "And hence it is, that the workmen are forced to let [glass] cool by degrees in such relentings of fire, as they call their nealing heats, lest it should shiver in pieces by a violent succeeding of aire in the room of the fire."[96]

Other methods served to protect vessels. Chymists recommended shaking containers to prevent contents from fusing with them. In a recipe for a ruby-colored tincture of antimony, the chymical practitioner Edward Lloyd noted that a bolt-head of the mixture used to make the tincture should be shaken, "very often, otherwise twill stick soe hard to ye bottom, yt you must be forc'd to break it."[97] Another ingenious way to minimize damage was to miniaturize apparatus. In the 1790s, English readers of Crell's *Chemical Journal* could learn of Swedish chemist Peter Jacob Hjelm's isolation of the metal molybdenum from its ore.[98] The process involved mixing relevant ingredients in a crucible, which was luted and placed in a furnace. However,

> In consequence of the violent and long continued heat to which they are exposed, the crucibles in these experiments sometimes crack and break asunder, in which case the calx of molybdaena is lost, and the labour thrown away. On this account it is proper to subject to these trials only very minute quantities, and in the smallest possible crucibles; but as it is difficult to manage such small crucibles by themselves, when a strong blast is applied, it was found best to make several assays at once.[99]

Experimental procedure needed to be altered to account for the difficulty of handling miniature crucibles, but the change in scale reduced breaks and losses.

The Interdependence of People and Things

Inspection, cleaning, and containing thus served maintenance. But these techniques were not applied only to household objects. There was much crossover between the maintenance of people and things. It is notable that the highly moralized early modern language of damage and maintenance was similar for both bodies and things. As the seventeenth-century natural philosopher Margaret Cavendish noted, old men shook because, "As in a decayed House, every Material is looser than when it was first built; but yet, sometimes an old shaking House will continue a great while, with some Repairs: so old shaking Men, with Care, and good Dyet, will continue a great time."[100] Both human and other bodies could be "offended" and "wear out," and instruments might "suffer" and be "injured" or "bruised" and their defects "remedied" or "cured." The hydrographer and printer Joseph Moxon explained that "the force that must injure a Tennant [joint], must offend it [a]cross the grain of the wood, in which position it will best indure violence."[101] This proximity of things and people recalls the animistic vocabulary of Renaissance "artisanal epistemology" described by Pamela Smith.[102] Sixteenth-century artisans, Smith argues, knew nature through their bodies, in the skills they developed to manipulate the material world, which they often spoke about in animistic terms, attributing an active, organic agency to materials. To create was to sound out the qualities of a material, to listen to what it had to say, and to negotiate with it in order to generate a new form. Smith argues that such an approach declined with the rise of experimental science, as experimenters sought distinction from artisans. Nevertheless, in speaking of maintenance, experimenters continued to use this animated vocabulary well into the eighteenth century.

One term of maintenance that applied equally to people and things was "preservation." Much insight into nature was gained by preserving vegetable and animal bodies, and these bodies might, in the case of foodstuffs, contribute to the health of householders. Early moderns devised an array of techniques to preserve plants, animals, and anatomical specimens, recording them in recipe books and printed oeconomic treatises.[103] Mary Bent's manuscript recipe book, compiled over many decades, explained how to preserve green plums, raspberries, walnuts, currants, gooseberries, and apples, among other things.[104] The printed *England's Happiness Improved* suggested fruit could be preserved by rubbing it over with spearmint juice. Salting fish kept it from tainting for an extra day, and fowl could be placed inside an old wine cask in a dry place to keep them from going bad.[105] "Preservation" applied across people and things in these works. *England's Happiness Improved* spoke of preserving foods and equally of preserving health with medicines such as "Plague-Water" and "Aqua Multifaria," a polychrest "stiled to be of many Virtues."[106]

Certainly, the preservation of the experimenters' body was as important an element of maintenance as the preservation of apparatus. Apparatus and experimenters faced common threats. One of the most pressing in the seventeenth and eighteenth centuries was fire. Between 1650 and 1700 alone some eighty-nine large conflagrations ravaged British cities, causing £913,000 of damage. Between 1730 and 1779 at least 940 fires destroyed 8,300 houses across England, Scotland, and Wales.[107] The threat from fire might be one reason why early moderns were so amenable to thinking of material culture as open-ended and incomplete, since possessions could easily be ruined and homes destroyed and rebuilt much more often than today. In larger houses, kitchens were set at some distance from other rooms to allay the threat of fire.[108] In smaller houses, however, people had to protect themselves by other means. Some slept with rattles by the bed to raise the alarm, so that Robin Pearson has described Georgian London as

> a city of insomniacs, with residents reacting within minutes to the smell of smoke or the sight of flames, leaping out of their beds and escaping into the streets dressed only in their nightshirts, knocking up neighbours, alerting watchmen and firemen, and carrying what they could of their linen and goods.[109]

Avoiding such dangers depended on a close interdependence between people and things. Maintenance was moral and depended on the cooperation of the whole family. Husbands were advised to keep order in the home to avoid risks. Dudley North explained that it was essential

> to observe set hours for publike Meals, and for going to rest; for by that means Servants know their proper times allotted to their businesse; meat is the better drest and served in, and night disorders are much prevented, whereby there comes not only great wast for the most part, but danger by fire.[110]

Certain kinds of experiments were additionally prone to risks that households might not face in other situations. Experiments might entail the disciplining of unknown and uncontrolled phenomena, and these could be dangerous. Corrosive liquids, poisonous fumes, hazardous substances, and risky effects could all threaten health. No doubt one of the most important early modern means to alleviate such dangers was prayer. North provided a prayer for families to recite each morning in his *Observations and Advices Oeconomical,*

> O Lord our most gracious God: We of this Family do acknowledge with all thankfullnesse thy many great Mercies unto us; but more especially thy freeing us from danger and temptation during this night past, and thy giving of healthful and pleasing rest unto us therein, whereby we are enabled to perform the duties of this day, and chiefly that of prayer unto thee.[111]

Should prayer fail, more material precautions served to avoid dangers. Families prepared for the eventualities of fire by recording numerous remedies for burns. A recipe collection signed by Frances Springatt included a burn ointment made with "a pound of brown hog's grease and a good quantity of sheep dung and four handfulls of alehoofe."[112] Mary Bent's recipe book described a plaster laid on burns and impregnated with a mixture made with beeswax, rosin, and sheep's suet mixed with oil.[113]

Another remedy for the dangers of experiments, as in the home generally, was just to be careful. The chymist Peter Shaw and Royal Society curator of experiments Francis Hauksbee reminded their readers that no one was "obliged to try dangerous experiments" and any chemical trials should be conducted with "prudence and discretion."[114] Furthermore, haste should be avoided: "Chemical Operations require a length of time, and a good deal of patience; to which the Chemist must by all means inure himself."[115]

Clothing was a consideration in protecting the body. Householders often wore old clothes when doing experiments. John Webster reckoned scholars should not "be ashamed to cast off thy fine clothes to work in a laboratory" if they wished to do more than just peddle "Aristotles dreames."[116] Alternatively, they might avoid doing experiments themselves and instead have "laborants" or assistants perform them. The countess of Pembroke kept a laborant to help in her chymical experiments at Wilton House in the seventeenth century.[117] Robert Boyle's laborant wore a calico apron for protection.[118] Boyle did not leave everything to assistants, however. His work diary noted that when producing

Figure 4.2 Some substances threatened damage to experiments, or to experimenters. Work diary 34 of Robert Boyle (in the hand of his assistants), damaged by a corrosive spill. Boyle Papers 25, p. 51, © The Royal Society. Such accidents prompted Boyle to publish, in case any more manuscripts were lost. See Michael Hunter, *Boyle: Between God and Science*, 222–24.

violent chymical reactions, one should "have a muffler with glasse eyes upon your face to avoid the fume" (fig. 4.2).[119] In the eighteenth century, Joseph Priestley used "opake Glasses for viewing Luminous Objects."[120] Benjamin Franklin laid down thick paper over glassware in electrical experiments "to save my eyes," and on one occasion when he did not, a glass vessel exploded, so that "several of the pieces struck my face smartly, and one of them cut my lip a little so as to make i[t] bleed."[121]

Despite the risks, Franklin did not desist from electrical experiments, which were often focused on preservation. His invention of the lightning rod was perhaps one of the principal improvements in "securing . . . Habitations and other Buildings from Mischief" in the eighteenth century.[122] But dangers also led philosophers to avoid repeating their experiments. Pieter van Musschenbroek famously declared, on his discovery of the electric shock from a Leyden jar, "I would like to tell you about a new but terrible experiment, which I advise you never to try yourself, nor would I, who have experienced it and survived by the grace of God, do it again for all the kingdom of France."[123]

Carriage

Householders thus devised many means to make material things endure in their homes, using good design, careful handling, secure storage, and preservation methods. But not everything could be kept at home. Houses were not static collections of possessions but nodes in multiple circulations of goods and people, and when things needed to travel, a different set of maintenance practices was required. Issues of carriage, the packaging and transport of things from place to place without their being damaged, were of much importance in early modern science. Perhaps the most significant scientific project of the eighteenth century, the discovery of longitude at sea, boiled down to an investigation of how to make sure a timekeeper would not lose time on a long voyage so that an accurate measurement of position would be possible. The clockmaker John Harrison invented a whole series of ingenious mechanisms to keep his chronometers in good order. As Alexi Baker has argued, efforts to maintain and repair instruments like the marine chronometer were highly productive of new knowledge and inventions in the eighteenth century.[124]

The longitude story is well-known, but it is really only a famous example of problems that faced any instruments made to travel in the seventeenth and eighteenth centuries.[125] Most traveled much shorter distances than marine chronometers, but still had to be kept from becoming deranged. The circulation of delicate instruments, like the "circulation of knowledge," was laborious and a source of great trouble.[126] Probably most carriage of instruments involved philosophers (or the porters they employed) walking from place to place and carrying them. This could be cumbersome. "As for the analysis of the Cheltenham mineral waters," Cavallo explained to his correspondent James Lind in 1788, "I have brought nothing with me for the purpose, considering the trouble of carrying about the necessary articles."[127] For longer distances, larger items were packed in trunks or wooden crates, such as those still extant in James Watt's workshop, perhaps used for his move from Glasgow to Birmingham in 1774.[128] Chemical powders and samples might be wrapped in paper for transport, and then deposited in jars once they reached the laboratory. When Benjamin Vaughan purchased an electrical machine in 1789, it was packed in a case with paper shavings to protect it.[129]

A mail service enabled things to be sent from one house to another. From the seventeenth century, packages could be handed over to a mostly unregulated profession of carriers who sent packhorse trains of carts and wagons led by drivers across England's rough and ready roads, all liable to flooding, robbery, and accident.[130] It was a precarious and expensive business. Prices for carriage varied. The instrument maker Edward Nairne sent a patent electrical machine from London to the physician John Fell's home in Ulverston, Cumbria, in October 1783. The machine cost £11 11s., and the "Packing Case" cost nine shillings, about 4 percent of the price of the machine. Carriage from London to Ulverston took thirteen days and cost 18s. 6d., or 8 percent of the price of the machine.[131]

Goods were often damaged in transit. Fell was relieved that his electrical machine arrived from London "without the least injury."[132] But others were less fortunate. One way to alleviate the troubles of carriage was to buy surplus goods. In eighteenth-century England, glassware dealers told customers to purchase 25 percent excess to cover breakage.[133] Another way was to experiment on improving carriage. Shaw and Hauksbee insisted in 1731 that chemistry offered the potential to improve carriage:

> Chemistry . . . teaches the best methods of packing and securing Goods of all kinds; reducing them to their least dimensions; and preserving them from the accidents of the weather, the Sea, and the like . . . Thus, to give an obvious instance, as Treacle is very subject to waste, on account of the property it has of penetrating and dissolving the

resinous substance of the Cask, this cannot be esteem'd its best kind of Futail; whence we are directed to look out for another.[134]

As in cleaning and storing, carriage required a combination of social and material techniques to succeed. People and possessions needed to be managed correctly. In May 1684 William Molyneux sent instructions to John Flamsteed on procuring a watch with a regulator and chain for him, which Flamsteed was to test before sending it on.

I desire you would seal it up carefully in a little round twopenny box, putting cotton or some soft thing about it, and leave it at Mr Pitts shop in St Pauls to be delivered to him that I send for it, and desire Mr Pitts to leave it so in the way, either in the Custody of his Wife or some Trusty servant as Mr Pullein, that if he himself should be out of the way, when I send my messenger to call for it, it may notwithstanding be delivered to him.[135]

Written advice also helped to avoid damage. In August 1792 Cavallo sent a new micrometer to Lind: "Herewith you will receive the micrometer, which is extremely delicate, and therefore it must be handled very cautiously. I have broke two before I could adapt one to the brass piece."[136] Discussions of accidents sometimes reveal the hidden labor behind carriage. Cavallo reported to Lind in 1784,

I am very sorry that the essence bottles were broken in their way to [you in] Windsor, notwithstanding, that I wrapped a great deal of paper round them, and gave strict charge of them to the man of the inn, where I myself delivered them.[137]

Transport further afield, for instance to Britain's colonies, depended on more extensive networks of support. In North America, thermometers and barometers were all imported from Britain during the eighteenth century and were regularly damaged. In Philadelphia, Franklin complained of three barometers sent to him by Peter Collinson in London, "two of which came safe, but the Ball of the third broke to pieces and the Mercury gone."[138] After his move to Northumberland, Pennsylvania, in 1794, Joseph Priestley relied on many individuals to transport chemical instruments to his home. Historians have identified William and Samuel Parker as his glassware suppliers, but to get glass from London to Pennsylvania involved many more persons and skills, necessary to negotiate proper packaging, rough roads, and customs inspections. When the network failed, Priestley enlisted friends and family to try to recover his goods, and when these reached him intact he was elated.[139]

Portability

As in the case of household maintenance, good design could alleviate problems of carriage. Some experimenters tried to design portable instruments whose reduced scale made them easier to handle and transport. Portability served transport and also helped make instruments accessible to potential users. In 1731, Shaw and Hauksbee advertised for sale a new "portable laboratory" that centered on a furnace to "facilitate and promote the Practice of Chemistry."[140] The laboratory is a good example of an instrument valued for its capacity to be put to many uses. Concerned that furnaces were difficult to procure, they offered a polychrest design capable of serving as many different functions as possible, in a form that made it easy to transport (fig. 4.3). This was achieved by reducing the laboratory in size so that it would "come into a small Compass; and may be commodiously carried, either by Land or Sea."[141] To manage this, the furnace was divided into a series of parts whose different combinations could serve different operations. The parts consisted of a cover to prevent heat escaping, a ring on which operations could be performed, a body to contain fuel, and a foot that supported the whole and contained overflowing substances. Several grates could be placed inside the furnace at different levels, and different combinations of these parts allowed the furnace to be "readily converted" from an apothecary's furnace to an assayer's furnace to a philosophical furnace for experiments to a furnace for work such as distilling, enameling, or glass staining. It could also be used as a cooker in the kitchen or as a heater. Other designs for portable furnaces followed in the eighteenth century.[142]

Another way to make instruments more portable was to soften brittle or easily damaged parts. For much of the seventeenth century barometers could not be moved in case the mercury they contained shattered their glass tubing. But in 1695 Daniel Quare added a soft leather bag at the base of the tube which could be compressed using a screw to push the mercury to the top of the tube, stopping any movement during travel. At the desired location the screw was turned to open up the bag, and a vacuum above the mercury reappeared. Fixed on a tripod, these portable barometers allowed figures such as the surveyor William Roy to make measures of pressure and altitude on mountains, while others used them to detect changes in pressure caused by leaking gases in coal mines.[143]

Making instruments small enough to fit inside protective containers also contributed to their portability. In 1786, the London physician

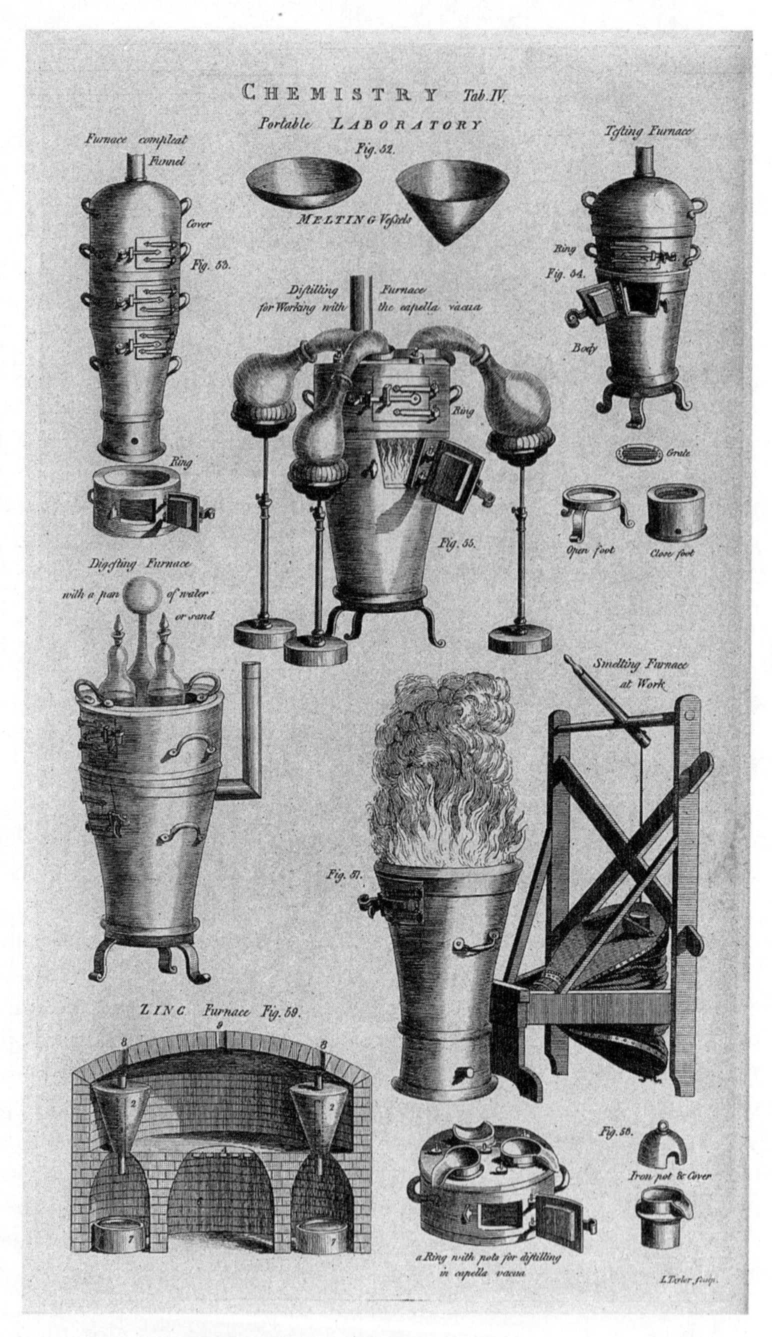

Figure 4.3 Various configurations of the portable furnace, based on Shaw and Hauksbee's design of 1731. Wellcome Collection.

Miles Partington advised John Fell on building a portable electrical machine. This consisted of a twelve-inch glass plate set in a mahogany stand with a Leyden bottle and Lane electrometer. The machine could be dismantled for travel, and relied on ingenious design and the labor of servants: "when the Bottle was taken away, the Stand turned up & took very little Room. The whole went into a flat Box; & the Servant carried it by the Handle on the Top in one Hand."[144] Finally, portability involved a trade-off between the capacity of an instrument to be moved and its capacity to produce effects. As Fell explained of his Nairne's patent electrical machine, "had it been larger, it would have been less portable, tho' more powerful."[145]

Conclusion

Historians and philosophers of science have made much of the role of accidents in experimental inquiry, but rarely with the material consequences in mind.[146] Mistakes and breakages could certainly be productive of knowledge. In the seventeenth century Boyle investigated the phenomenon of spontaneous cracking in glass objects such as telescope lenses, looking glasses, and a "thick Glass Cup." Of the latter he wrote, "though it did not so crack as to fall to pieces . . . yet it was flawed with such a multitude of little cracks, that at a distance it looks . . . white."[147] He then tried to explain the "intestine motion in glass" that might lead to such an effect and proposed that "saline corpuscles" slowly moved to the surface of the glass and in doing so formed weaknesses that became cracks.[148] In the 1730s, Hauksbee and Shaw also insisted that much could be learned from accidents: "an intelligent Operator presently gains light from his own Errors, and instructs himself by his miscarriages."[149]

Even so, as this chapter has shown, much trouble, effort, and expense went into the maintenance of experimental objects in the home. If things were to be made use of to best effect, then they needed to endure, and this required maintenance. Inspection, cleaning, careful storage or packaging, preservation from injury and decay, good design, and careful handling all served this purpose. Providing solutions to these problems, Baconian "polychrests" provoked experimentation and ingenuity. Material maintenance was not peripheral but had substantive effects on experimental agendas and techniques. Boxes and cases formed fundamental parts of some instruments, while potential threats prompted experiments in preservation. Containers prevented contaminants from causing damage and decay, and might themselves become subjects of experimental inquiry.

These material practices were also social, and early moderns used a common language to speak about injury to things and to people, while failures to maintain were identified with failures of morality or character. A focus on innovation in the sciences has tended to obscure these practices and values.

So while experimenters did make use of mistakes and accidents, they did not wish to see good instruments get broken. Accidents were for the most part to be avoided, and if lessons were missed, then so be it. Shaw and Hauksbee assured would-be chemists that accidents in chemistry were rare, "And in case any uncertain Operation were to be gone upon, a little common prudence will provide against contingencies; so that the bursting of Glasses, or the unexpected eruption of Fumes, shall occasion no farther mischief."[150] While an occasional accident could be informative, most of the time it was a good idea just to be careful.

FIVE

The Broken World: Repairs and Recycling

Despite the best efforts of philosophers to keep things well maintained, accidents did happen. In September 1794, writing from his home in Little St. Martin's Lane, London, Tiberius Cavallo reported to his friend James Lind,

> In unpacking my trunk after my arrival in town on Thursday last, I had the misfortune of dropping the box with the doubler of electricity, upon the ground, which broke almost all the pieces of glass in it, but it is now repaired, and costs me almost a whole day's work.[1]

Such experiences were frustrating, but early moderns did not brush broken things under the carpet. While a broken object might lose its value for experiment, this was usually only temporary and could be rectified in various ways. Some sought out repairs or recycled damaged goods, while others made a shift of broken things as best they could, reflecting the values of thrifty oeconomy. "A cracked [Leyden] jar may be made to receive a charge," explained the instrument maker George Adams, "by taking away the external and internal coatings which were over the crack, so as to leave a space of about one-fourth of an inch between the crack and the remaining tinfoil."[2]

Histories of repair have long been obscure but are increasingly coming to light as historians move away from a fetishization of novelty and innovation toward studies of circulation and use.[3] Sara Pennell critiques consumption histories for overlooking episodes of breaking and repair in households in the eighteenth century.[4] Breakages, Pennell writes, were "a fact of domestic life" and as such were exploited for all their "possibilities and consequences."[5] David Edgerton has noted that repair and maintenance "are the most widespread form of technical expertise . . . a great deal [of which] takes place outside the formal economy."[6] Simon Schaffer has explored the ambivalence of states of repair in the early modern sciences. It was not obvious that instruments or experiments were working properly, and this status was negotiated. Schaffer reminds us that repair work was simultaneously both social and material work: "States of disrepair refer simultaneously to tools and humans that interact with them and each other."[7] The need for repairs also prompted philosophers to investigate the nature of materials and to collect recipes and techniques for mending them. This contributed to a sense of the "incompleteness" of objects, so that, as Alexi Baker has argued, "most early modern instruments were in near-constant physical flux . . . which sometimes impeded or entirely prevented instrument usage and maintenance."[8] Baker concludes that repair was then

> a central influence upon the production and use of instruments, the conduct and communication of related activities including research and observation, and efforts toward the standardization of practice and technology around the globe.[9]

Truly broken things could still be recycled. The term "recycling" was not in use in the early modern period, since objects were understood to be undergoing changes of use on a regular basis. But the term helps to identify practices that featured in thrifty science. Ariane Fennetaux, Amélie Junqua, and Sophie Vasset have proposed that recycling was "a fundamentally relevant operative process . . . that ran through the whole fabric of eighteenth-century society."[10] Following pioneering work by Donald Woodward, Fennetaux, Junqua, and Vasset equate recycling with a "cyclical process of valuation, devaluation, and re-evaluation" of things as they became redundant, were discarded, then reworked into serviceable materials by a variety of trades.[11] This chapter shows how experimenters also participated in "recycling." While previous chapters have been mostly focused on experiment inside the home, this chapter moves outside the house and into the streets. Repairs and recycling could sometimes be done at home but usually involved larger networks of operators within which early modern dwellings were embedded and whose trades and skills they depended on to function.[12]

Repairs in the Home

If something broke in early modern England, there were a variety of options available as to what to do with it. Breaking constituted a loss of value, in that some or all of the properties that made an object functional or desirable were lost. These might be physical properties such as integrity, the capacity to produce some effect, to contain or carry some substance or to perform some task. In such cases, one option was to keep an item and make use of it in a limited capacity, or put it away on the basis that it might prove serviceable in the future. Then as now, such items might never end up being put into use, but remain in storage or on display in the home until the death of their owners. One eighteenth-century inventory of a deceased's household thus recorded "one pair of broken Sconces . . . Thirteen China Dishes, some of them broken . . . some odd cracked Coffee cups . . . some broken China . . . a broken Chimney Glass."[13] Another option was to have the item repaired, in which case value might be restored by recovering the properties lost through damage. Repair was a common component of household labors in the early modern period and was often preferable to replacement if something broke.[14] The term "repair" had many synonyms, such as mend, fix, "new-make," or "vamp up," originally referring to the mending of shoes or stockings but then applied more generally.

In the home, clothes, shoes, kitchen utensils, furniture, locks, and buildings were all regularly repaired. The medieval *Booke of Thrift*, reprinted in the sixteenth and seventeenth centuries, reminded people of the routines of repair work, "Mending of houses, walles, hedges and ditches if neede be, ought to be allowed according vnto season."[15] In the eighteenth century, lodgers were obliged to keep property

> well and sufficiently repaired, upholden, sustained, cleansed, scoured, fenced, stocked, planted, weeded, gravelled, rolled, preserved, and kept in good order and repair . . . together with all the household goods, furniture, implements of the household, and other things.[16]

The Royal Society periodically ordered repairs for the meeting rooms and repository at its quarters in Gresham College and Crane Court, appointing a surveyor to assess what needed to be done (usually whitewash the walls, fix the floors, mend broken windows).[17]

Mending things required various levels of skill and experience and was divided up along class and gender lines. In less well-to-do households, people might take care of repairs themselves, with women supposed to mend clothing and linen, and men taking care of hardware. Since all these things might be used in experiments, both sexes effectively contributed repairs that enabled them to happen. The Maine midwife and diarist Martha Ballard recorded numerous occasions when she mended breeches, stockings, trousers, gowns, coats, mittens, and quilts, while her husband, Ephraim Ballard, mended a teapot, a sled, and a plow.[18] Repair work was hard work. As Ballard wrote, "I have been at home. the hoop of my Soap Barrels broke and let my Soap out. I had a great fatague [*sic*] in geting it up; lost a good deal."[19] Friends could help out. In London, Robert Hooke recorded in his diary that he "Returnd Mrs More china cover mended."[20]

Repairs could serve aesthetics as much as functionality for early moderns. Repaired objects were not necessarily considered inferior to undamaged goods and might even be displayed as signs of improvement and progress.[21] Several of William Hogarth's family portraits of the 1730s showed a repaired teapot as a feature of the elegant interiors they depicted.[22] An acceptance or approval of fragments and broken things, or what art historians call *Ruinenlust* (an appetite for ruins), was also apparent in the sciences.[23] This might be epistemic (an interest in accidents and breakages as opportunities for discovery) or material. When Nehemiah Grew wrote up a catalog of the Royal Society's repository in the 1680s, he included exhibits that were incomplete or broken, suggesting a tolerance for partial objects. These included a headless hummingbird, a dodo's leg, a crocodile with its tail missing, and a "great gogle-eyed beetle . . . His Horns rooted between the Eyes and the Snout; but they are here broken off."[24] Perhaps the Society's collection was just poor, in a state of bad repair, or partial items served when whole ones were unavailable. But some specimens were unlikely to have been rare or hard to replace, such as a stag beetle "with the Head broken off."[25]

Many repairs might take on the character of a shift or improvisation, especially if they needed to be performed quickly. In 1670, Robert Boyle improvised a repair after noting the way a frog inside the evacuated receiver of an air pump

> began to swell much upon the withdrawing of the Air, and seemed to be distressed . . . he held out half an hour, at which time it was remarkable, that the Receiver, though it had held out against the pressure of the outward air, during that space of time, notwithstanding that a piece of it had been cracked out, and was mended with a cloth

deep'd in Cement, yet at the end of the half hour, the weight of the outward Air suddenly beat it in, and thereby brought the imprisoned Frog a reprieve.[26]

More common than such hurried repairs were those that took time and some skill to perform. In such cases, labor from outside the home was needed.[27] Domestic life depended on this integration into larger networks of craftspeople and traders. The managers of rented properties employed carpenters, smiths, gardeners, and bricklayers to make repairs on their property.[28] Amanda Vickery records the accounts of Kent gentlewoman Anne Brockman in the first decade of the eighteenth century and John Arderne of Harden in 1739, who paid various artisans to retin worn pans, solder or patch metalware, add new handles and legs to pots and pans, clean clocks and irons, and grind scissors, knives, and scythes (fig. 5.1).[29] Clockmakers returned periodically to homes to repair and clean clocks.[30] A brazier visited the Fox Strangeways family at Melbury House near Dorchester, getting paid five shillings for "tinning and mending" their distillation apparatus in 1749.[31] Other traveling artisans included cobblers, who repaired shoes, and chair menders, who rewove the canes on worn seats.

Many of these laborers were poor and depended on household trade to survive. Martha Pillah, condemned to death in 1717 for stealing six guineas and fifteen shillings, was eighteen years old and lived in the parish of St. Margaret, Westminster. After serving as an apprentice to a tailor,

Figure 5.1 Anon., People approach a knife grinder to have knives sharpened. Undated etching. Wellcome Library, London.

"she work'd for herself, whose chief Business then was, the making and mending Men's Cloaths."[32]

If particular expertise was required to get instruments repaired, then householders might need to travel further afield, visiting the shop of the relevant artist. While the Ballards did most repairs at home, "mr Ballard went to Capt Coks^{s} to get a Surveyers Compass mended."[33] A growing number of traders offered specialized repair services in eighteenth-century London. John Oliver traded from Grub Street, where he made and repaired "all sorts of Glaziers & Plumbers Tools at the Lowest Prices."[34] Edmund Morris was a "China Rivetter" located in Grays Inn Passage, who "Mends all Sorts of China with a peculiar Art . . . so as a Rivetted Piece of China will do as much Service as when New."[35] Instrument makers also offered repairs. London became a global center for the instrument-making trade in the eighteenth century. Mathematical and optical instrument makers flourished as they supplied precision instruments to the Royal Observatory in Greenwich and to the Navy, and makers congregated alongside clockmakers, carpenters, and metalworkers in the Clerkenwell area.[36] As Richard Dunn has noted, much of their work involved cleaning and repairs.[37]

Robert Hooke, like many other philosophers, regularly set out from home in search of repairs. At his quarters in Gresham College, Hooke had his servants mend his clothes, or took them to a tailor to be repaired. He had a Mr. Scarborough fix his chimney. He took mathematical instruments for repairs to the clockmaker Thomas Tompion on Fleet Street.[38] John Evelyn similarly scoured London and Paris in search of artisans to do cleaning, repairs, and varnishing. He picked up useful recipes along the way. French coachmakers, for instance, told him of a "black putty wherewith they stop cracks in Ebony or any black polished works . . . made with a part of Rosin bees wax & lamp black."[39]

Despite the frequency of repair work, instrument makers might not acknowledge their services on trade cards or in catalogs, where, as Alexi Baker has argued, they presented instruments as "idealized and unproblematic."[40] Trade cards showed instruments as complete objects, designed and constructed and ready to buy. The absence of references to repairs on trade cards may simply have been a convention for lack of space, or else signals that repair work was so taken for granted that it did not need advertising. The printed trade advertisement of Charing Cross instrument makers Jeremiah and Walter Watkins, for instance, did not advertise repairs, but when they were used as bills for work done, repairs were recorded on them (fig. 5.2).

How, then, were repairs effected? Art conservators have a good sense of

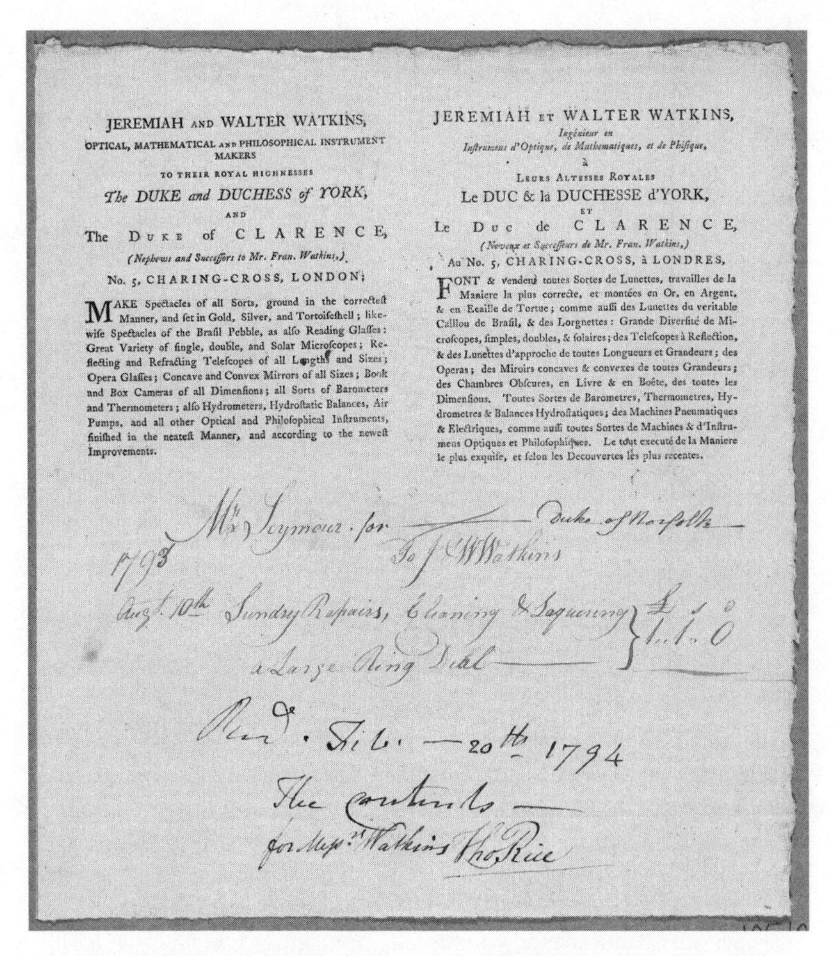

JEREMIAH AND WALTER WATKINS,
OPTICAL, MATHEMATICAL AND PHILOSOPHICAL INSTRUMENT MAKERS
TO THEIR ROYAL HIGHNESSES
The DUKE and DUCHESS of YORK,
AND
The DUKE of CLARENCE,
(Nephews and Succeſſors to Mr. Fran. Watkins,)
No. 5, CHARING-CROSS, LONDON;

MAKE Spectacles of all Sorts, ground in the correcteſt Manner, and ſet in Gold, Silver, and Tortoiſeſhell; likewiſe Spectacles of the Braſil Pebble, as alſo Reading Glaſſes: Great Variety of ſingle, double, and Solar Microſcopes; Reflecting and Refracting Teleſcopes of all Lengths and Sizes; Opera Glaſſes; Concave and Convex Mirrors of all Sizes; Book and Box Cameras of all Dimenſions; all Sorts of Barometers and Thermometers; alſo Hydrometers, Hydroſtatic Balances, Air Pumps, and all other Optical and Philoſophical Inſtruments, finiſhed in the neateſt Manner, and according to the neweſt Improvements.

JEREMIAH ET WALTER WATKINS,
Ingénieur en
Inſtrumens d'Optique, de Mathematiques, et de Phiſique,
à
LEURS ALTESSES ROYALES
Le DUC & la DUCHESSE d'YORK,
ET
Le DUC de CLARENCE,
(Neveux et Succeſſeurs de Mr. Fran. Watkins,)
Au No. 5, CHARING-CROSS, à LONDRES,

FONT & vendent toutes Sortes de Lunettes, travailles de la Maniere la plus correcte, et montées en Or, en Argent, & en Ecaille de Tortue; comme auſſi des Lunettes du veritable Caillou de Braſil, & des Lorgnettes: Grande Diverſité de Microſcopes, ſimples, doubles, & ſolaires; des Teleſcopes à Reflection, & des Lunettes d'approche de toutes Longueurs et Grandeurs; des Operas; des Miroirs concaves & convexes de toutes Grandeurs; des Chambres Obſcures, en Livre & en Boête, des toutes les Dimenſions. Toutes Sortes de Barometres, Thermometres, Hydrometres & Balances Hydroſtatiques; des Machines Pneumatiques & Electriques, comme auſſi toutes Sortes de Machines & d'Inſtrumens Optiques et Philoſophiques. Le tout executé de la Maniere le plus exquiſe, et ſelon les Decouvertes les plus recentes.

Mrs Seymour for —— Duke of Norfolk
1793 To J & W Watkins
Augt. 10th Sundry Repairs, Cleaning & Lacquering a Large Ring Dial —— } £ 1. 1. 0
Recd. Feb. — 20th 1794
The contents ——
for Messrs Watkins Tho. Rice

Figure 5.2 An invoice from Watkins, instrument makers, for £1 1s. 0d. to Mrs. Seymour, for "1793 Augt. 10th Sundry Repairs, Cleaning & Lacquering a Large Ring Dial." Trade card of Jeremiah & Walter Watkins, scientific instrument makers, 1794, British Museum, Heal, 105.109. © The Trustees of the British Museum.

the repair techniques of the past. Jonathan Thornton suggests that early modern artisans desired their repairs to be "as strong, permanent, and neat as the available technology would allow."[41] Typically, practitioners began a repair using the material from which the object to be repaired was made and which most likely they had a particular expertise in working with. This made repairs difficult to see, but there was no expectation that they should be invisible, as Hogarth's paintings make apparent.[42] Repairs to wineglasses were visible since the top and base of the glass were held together with a lead or copper wire brace woven around them.[43] Other

Figure 5.3 A Derby porcelain scalloped dish, c. 1800–1825, repaired with rivets. Photograph by the author.

repairs required making use of substitutes. China or stoneware could not be repaired with new china or stoneware parts, so metal rims were added to broken edges ("rimming"), pewter or lead plugs filled holes or chipped edges, while broken spouts or handles were replaced with metal or ivory ones riveted on.[44] The original makers might repair brass or wooden items, but others demanded specialized skills. China, for example, was riveted by a specialist china riveter, who usually doubled as a retailer (fig. 5.3).[45]

Repairing mathematical, optical, or philosophical instruments involved various skilled practices, though these were scarcely articulated in print before 1800, perhaps because they were secrets or perhaps because they were not considered worthy of recording. Richard Dunn has used bills to show how the mathematical and optical instrument maker James Ripley repaired compasses in the early nineteenth century. Ripley cleaned and remagnetized needles ("retouching" them), replaced worn parts and cards, and mended cracks and damage to the cases. He also lacquered telescopes and replaced their broken lenses and caps. Repairs to a compass cost between two shillings sixpence and three shillings sixpence, while those to a telescope were six shillings and sixpence.[46] John Wilkes's *Encyclopaedia Londinensis* (1814) described the hardware needed to repair chemical apparatus. Wood and metal items could be mended by fixing them in a lathe or vice to be manipulated with various tools such as files and rasps. Small metal objects such as screws could be made with a hammer, anvil, screw plate, and taps, and glassware was bent and sealed with a glassblower's lamp and bellows.[47]

Repairs as Social and Material Judgments

Seeking out repairs was thus a social practice, and repair work contributed to social order in a variety of ways. Acts of repair manifested rank and distinction. The genteel distinguished themselves by their delicate handling of fragile items, blaming servants for clumsy mishandling and breakages. China earned its luxurious status in part by its easy breakability, prompting more oeconomical householders like Samuel Johnson to scoff, "I am not yet so infected with the contagion of china-fancy as to like anything at that rate which can so easily be broken."[48] Problems with instruments offered occasions to manifest social distance. Joseph Priestley thought it a defect of the Hauksbee electrical machine that its handle could not be turned by its operator: "A servant is therefore necessary who must fit to his work."[49] Once again, servants could be instruments. A common way to say something was damaged in the eighteenth century was to refer to it as "deranged," implying an aberration from some correct state. Work and materials might also be "put out of order," which again implied some proper order and the action of an agent to derange it. Social trust and material order went hand in hand in derangements. When "one of the servants" of Pennsylvania astronomer John Ewing looked through an observing instrument out of curiosity, he "so deranged it that no confidence could be placed in the observations."[50] Such language recalls the vocabulary of injury and preservation used for both people and things discussed in chapter 4. The language of repair could also serve distinction. Artisans developed their own specialized vocabulary for repairs and looked down on more common terms. The printer and globe maker Joseph Moxon explained that wooden boards, "if the Sun shine fiercely upon them, it will dry them so fast, that the Boards will *Tear* or *Shake*, which is in Vulgar English *Split* or *Crack*."[51]

As the market for specialized instruments grew in the eighteenth century, so experimenters were obliged to rely more on instrument makers for repairs. As Schaffer notes, scholars might resent such reliance on artisans.[52] To justify independence from instrument makers, experimenters blamed them for poor workmanship rather than accept that their own errors might make instruments defective. Joseph Priestley reckoned the "modern constructions" of electrical machines enforced such dependence, since they were "liable to accidents, which electricians in general cannot easily repair; and I would wish philosophers to be as independent as possible of all workmen."[53] Schaffer has also shown how decisions about whether or not to rely on expert repairs depended on social and material judgments.[54] Modes of repair were decided on collectively. As Tiberius Cavallo wrote to James Lind, "Herewith you will receive the electrometer repaired in the manner we fixed together . . . Let me know how it answers."[55] Decisions whether or not to opt for repair depended on context. Astronomers used repaired instruments to make observations. One of the largest telescopes of the late seventeenth century was fitted with a lens that had broken in half and then been repaired.[56] While observing the transit of Jupiter and a star in Virgo from his Derby home in April 1672, John Flamsteed held the lens of his telescope "over the flame of a candle to smeare it as I used to doe frequently," but "it cracked in a veine through the middle & a bit of the breadth of a great pins head burst from it."[57] As with Nehemiah Grew's broken repository exhibits, Flamsteed showed a tolerance for broken things and reckoned the lens could still be serviceable: "I have since tried it on the moone and can not say that it performes any worse then it did."[58] After some consideration, however, he decided to find a replacement, since the observation he needed to make on this occasion had to be competitive with those of the best astronomers.

The Broken World: Repairs at a Distance

Not everyone had access to the makers of the metropolis, and A. D. Morrison-Low has argued that repair work was especially valued in the English provinces, where makers were fewer, instruments were rarer, and skilled repairmen could make a good living maintaining instruments. This was particularly true in port towns where octants, sextants, and compasses were regularly renovated and resold.[59] Similarly in Britain's colonies, repair work could flourish. Ports in North America, like those in Britain, were home to many repair services for navigational instruments.[60] The trade card of New York mathematical instrument maker Adrian Br. Hicks announced "Quadrants, Compasses, &c. repaired in a neat workmanlike manner. If any of his work should not give entire satisfaction, he will require no compensation."[61]

As discussed in the previous chapter, it would be wrong to see provincial or colonial repairs as evidence of an inferior scientific culture, since repair work was ubiquitous. But certainly skills were unevenly distributed. In the early seventeenth century, even Oxford was considered to lack artisans capable of keeping instruments in good repair.[62] Distance meant disintegration: the further one traveled, the more difficult it could be to effect repairs. The Pennsylvania astronomer and surveyor David Rittenhouse was forced to make shift when he took his instruments to a specially prepared observatory to measure the longitude of Wilmington, Delaware, in July 1784:

> from the first to the 18th [July] the Clock went something irregular, the Bob striking the Case at times notwithstanding our frequent attempts to fix it. But on the 18th at noon by cutting quite thro' the Case & pasting paper over the Holes this inconvenience was effectively removed.[63]

The previous chapter noted the risks of early modern carriage. When things broke during travels, it was hard to find skilled persons to repair them. In 1751 Governor Jonathan Belcher wrote to Benjamin Franklin from New Jersey complaining that electrical apparatus sent to him was ruined: "the Glass Globe was broke all to pieces . . . I have tryed to get another at New York without Success. Can you Sir, put me into any method to repair it?"[64]

One solution for American philosophers was to send instruments back to Britain to be fixed, but this could prove disastrous. In 1784 the Charleston Library Society sent its Manigault telescope to London for repair. It took nine years and several instrument makers before the instrument was finally returned in good order.[65]

Repairs and Knowledge

Headaches over repairs in North America probably helped to fuel ingenious substitutions of the sort discussed by Bedini in the previous chapter. Repair work was thus more than just a support activity for scientific inquiry. Repairs were constitutive of new inventions and knowledge and lay at the root of many innovations.[66] Perhaps the most famous case was James Watt's steam engine improvements that followed his repairs to the dilapidated Newcomen steam-engine model of Jonathan Sisson in the winter of 1763.[67] Watt's improvements are sometimes presented as the work of a "lone genius," but they reflected the domestic setting of his investigations.[68] Watt took much from his home, where he lived with his family, who were all involved in ingenious enterprises. Watt shared his home with his wife, Anne (née McGrigor), and children Gregory and Janet, or Jessy. Gregory had "precocious talents," going on to become a geologist.[69] When Jessy died prematurely aged fifteen in 1794, Watt recalled her abilities, "her little works of ingenuity, her books and other objects of study, serve as mementos of her who was always usefully employed even to the last day of her life."[70] Watt claimed to draw inspiration for the separate condenser from kitchen observations: "I saw no boyler so perfect . . . as the Common tea kitchen (an invention for which we are beholden to the Chinese) here the fuel is always in Contact with the sides of the boyler Containing the water."[71] His first model of the condenser was made with what was probably a defunct enema syringe, a medical object.[72]

Other repairs led to innovation. In the 1770s, Joseph Priestley found that lutes used to seal tubes inserted through corks into glass vessels in experiments on air were "apt to give way, and to be the occasion of very disagreeable accidents."[73] To avoid this, Priestley designed a glass stopper and tube with no seals, made for him by Fleet Street glassmaker W. Parker and Sons. The stopper worked well and avoided another problem of lutes interacting with vapors in experiments. Priestley concluded that the stopper was "of most admirable use. For, in experiments with air, where the greatest possible accuracy is required, lutes are by no means to be trusted."[74]

Another way repairs generated knowledge was through attempts to articulate mending skills and recipes in text. As Sara Pennell notes, families recorded recipes for repairing broken glass in manuscript recipe books from the late sixteenth century, with others for porcelain and fine earthenware (called "chiny," "cheny," or "china") appearing later.[75] The Lowther family's receipt book recommended boiling isinglass in brandy until clear "to glew china to gather."[76] Chymists similarly recorded recipes and published them. The apothecary and chymist Robert Dossie detailed techniques of repair in several works intended to improve the state of the arts. In his 1758 *Handmaid to the Arts*, for example, Dossie described cements, pastes, and glues for joining together paper, silk, leather, glass, marble, and china.[77] Common glue was made by putting old materials into service again, boiling gelatinous scraps of coarse leather or animal hide in water. Isinglass, mastic, beeswax, and resin offered additional strength. Cements could be made for visible or invisible repairs depending on preference and varied from fine mixtures for mending broken china to coarse cements for building work.

Seventeenth- and eighteenth-century householders thus gave much importance to repair work. This significance was further reflected in the common habit of discussing both the sciences and nature in terms of states of repair. The "new science" was often discussed as a project not of innovation but of renovation. Restoration natural philosophers claimed to be much preoccupied with the physical, spiritual, and political recovery of humankind. Bacon, Hooke, and others reckoned experimental philosophy might remedy the deficiencies of Adam's diminished senses by the use of instruments. In *Micrographia*, Hooke explained, "by the addition of . . . artificial instruments . . . there may be, in some manner, a reparation made for the mischiefs, and imperfection, mankind has drawn upon it self."[78] Repairs could thus be highly moralized, evoking the recovery of humans from original sin. Robert Boyle compared the integrity of his instruments to the integrity of English society after the Civil War. He complained when an earthen furnace was delivered to his home in Stalbridge House in Dorset. It had "been brought to my hands crumbled into as many pieces, as we into sects."[79]

Perhaps the most famous discussion of repairs in the eighteenth century was in the Leibniz-Clarke correspondence. From 1715 to 1716, the Hanoverian court philosopher Gottfried Wilhelm Leibniz debated the merits of his system of the world with Isaac Newton, who was defended by the Reverend Samuel Clarke. Clarke insisted that God manifested his will in the world through periodic interventions to set right a solar system whose orbits decayed through gravitational irregularities. Leibniz, who imagined God as the most perfect of beings, reckoned such interventions absurd. Using the analogy of God as a clockmaker, Leibniz insisted that God would surely design a universe that would never break down and need repairs. "Tis true that every particular Machine of Nature, is, in some measure, liable to be disordered; but not the whole Universe, which cannot diminish in Perfection."[80] Repair work was a manifestation of divine will or a sign of poor construction. It is notable that Leibniz's arguments appeared as he was engaged in efforts to get one of his own mechanisms, a calculating machine, into good repair. These efforts Leibniz found "tortured."[81] Shifty artisans assisting took off with all the tools needed for the construction. Another dismantled a model of the device and used the parts to repair a clock. Leibniz was never able to properly mend a contrivance for carrying tens. No wonder that in Leibniz's world, perfection was best manifested by an unbreakable mechanism.

Recycling

Newton's universe was a "cyclical cosmos" in which the world was perpetually worn down, broken, and replenished by God using vital spirits.[82] Newton's everyday world valued such cycles too, since daily all kinds of materials that could no longer be made use of in the home were brushed out of the house as "refuse," thrown onto the wagons of scavengers, or broken down to be remade as new materials. Newton's contemporary John Beale, writing in 1666, made an explicit connection between the circulation of old materials and the circulating spirits that brought life to things in nature:

> I see rags, leather, mettalls, all things that are caste out upon the face of the Earth in time resolve to Earth, & then to invigorate liquids to aspire into all the flourishing beautyes of the fields, & gardens, This travayle of Spirits will bring us to . . . the Elysian groves.[83]

If objects could not be repaired or put into service while broken, early moderns dispatched them to a variety of trades where, to use the modern term, they were "recycled." It is important to recognize the limits of the term "recycling." Besides the obvious point that recycling today carries an unprecedented environmental meaning, chapter 3 proposed that in the era of oeconomy, things were *always* seen as pregnant with different uses, so the idea that there was a special stage when they would be given a different use did not exist. This is why early moderns had no term for "recycling."[84] For convenience sake "recycling" will be used here to signal a certain set of practices, but it should be kept in mind that it is an anachronism.

Materials might retain valuable properties despite being considered no longer serviceable for some purposes. These properties could be recovered by collecting and breaking down the material before reforming it as a new raw material or product. As chapter 1 noted, early moderns engaged in the widespread collection of old materials considered no longer serviceable, including rags, cinders and dust, dung, straw, and rubbish. Inside the home, food was routinely recycled. "Scraps," the remains of a meal ("leftovers" is a Victorian term) served to make sausages, pies, and other fare or were fed to animals. Soot and cinders from the fireplace and dust from the floor were swept across stone (later wooden) floors with besoms and brooms and outside into piles.[85] Such things were, as Mary Douglas famously observed, dirt, or "matter out of place," and were removed from the house.[86] But there was little that was not reused. Today, when waste pits are excavated by archaeologists they typically find only fragments of broken ceramic vessels and even these were sometimes put into service as fill for ditches or foundations.[87]

From the fifteenth century, parishes and wards organized "scavengers" and "rakers" to clear the streets of dust and dirt outside people's homes.

Nightsoilmen cleared out cesspools filled with the contents of privies which they carted out to fields beyond the city walls to be used as fertilizer. A series of acts of Parliament between 1662 and the 1760s regulated the laws regarding the paving and cleansing of streets in London and Westminster. On pain of a fine householders were responsible for keeping the lanes and alleys surrounding their homes in good repair. From 1690 they were ordered to keep the street swept, with scavengers taking away the "Dust Dirt Ashes Filth and Soyle" using carts, carriages, and dung pots. Scavengers rang a bell or clapper to warn of their odorous approach.[88] Of course, there were constant infringements, and many noted the disgusting state of streets in London. From 1728 justices of the peace appointed "Surveyors of the Streets" to report on miscreants.[89]

Materials ejected from the home then circulated through various sites and trades, where much could be transformed into raw materials for new goods, such as bricks or paper. Dunghills grew up in locations inaccessible to the scavengers' carts, filled with rotten wood, moldy fruit, fish skins, broken china, soot, dead animals, and dust. These might be picked over for serviceable items. Dung, straw, and refuse were taken to holding grounds called laystalls. Straw served to thatch roofs and was combined with hair left over from animal hides treated by tanners to serve as a binding agent in the plaster on houses.[90] Urine was collected and sold to the makers of alum and leather (twenty-four tons of "London urine" was shipped to Newcastle on returning coal vessels in 1612).[91] Brick dust was used as a scouring agent. Others gathered old rags and clothes to sell in the Rag Fair in Rosemary Lane, a dangerous place northeast of the Tower of London notorious for robberies.[92] Fripperers sold old clothes "new vamp'd" on the street or in shops.[93] Carpenters used dried fish skins to make the early modern equivalent of sandpaper (a nineteenth-century invention). Tanners used the hair scraped off animal hides for stuffing or string and boiled down the hooves for glue.[94]

Scholars proposed elaborate schemes to recycle materials as fertilizer. At the turn of the seventeenth century, as Ayesha Mukherjee has shown, Sir Hugh Platt promoted the use of "seaweed . . . dregs of beer and ale, soot of chimneys, saltpetre residue, shavings of horn . . . dead cattle or stray cats and dogs . . . as well as drained water from rubbish heaps" to serve as manures. Platt believed such materials fed the earth with a generative virtue that enabled plant growth and would enhance harvests to prevent famine.[95] In 1644, Hartlib's collaborator Gabriel Plattes published a pamphlet advertising a new work called *The Treasure House of Nature unlocked, and set wide open to the world*.[96] The *Treasure House* was never published, though Hartlib reprinted Plattes's summary. Plattes proposed that the good use of old materials could enhance manures and enrich the commonwealth. Since linen rags, old shoes and leather, horns, hoofs, and hair contained more "vegetable spirit of the world" responsible for fertility than common dung, they could increase the yields of soil if carefully collected and turned into fertilizer. Blood, urine, and soot collected from butchers, poulterers, and chimney sweeps could serve the same purpose. The Puritan Plattes viewed this project as a moral imperative, stressing the general importance of making the best use of materials:

> If [a]ny one should be seen to cast away good bread, when so many poor people want it, then all the world would cry shame upon him; but why should not the casting away of any of these materials fondly be reputed a more heynous sin, when as they will produce divers times their weight of as good bread as any Prince eateth.[97]

He went on to list many examples of materials that could be put to better use. Thus "food of all kind spoyled for want of looking as musty Corn, mouldy Cheese, stinking Flesh and Fish" could all be made use of, while old clothes deserved longer lives.[98] Servants could "get themselves every year a suit of cloathes, if they buy them at the second hand, and shew their love to the Common-wealth."[99]

While Platt and Plattes imagined as yet unrealized circulations of materials, recycling continued in practice in many areas of activity. New buildings were made from old stone, bricks, and timber.[100] Christopher Wren's Royal Observatory at Greenwich was built in 1675 using old iron, lead, and bricks from a demolished fort at Tilbury in Essex. The timbers came from an old gatehouse in the Tower of London, and the observatory was paid for from funds raised through a sale of spoilt gunpowder.[101] Wren and Hooke also contributed much to attempts to improve the circulation of waste in London, overseeing the construction of new sewers and the siting of latrines and laystalls in the city after the Great Fire.[102]

Householders also engaged in recycling to serve experimental inquiries, breaking down and making use of unwanted materials to produce new items or serve expedient ends. The seventeenth-century chymical author Alethea Talbot made use of spoiled organic materials such as the juice of old leeks, old grease, and stale ale to make medicines. She burnt "six soles of old Shooe" whose ashes she mixed with copperas, alum, and brass coal to produce a powder for treating fistulas.[103] Dung was also very serviceable. Talbot recommended a drink made with a mixture of dung and beer, nutmeg and treacle as a treatment for pleurisy, a recipe she borrowed from Lady Fortescue.[104] Other chymists immersed substances in horse dung to achieve dissolution through the heat generated.[105] In 1666 the physician Daniel Coxe placed glass boltheads of Glauber's Sal Mirabile in a fresh dunghill "to Putrify" them, but the heat broke the glass and ended his experiments.[106] Ashes and cinders were also valuable. Thomas Smart imparted a recipe to Boyle to make a hard alloy using tin calcined with "the (old) cinders of Iron oare."[107] Broken stoneware, earthenware, or porcelain could be of service in the laboratory. Smart told Boyle of his method for making new crucibles out of the powdered remains of old ones, and commonplace books recorded recipes to mend vessels using powdered crucible.[108] Priestley used pieces of broken crucibles to hold materials for heating with a burning lens.[109]

Precious and base metals were also recycled.[110] Metal goods were routinely collected and melted down. As warden and master of the Mint, Newton presided over England's "Great Recoinage" of 1696 to 1699, calling in old coins to melt them down to make new ones.[111] Plumbers recast old lead to make roofs, and artisans kept stocks of old iron, brass, and copper for reuse. Donald Woodward estimated that some ten percent of all iron used in seventeenth-century England was reused, so undoubtedly many mathematical and philosophical instruments were made with recycled metal.[112] When he died in 1703, Robert Hooke's quarters in Gresham College contained, "Eighty eight pound of old Brass, 12 pounds ditto, 4 hundred 3 quarters & 14 pounds of old Iron, Sixty tenne pounds of old Lead."[113]

The trade in "old brass" was international. In eighteenth-century England, instrument makers used imported fine brass from Germany (known as "Dutch" brass), brass made with copper ore from Staffordshire, or a cheaper brass containing old brass or "shruff" from Birmingham.[114] John Harris's *Lexicon Technicum* explained that shruff was "a Collection of Pieces of old Brass" that served instead of copper to cast new brass.[115] Old materials could be of a better quality than new. In 1738 Cambridge professor of astronomy Robert Smith's *Compleat System of Opticks* recommended a mix of "six parts of good shruff brass and one part of tin" to make the white metal speculum for a reflecting telescope because it was harder and whiter than brass made from three parts copper and one part tin.[116] In other cases, the presence of old metals in instruments posed problems. In the 1750s, the Royal Society fellows Henry Baker and William Arderon detected magnetism in brass, but were unsure if the brass had been melted down with old iron, which would account for its magnetic virtues.[117]

Another material to be widely recycled was glass. Glass manufacture in England in the seventeenth and eighteenth centuries centered on two types. Flint or crystal glass contained lead and served to make fine drinking vessels, lenses, and prisms. Crown glass was used to make windows and bottles and in the manufacture of barometers, thermometers, chemical glassware, and receivers for electrical machines and air pumps. Much glass was recycled, since glassmakers added "cullet" or shards of broken glass to melts in the manufacture of new glass. Remains of cullet have been identified by archaeologists at former manufacturing sites in locations such as Vauxhall, London, and Bristol.[118] Traders and artisans using glass also recycled plate glass. The *Plate-Glass Book* (1757) explained how to divide up a broken pane of glass into rectangles of a maximum value.[119] Experimenters also put broken glass to service. Robert Hooke made single-lens microscopes by melting a "clear piece of a broken Venice Glass" into a fine thread and taking off a tiny globule.[120] Francis Hauksbee used pieces of a "broken Looking-glass" pressed together to study capillary action.[121] The archaeology of the site of Joseph Black's chymical laboratory in Edinburgh revealed "black glass wine bottles, and assorted bottle bases containing residues" of chemical substances.[122] Broken glass was also used in instrument making. Optical instrument makers evidently kept pieces of broken glass to use in their workshops, as may be seen on the top shelf in Johan Zoffany's 1772 portrait of the optician John Cuff (fig. 5.4).[123]

Rags and scraps of fabric were another material widely recycled and put into service in the sciences. Rags served to strain off residues in chemical experiments and to protect the bulbs of thermometers when they were immersed in boiling water.[124] To filter distillates John Evelyn used a basket containing ashes and straw with a "linnen cloth tacked in it."[125] Alethea Talbot used "old flaxen cloath" as an ingredient in a medicine for "a man that is burned with a Harlot."[126] Christopher Wren made an unusual use of rags to investigate sympathetic powders, which were supposed to heal at a distance. His instruments were "a servant, who had grievously cut her finger; and a rag rubbed upon the wound being dressed with calcined vitriol, and put into the maid's bosom, her finger within a short time was cured."[127] Rags also served for the first litmus test, made with "linen rags or bits of cambric [a linen cloth], which had been tinged with an infusion of litmus."[128] Joseph Black explained that "A little bit of these, when touched with a liquor that is in the smallest degree acid or alkaline, has its colour changed from the purple to red or blue."[129]

The main use of rags was in making paper. In the seventeenth century London became a center for rag collecting to supply paper mills. The work was done by poor stragglers for a few pennies a day.[130] Scholars of course relied on paper, and scholarly efforts to improve paper-making in the seventeenth and eighteenth centuries were substantial.[131] John Ev-

Figure 5.4 John Cuff, the optician, by Johan Zoffany, oil on canvas, 1772. Royal Collection Trust. © Her Majesty Queen Elizabeth II 2017. RCIN 404434.

elyn was the first person to record English paper-making practice in Surrey in 1678: "First they cull the raggs (which are linnen for White paper, Wollen for browne) then they stamp them in troughs to a papp."[132] An intimate relationship existed between rags and paper in the period. Chloe Wigston Smith has stressed how much text and textiles were connected, through, for example, hybrid techniques such as printing on fabric.[133]

Scholarly works printed on paper made from rags might be the source of illustrations printed on fabric for dresses that would one day become rags again.[134] Paper, as chapter 3 indicated, was then made serviceable in a great variety of ways.

Conclusion

Alongside maintenance, repairs formed an important aspect of household and experimental life in the seventeenth and eighteenth centuries. While there was a tolerance and sometimes a use for broken things, fixing and mending were preferable if possible, to increase the serviceable life of goods. Even when things could not be saved or repaired, they might still be "recycled." Repair and recycling skills were distributed across the household. Men, women, and servants kept domestic and experimental possessions in states of good repair. Repair and recycling work also necessitated sending goods out to circulate among tinkers, chair menders, instrument makers, scavengers, and traders in old rags, glass, and metal. Sometimes the circulations associated with repairs and recycling were international in scale, and making goods travel over long distances could be laborious. In all locations, repair work was also social work, serving hierarchy and distinction, and contributing to class and gendered divisions of labor. Repairs and recycling were also productive, generating new investigations, inventions, and natural knowledge. Early moderns recognized the ubiquity and significance of repairs and recycling in the metaphors they used. These articulated natural and divine action in terms of repairs, and imagined transformations in the sciences not as innovation but as renewal and reparation. This view reflected a widespread use of repaired and recycled materials in experimental culture.

SIX

Secondhand Science

Practices such as maintenance and repair took householders and experimenters outside the home in search of assistance from skilled cleaners, menders, and makers. This chapter and the next expand this geographical scope to consider wider circulations that served thrifty science, in practices of exchange among family and friends, modes of inheritance, secondhand markets, and, in the next chapter, auctions. These practices demonstrate how many options were available to experimenters of the seventeenth and eighteenth centuries to obtain scientific goods without buying them new or having them made by an instrument maker. Thrifty science was often "secondhand" science.

Historians of recipes have done much to trace networks of exchange inside and between families in early modern England.[1] Exchanging culinary, medical, chymical, and other recipes served social cohesion in the household, established new relationships, and cultivated the credit of participants. This chapter builds on the work on recipe exchange to explore a variety of means by which experimental goods were swapped and traded in scholarly networks in seventeenth- and eighteenth-century England. The first two sections examine how householders might come into possession of experimental goods through exchanges of gifts or "presents" and by borrowing and lending among friends and acquaintances. The third section follows scientific goods as they passed down the generations through inheritance and donations. Like recipe books, early modern instruments might endure for many decades in their serviceable life, a fact reflected in the bustling markets for secondhand instruments and specimens examined in the following section. Finally, a case study of the correspondence between the electrician Tiberius Cavallo and his friend the physician James Lind makes apparent the various forms of exchange that were entailed in thrifty science, with assorted goods being passed around from what Cavallo called his "Rubbish Box."

Presents

Another way that people extended the serviceable life of things in the seventeenth and eighteenth centuries was to give them to someone else when they no longer needed them. Possessions could be passed on to other people by giving them as presents, bequeathing them in wills, lending and borrowing, or by selling things on through secondhand dealers and auctions. These practices were common, so that buying things new or having them made was only one of many possible means to acquire goods. As the auction sale catalog of the duchess of Portland's natural history collections noted in 1786, "Every Subject here recorded came into her Possession, either by Inheritance, the Assistance of those who were honored with her Friendship, or by her own Purchase and Industry."[2]

Gift economies have been well documented by economists, anthropologists, and historians of science, with Galileo's presentation of the moons of Jupiter to the Medici being perhaps the best-known example in early modern science.[3] Gifts have often been discussed as a means of social advancement, given and received between patrons and clients of greatly differing status. More recently, scholars have focused attention on exchanges between people of more similar rank and status. Ilana Krausman Ben-Amos has argued that "horizontal" exchange among friends and neighbors was common in early modern England.[4] Recipes were routinely given and received between family members, relatives, and acquaintances. Household books belonging to women such as Lady Ann Fanshawe, Elizabeth Godfry, and Lady Joanna Singen recorded medicinal, culinary, and domestic recipes and the persons who had donated them. Elaine Leong reads these books as "ledgers of social credit and debts," revealing networks of exchange that served not just to impart knowledge but also to cultivate social standing among participants in the network.[5] As chapter 1 noted, these recipe collections recorded a variety of experimental inquiries. Householders, according to Leong, modified and customized processes and ingredients to generate novel recipes that might earn credit from those to whom they were imparted. Householders competed to generate the best recipes and hence earn the highest evaluations among their peers. Household manuscript books, surveyed by one's rivals and allies, served as archives of social relations and the social power these entailed.

Householders exchanged not only medical, culinary, and household recipes but also optical instruments, natural history specimens, and various apparatus. There was no hard distinction between these practices: they might all be seen as a form of experiment. Some of these ended up designated as "natural philosophy" in male publications and academies, while others remained in manuscript recipes, correspondence, or memory. Prior to this gendered division of labor, experiment belonged to both men and women, to husbands and wives, to friends and neighbors, and experimental things routinely circulated amongst them. In 1666, for example, Ezreel Tonge sent Robert Boyle a book, an "excellent piece of good huswiferie," containing recipes for preserving meat that Tonge reckoned should be "referred unto tryall."[6] Such exchanges formed an intrinsic part of good oeconomy, cultivating social relations among people within and beyond the household. For Joseph Banks and his sister Sarah Sophia, gift exchanges of plants and other items were occasions to build collections and initiate conversations. As Banks told one correspondent, "Plant [this root] in your Garden or in that of Some friend that we may when it Comes into fructification Conferr about its right name."[7] Scholarly exchanges of recipes, seeds, books, or instruments were part of a more general culture of informal exchange in early modern England. To get by in a city like London lodgers needed to trust their landlords, families their servants, and neighbors one another, cultivating, according to Ben-Amos, "intense contacts forged on the basis of mutual dependencies and the indispensability of trust."[8] Exchanges of "an immense array of reciprocal favors" served this end, including the exchange of goods that might be made use of in experiments. Ben-Amos lists "coal, cloth, livestock, salt, grains, baskets, and brooms . . . casks, tools, utensils, beddings, clothing" among the many material objects that passed between neighbors and friends as they visited one another's houses.[9] Things were either given as "presents" (a term more common in early modern England than "gifts"), bought and sold on credit, or borrowed and lent. Material objects were exchanged hand in hand with advice, gossip, news, plans, and other verbal information. Such exchanges continued throughout the seventeenth and eighteenth centuries. Elizabeth Yale has recorded the rich culture of exchange among early modern British naturalists, who shared and gifted a myriad of letters, books, notes, drawings, leaves, flowers, seeds, rocks, and coins. The result, as Yale shows, was that exchange was constitutive of the very nation in which it occurred, since the combination of these items shaped a sense of "Britain" as a coherent and knowable entity.[10]

Practices of exchange might have little in the way of disciplinary boundaries. In the late eighteenth century the secretary of the Royal Society Charles Blagden routinely went about London visiting the homes of numerous lady and gentleman friends, carrying with him assorted philosophical items to show or to give as presents. New Years Day 1795 was typical, when Blagden, who resided in Fitzrovia, recorded in his diary how he breakfasted with Joseph Banks at Soho Square, then visited Lord Lucan in Mayfair, to whom he showed the drawings from William Hamilton's paper on Vesuvius. He met Viscount Bulkeley there. After calling on Matthew and Elizabeth Charlton Montagu in Marylebone, he visited Henry Cavendish in Bedford Square before dinner, giving him a copy of another treatise on Vesuvius. Along the way, material exchanges offered opportunities for Blagden to converse with friends, hearing and forming opinions with them concerning men and women prominent in science and society ("Major Rennel character of Mr Wilberforce, harsh/just as he deserved . . . Cambden poor weak character . . . agreed about L[ad]y Spencer").[11] Like the "domestic experimenters" promoted by Hartlib in the seventeenth century, Blagden was enthusiastic to communicate knowledge about nature with other men and women but had little compulsion to publish this intelligence, preferring to collect and collate it in his extensive diary.

Much present-giving occurred at a greater distance, between householders and scholars located in different towns or cities, who wrote to one another and posted books and goods using the somewhat unreliable carriage system described in chapter 4. As in the case of patrons and clients, present-giving through the post might offer a chance to initiate, test, or solidify social relationships. In the first decades of the eighteenth century, the Yorkshire physician Richard Richardson managed to secure election to the Royal Society after sending a variety of dried plants, curiosities, and dishes of potted woodcocks to Sir Hans Sloane in London.[12]

Networks of exchange varied depending on the nature of what was being exchanged and who was exchanging it. A recipe on a scrap of paper was more mobile than cumbersome instruments, rare items needed more care than mundane ones, and different persons might be given varying levels of access to things, depending on their status. Material and social considerations were invariably intertwined in what could often be prolonged series of exchanges over great distances. Class, gender, and wealth informed present-giving. In the 1770s, the artist and botanist Mary Delany lamented the increasingly "enormous price" of shells and corals and noted, "were it not for the Duchess of Portland's bounty, I have small chance of additions to my collection."[13]

Presents often enabled and encouraged new scientific inquiries. The London botanist Peter Collinson famously sent a glass tube to Benjamin Franklin in Philadelphia in 1745, prompting Franklin's lifelong investigations of electricity: "Your kind present of an electric tube, with directions for using it, has put several of us on making electrical experiments."[14] By the late eighteenth century, instrument makers also used presents as an opportunity to promote commodities. Cavallo was delighted when the instrument maker Edward Nairne "made me a present" of his new patent electrical machine and medical apparatus.[15] Thrifty experimenters anticipated the serviceability of exchange. Cavallo spoke of having a "rubbish box" of assorted instrument parts that he was willing to send to Lind when needed, and in turn he reminded Lind to send him items from his "Broker's room" or office.

> Together with this letter you will receive a little provision for your collection of mechanical prints, viz: the plates of Mr Hatchet's new coach of safety . . . I shall, whenever I can, take care to provide duplicate bills or prints for you; but remember that I am also a collector of the same species as well as of minerals, and therefore if rummaging in your Broker's room, you find any duplicates, remember N. S. [*nous*—us] in Little St. Martin's Lane.[16]

These kinds of horizontal exchanges also bore on the more vertical giving and receiving of gifts between scholars and patrons. Scholars might pass on broken instruments to great patrons. In August 1674 Johann Hevelius asked Henry Oldenburg to send him a new microscope, since the wooden rings holding the lenses on his existing one had shrunk "so hard . . . that the lenses cracked a little near their edges."[17] Hevelius went on to explain that "I have lately presented the one I owned to a certain grand patron of letters."[18] Perhaps the gifting of broken instruments made little difference to a patron since he or she would not be using them personally. Perhaps it reflected *Ruinenlust*, a general tolerance for broken objects.[19]

Lending and Borrowing

Another way to circulate used goods was in practices of lending and borrowing. Good Christians were expected to lend freely to their neighbors.

Newton's rector Samuel Clarke evoked the Gospel of Luke, "To him that begs any thing of you, give freely; and to him whose Wants oblige him to borrow, be always ready to lend, and never rigorous to exact it of him again."[20] At the same time, thrifty householders were advised to be self-sufficient. Gentleman farmers, for instance, should "Have all necessary implements of your own; never borrow, if possibly you can help it. You lose time; and if any accident happens to that you borrow, you may find insuperable difficulties."[21] In practice, status and power might be negotiated through acts of lending and borrowing that operated like a gift exchange. The owner of a large number of goods could lend more than the owner of a small number, and so gain prestige through this capacity. It was common for the wealthy or nobility to make their collections accessible to more lowly scholars, as Henry Cavendish did with his library in eighteenth-century London.[22] The powerful also had a habit of borrowing things that they never gave back, leaving their clients at a loss because it was not appropriate to ask for their return.[23] Much lending and borrowing, however, happened between persons of more equal status. Sharing helped to build relations between individuals, made things more accessible, and saved the money and time needed to find and purchase new goods.

The earliest entries in Robert Hooke's diary refer to his borrowing a paper by the microscopist Malpigi from fellow experimenter Richard Waller, and Hooke regularly borrowed and lent books.[24] Historians of reading have examined the lending and borrowing of books in early modern England.[25] Borrowers not only read books but also copied or abstracted them to make a permanent record of the borrowed item. In his diary Hooke noted, "Sir R. Southwell till 11 . . . I lent him my last two lectures on Dr. Burnet's *Archaeologia* to coppy."[26] Lending was both a social and a material practice, entangling judgments of value, changes in social status, and material acts. Lady Ranelagh sent a copy of a book by the Hartlib circle member Gerard Boate to her sister-in-law with the stipulation that it must not be shared with physicians in case the secrets therein were lost to Boate's son:

> I send yu Dr Botes Booke to write out for yr owne use but must beg yu yt noe Dr nor Apothecary may have any thing out of it because he has a sonn of his owne yt is studying towards being of his fathers profession, & therefore for him I would reserve ye assistance of his fathers Experim[en]ts.[27]

Borrowers might travel long distances to collect a promised volume. Some unfortunate individuals agreed to carry books on behalf of a lender to their recipient if they were traveling to see them. Wealthy borrowers might have parts or all of a work transcribed while on loan. Libraries were also places where books were shared, although institutional lending from libraries was limited before the eighteenth century, when chains began to be removed from expensive volumes, previously viewable only by visiting the library.[28]

Instruments were usually loaned between individuals and then used for observational or experimental work. Sometimes individuals had no access to expensive or specialized instruments and approached others to help them. In 1748 Buffon sent his writing on generation to John Turberville Needham, then residing in Paris, in the hopes of using one of Turberville's microscopes, which were "infinitely superior to my own," to study infusorial life, which Needham agreed to.[29] Lending was often reciprocal, in that the borrower agreed to provide some service in return for the opportunity to borrow. Parties negotiated to maintain what was felt to be a fair equivalent of contributions. Reciprocity might involve acknowledging a loan in public, for example in print. When William Derham sought to compare the rates of pendulums oscillating in air and in a vacuum, he at first "recommended the Experiment to many ingenious persons who had better opportunities of trying it than I."[30] But Derham was able to borrow the necessary instrument from Francis Hauksbee. He acknowledged the loan in the *Philosophical Transactions*, noting "Mr Hauksbee's generous kindness, in lending me one of his Air-Pumps (which I shall always gratefull acknowledge)."[31] Acknowledgment also offered opportunities to advertise links to persons of status. In the preface to his *Essay on Electricity* (1787), George Adams noted that he was "particularly obliged to Sir Joseph Banks, for his politeness in lending me 'Les memoires de l'Academie de Berlin' for 1780, at a time when I could not procure them elsewhere."[32]

If loans extended too long, were not returned, or entailed damage to property, there could be tensions. Lenders needed to be appraised of who possessed their loans. Joseph Priestley wrote to John Vaughan in May 1796,

> I forgot to tell you that Mr. [Benjamin Smith] Barton had lent me the copy of Dr Bancroft's book on colour, and that, intending to make some experiments in persuance of this, I took it with me. I hope you will excuse the liberty I took. If you want it I can return it.[33]

Borrowers promised to take care of loaned things and return them in the state they were borrowed in. William Derham promised Hans Sloane that a book he had borrowed would "be used wth as much safety as if it lay in yr Library, & soon returned."[34] If an item was damaged during a loan there could be embarrassment or dispute. At the very least an apology was in order. Benjamin Franklin lent an electrical machine to Jonathan Belcher in New Jersey in 1751. The glass receiver was smashed in transit, but Belcher informed Franklin that he had made the best of what was left and was now sending it back: "I now return your Apparatus with a great many thanks and am very sorry for the mischance it met with."[35] Sharing worked better over short distances. The watchmakers of Clerkenwell arranged themselves so that "no one man had a complete set of tools belonging to his own department . . . the system prevailed of borrowing from one another, and of lending in return such tools as might be possessed."[36] Such arrangements meant that lenders and borrowers needed to live and work in proximity to one another, and probably practices of sharing contributed to the concentration of instrument makers in particular parts of London such as Clerkenwell, Fleet Street, and the Strand.

In an ethnographic study of cooperation among farmers, B. Lisa Gröger has suggested that lending and borrowing vary with the scale of objects.[37] Small, cheap things might not be lent, because they are easy to obtain, while very large bulky things are not lent because they are hard to move. Most lending and borrowing occur around objects of relative scarcity and easy mobility. This may explain why recipes, seeds, books, and instruments were most commonly lent and borrowed. As William Derham wrote to Hans Sloane in January 1707, "I have here sent you the 3 books of Mr Boyles (wth my many thanks) wch I borrowed, & wch are all I had except Hakewill, wch is too large to be so soon dispatched."[38]

Inheritance and Bequests

Advice on husbandry from 1697 proclaimed, "One hath well said, That Frugality is the Left-hand of Fortune, as Diligence is the Right. And the witty *Italian* hath this common Proverb, *That if Frugality be a Vice, it never disinherited any Man*."[39] By taking good care of one's possessions, one might hand them down to the next generation. As Leong has noted, "Bequest and inheritance . . . played a pivotal role" in the careers of books of medical and culinary recipes.[40] Leong records how a seventeenth-century book of recipes belonging to Rhoda and Ferdinando Fairfax was inherited by their daughter Ursula, whose own stepdaughter recorded her name in the front flyleaf. The book remained with the stepdaughter's family for two generations and was eventually auctioned in the twentieth century to Henry Wellcome.[41] Other household books were mentioned in wills, and in addition to being repositories of practical advice, books became treasured family possessions. Inheriting a book gave the person receiving it a responsibility to take care of it, signaled the aspiration of the donor that their inheritors would make use of the book, and acted to reproduce knowledge, techniques, and practices across generations.[42]

The same social and material considerations applied to other experimental goods passed down in wills and bequests, so that along with recipe books a variety of instruments, specimens, and household possessions that might serve for experiment were inherited by early modern families. When he died in 1691, Robert Boyle left, for example, his "best microscope and . . . best loadstone" to Robert Hooke.[43] Inheritance constituted a significant means of giving and receiving scientific materials in the early modern period, and acts of inheritance marked lines of descent and affiliation between scholars, manifested a record of friendships and allegiances, and provided the means for experimental investigations to continue across generations. Distribution of goods might follow a variety of patterns. The will of botanist and artist Mary Delany, who died in 1788, left to her nephew, Bernard Dewes,

> the deal cabinet of shell fossils, and any other fossils that the Duchess of Portland does not choose; but any duplicates to be given to Master Daniel Sanford, with the little flat case . . . in the closet in her bedchamber; Bernard Dewes also to have the drawers with corals "if he chooses it."[44]

Delany prioritized her friend and benefactor the duchess of Portland, then gave priority to Dewes, and then Sanford, a cascade of inheritors. Benjamin Franklin, alternatively, allied each possession with a particular beneficiary.

> I request my friend, Mr. Duffield, to accept moreover my French wayweiser, a piece of clockwork in Brass, to be fixed to the wheel of any carriage; and that my friend, Mr. Hill, may also accept my silver cream pot, formerly given to me by the good Doctor Fothergill, with the motto, Keep bright the Chain. My reflecting telescope, made by Short, which was formerly Mr. Canton's, I give to my friend, Mr. David Rittenhouse, for the use of his observatory.[45]

Delany's and Franklin's bequests honored friends and institutions, recorded supporters, and passed previous gifts to new owners. The same was true of the will of 1799 made by James Beattie, professor of moral philosophy and logic at Marischal College, Aberdeen. Beattie recorded how bequeathed books were intended as "memorials of me to other friends."[46] Presents were passed on and items returned to those who provided them. Beattie's will noted, "To Rev. Dr William Laing in Peterhead, to whom as a friend and as a physician I have often been obliged, I bequeath all my music books . . . and the telescope which he made for me."[47]

Bequests involving experimental goods often followed patterns of inheritance seen with other goods. It was a common practice for gentlemen to leave their clothes and linen to servants, and this carried over to laboratory assistants and technicians. Robert Boyle left his "wearing Apparell and linen" to his "servants" and laboratory workers Robert St. Clair and John Warr.[48] Widows of artisans inherited their husbands' workshops and often carried on their business, which, as Londa Schiebinger has shown, was a practice also taken up by astronomers, enabling women to undertake astronomical researches.[49] Public lecturers also followed this practice. The Dutch natural philosopher Charles Diller gained fame in the 1780s for his "philosophical fireworks" made with different kinds of inflammable airs, which he performed in the Hague, Paris, and London.[50] In his will Diller bequeathed his "wearing apparel linen and woolen of every sort" together with two silver buckles to the two assistants in his London shows.[51] Diller also bequeathed all the "machinery and other materials in my said profession" to his wife, Margaret, on the understanding that his musician friend Victor Gonetti would work with them for the benefit of Diller's son.[52] Gonetti subsequently displayed philosophical fireworks in Bristol and Birmingham.[53]

Bequests also served institutions. Fellows and patrons bequeathed monies for commemorative lectures, prizes, or purchases, and specimens, collections, and instruments. In 1701 Lady Sadleir left monies to the Royal College of Physicians and the Royal Society to establish the Croonian lecture in memory of her husband William Croone.[54] Alexander Marr has shown how Sir Thomas Bodley managed to accumulate some 8,700 volumes between 1598 and 1605 for his library in Oxford, giving public recognition to donors in return for support.[55] The benefactors' list blurred social boundaries by including lesser scholars together with great patrons.[56] Benefactors sometimes bequeathed instruments such as globes, armillary spheres, and quadrants with the books describing their use.[57]

The status of donations and bequests to institutions needed to be negotiated, depending on the circumstances of the parties involved and the quality of the materials received. In 1782, James Manning described the initial collection of Rhode Island College library as "about five hundred volumes, most of which are both very ancient and very useless, as well as very ragged and unsightly."[58] Damaged goods could be repaired. James Watt was employed by the University of Glasgow to "repair and put in order" astronomical instruments bequeathed by Jamaica merchant Alexander Macfarlane, which became the foundation for the university's Macfarlane Observatory, opened in 1757.[59] Or items might be given the status of relics. In Philadelphia in 1836, Joseph Hopkinson presented a battery of fifteen Leyden jars formerly belonging to Benjamin Franklin to the American Philosophical Society. Hopkinson explained that Franklin had left the battery to his father, Francis Hopkinson. The society accepted the gift and recast the instrument into a relic of Franklin, adding a brass plaque and replacing several broken or missing jars with facsimiles.[60]

Through present-giving, sharing, and bequests, the lifetime of books and apparatus might stretch over several generations. In 1691 Constantine Huygens donated a 123-foot telescope to the Royal Society, which periodically lent it to fellows over the next century. Borrowers included James Pound and James Bradley in 1728, William Derham in 1741, Lord Macclesfield in 1748, and Henry Cavendish in 1785. Borrowers reshaped the instrument, altering, breaking, and adding parts during loans that could last for several years.[61]

Instruments for demonstration lectures had similarly long lives. After the death of the natural philosophy lecturer John Horsley in Morpeth, Northumberland, in 1732, his lecturing instruments were purchased by Caleb Rotheram, who ran a dissenting academy in Kendal in Cumbria, after whose death they were purchased by John Holt, lecturer in natural philosophy at the Warrington Academy in Lancashire, where they were also used by Joseph Priestley. Then they were presented to Hackney New College in London in 1786, and then to the library of Dr. Williams in London's Cripplegate, where they still remained in 1821.[62] New institutions were happy to purchase old instruments. In 1673 James Gregory traveled to London to buy stock for the planned observatory at St. Andrews. Purchases included a large secondhand mariner's astrolabe made in 1616 and instruments by the Elizabethan maker Humfrey Cole, made some half a century earlier.[63]

As instruments passed between generations, their long endurance and wide dispersal might place them in quite different contexts, prompting the need for maintenance, repairs, and shifts to adapt them to novel conditions or uses. In the 1770s, the Irish physician John Longford in Cork purchased a secondhand clock at auction to make astronomical observations in his home. When Longford obtained the clock, he replaced parts and reworked the clock's setting to ensure it was sufficiently accurate for making observations:

> My clock was bought three years hence at an auction, among the collection of clocks of some gentleman . . . It had a common pendulum with a heavy bob. I got a wooden pendulum and a new crutch applied to it exactly according to Mr. Ludlam's directions . . . For four months after it was put up it went irregularly, upon which I fastened it to the wall with screw bolts and large washers. Since that time, May 1772, it never lost more than 2″ per diem, or gained more than 2″.5.[64]

Langford's observations, printed in the *Philosophical Transactions*, subsequently enabled the astronomer royal Nevil Maskelyne to calculate the longitude of Cork.

As in the case of recipe books, the long life of instruments might be reflected in inscriptions used to signal a change of ownership or to identify present users with illustrious forebears.[65] An octant made in 1784 now in the National Maritime Museum carries the name "R. Maxwell 1797" superimposed over the erased name "Nixon Hill 1784."[66] Instruments gained "pedigree." Referring to the celebrated instrument maker Jesse Ramsden, another instrument in the same collection, a sextant, bears the inscription "This Sextant was presented to Admiral Sir John Ommaney [*sic*] by Ramsden himself in 1792 being graduated by his hand and afterwards given by Sir John to Capt. Blackwood in 1851."[67] Such pedigree might inspire trust in instruments, while their continuing use could inspire innovations.[68] As Henry Charles Englefield wrote in the *Philosophical Magazine* of a new form of transit instrument, "An old transit instrument formerly belonging to Mr. Aubert, and at his sale purchased by Mr. Walker, gave me the first hint of the instrument now to be described."[69]

The Trade in Secondhand Goods

Philosophers not only presented, lent, and bequeathed materials to one another but also purchased them secondhand from friends or via traders and auctioneers. The term "secondhand" originated in the fifteenth century and by the eighteenth signified, according to Johnson's *Dictionary*, a "Possession received from the first possessor."[70] Historians have drawn attention to the great variety of secondhand circulations in this period. Young families setting up house visited house sales and auctions to pick up cheap domestic goods. Clothes, furniture, fixtures and fittings, books and paintings could all be purchased secondhand.[71] Not everyone was impressed by secondhand goods. Pennell records a father's response in 1703 to his son's interest in buying used bedding and brass: "buy new and it will last, for you or we may buy at sales which may seem new, yet be half worn."[72] Secondhand goods might carry a sense of being inferior or fraudulent, but as Jon Stobart and Ilja Van Damme note in a study of the trade, early moderns "from all social groups bought a wide variety of second-hand goods."[73]

An extended network of secondhand dealers operated in early modern England.[74] They ranged from impoverished criminals to genteel artisans. The illiterate Samuel Badham kept a stall selling secondhand shoes and doing odd mending jobs in Southwark, one of various trades he tried out before he was convicted of murder in 1740.[75] A more upmarket contemporary, the goldsmith Phillips Garden, sold secondhand silverware and watches from a smart shop at the Golden Lion in St. Paul's Churchyard and worked for various aristocratic clients. Like the families in Hogarth's family portraits who were proud to show off repaired goods, these retailers happily advertised secondhand items on their trade cards (fig. 6.1).[76]

Goldsmiths such as Garden, Thomas Chesson, John Hopkins, Henry Morris, and John Briscoe sold secondhand silver plate, jewelry, and watches.[77] Other secondhand goods available in eighteenth-century London ranged from coaches and chariots to interior shop fittings.[78] The Frying-Pan and Key on St. Margaret's Hill in Southwark sold "all Sorts of Smiths Tools, viz. Anvils, Bellows, Vices . . . both New and Second-Hand . . . at a very reasonable Rate."[79] Shops often sold used stock alongside new, so that as Ian Mitchell has suggested, "there was no rigid distinction between first- and second-hand circulations of goods. Many retailers were content [to] take old items in part exchange for new ones and would sell second-hand goods alongside unused ones."[80] This is again indicative of how early moderns viewed objects as "incomplete," with the assumption that goods would have several owners and uses over time. In some cases secondhand goods were sold as cheaper alternatives to new, so that tailors, for example, kept a stock of old garments for less well-off customers. But, as Mitchell indicates, "this was untypical of second-hand markets in general."[81] The decision to buy secondhand might be motivated by poverty but not always—it could equally depend on the provenance and quality of goods on offer. The wealthy bought secondhand too. The goldsmith John Hopkins dealt in secondhand plate in a shop in Fleet Street in the 1730s, where he

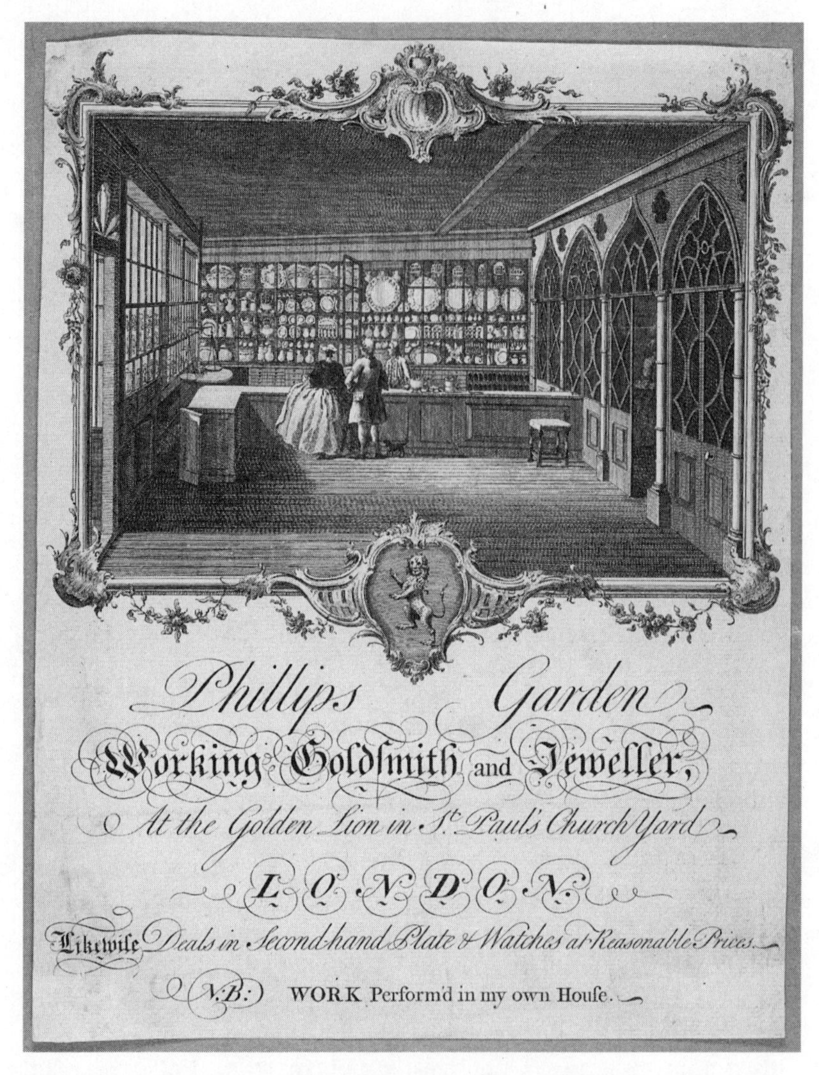

Figure 6.1 Trade card of Phillips Garden, "Working Goldsmith and Jeweller, at the Golden Lion in St. Paul's Church Yard, London, Likewise deals in Second-hand Plate & Watches at Reasonable Prices, Work Perform'd in my own House." British Museum Heal, 67.156. © The Trustees of the British Museum.

also sold new plate. He explained to genteel customers that "Nothing engrav'd with Coats of Arms, &c. will . . . be expos'd to Sale before the Engraving be entirely taken out, so that it shall not be known the same was ever engrav'd, which is presum'd will be most pleasing to Buyer and Seller."[82] Sellers reassured customers that used goods were "none of them worse for Use."[83] Goods could also be purchased with a guarantee. "Preston's Second-Hand Musical-Instrument Warehouse" at Exeter Change explained that "Every Instrument will be warranted genuine, and sold conditionally, if not approved on three months trial, to be exchanged, or the full money returned, deducting the hire."[84] Such securities were needed as increasingly in the eighteenth century certain traders in rags and old iron garnered the reputation of dealing in stolen goods.[85]

Used books, of course, were regularly sold or exchanged among the learned, assisted by a variety of booksellers.[86] Once bound in leather, books were durable objects, with many early modern books still in circulation today. Long lists of successive owners' names inside front covers testify to the way books passed through many hands over multiple generations. Like other goods, used books were typically included in booksellers' inventories in the seventeenth and eighteenth centuries. Dedicated secondhand or antiquarian traders appeared only once buyers made a sharper distinction between used and new books in the early nineteenth century (when, as chapter 8 will argue, the notion of "incomplete" objects was being transformed). Until then, as John Feather points out, many catalogs of booksellers in the eighteenth century were "dominated by the second-hand books."[87] Used books were available in London, traditionally the center of the book trade, and in the provinces, where the book trade expanded rapidly in the eighteenth century, catering to a growing reading public and a rising demand for stationery, which booksellers also supplied. By the 1790s, towns such as Nottingham, Sheffield, Leeds, Hull, York, and Manchester boasted more than ten bookshops each.[88]

Scientific instruments were also traded secondhand. Exchange was constitutive of community, and many transactions occurred between, and helped to create, friends and acquaintances. Tiberius Cavallo wrote to Lind in 1791,

> you told me, that a gentleman of Windsor . . . had a telescope of Dolland, which he wished to sell. There is now an acquaintance of mine who wishes to have a good telescope of Dolland, and therefore if this Windsorian telescope is still to be had, I beg you will inform me.[89]

Instrument makers dealt in secondhand items, selling both old and new instruments together. In the 1750s, the instrument makers Thomas Heath and his apprentice Tycho Wing placed advertisements in the *Public Advertiser* offering "Most Money for all Sorts of Second-hand Mathematical, Philosophical and Optical Instruments."[90] Notices of sales then followed, "To be Sold cheap, Being Second-hand, by Heath and Wing, Mathematical and Optical Instrument-Makers, near the Savoy, A Large double Sperit Level, fit for long Levels, an eight Inch Theodolite, and a plain Table compleat, and several cases of Drawing Instruments."[91] Other advertisements announced the sale of a secondhand orrery, telescope, air pump, electrical machine, and microscope at their premises near the Savoy on the Strand.[92] The brothers and instrument makers Samuel and William Jones also advertised secondhand goods for sale in the 1790s.[93] Lecturers engaged in the trade. James Hull, who styled himself a "Teacher of Geography," offered instruction to men and women in the use of globes and maps in private lectures at his premises in Half-Moon Passage in Aldersgate Street, north of St. Paul's Cathedral, where he also dealt in "all Sorts of mathematical Instruments, bought and sold, both new and second Hand."[94]

Provincial and colonial scholars asked metropolitan colleagues to purchase secondhand apparatus on their behalf. In 1706, writing from Derby, the Rev. Henry Greatorex requested John Flamsteed to buy him "a good second hand Telescope with some few instruments necessary for a right use of it" with the promise of reimbursement.[95] Shifts could compensate for wear to used goods. In 1789, Richard Price purchased a set of instruments in London for Ezra Stiles in Yale College. Among those from the firm of Nairne and Blunt were "An electrical machine and apparatus with all the latest improvements—Sold cheap because second hand but altered and improved."[96]

At Your Service: Tiberius Cavallo's Rubbish Box

An extended example serves to further illustrate the circulation of secondhand scientific goods and the ways various practices of exchange, maintenance, and making use might affect them over a period of time. Tiberius Cavallo grew up in Naples before emigrating to London, probably in the early 1770s. After publishing an extensive treatise on electricity in 1777, Cavallo was made a fellow of the Royal Society, after which he served as the Bakerian lecturer for a decade from 1782.[97] By this time Cavallo was living on Fairy Hill near Eltham in Kent, from where he regularly sent letters to James Lind, not the Lind of scurvy fame, but a Scots physician and astronomer who accompanied Joseph Banks on his voyage to Iceland in 1772. Lind set up practice in Windsor, about seventeen miles west of London, served as a physician-in-ordinary to the royal household, and busied himself with experiments, inventions, and amusements. Between 1782 and 1809, the two men corresponded, and their letters evince a busy trade of information, gossip, books, and apparatus. Cavallo also sought to assist in the buying, selling, exchange, and alteration of used instruments. In November 1792, Cavallo wrote to Lind to ask him about an old quadrant:

> An acquaintance of mine wishing to construct a quadrant on a different plan from those commonly used, I have proposed to him your quadrant, with which, altered &c., he may perhaps try the scheme at a cheaper rate. In short he will not pay more than about 35 pounds for the old quadrant, provided Mr. Haas, who must make the alterations, thinks that he can make use of most of the old parts. Therefore if you think of parting with that useless encumbrance, which I would by all means advise you to do, let me know it, and I shall send Mr. Haas down to Windsor in Sunday the 2d of December to inspect the quadrant, &c.[98]

Although the design was new, Cavallo's friend chose to have an instrument maker (in this case the German émigré Jakob Bernhard Haas) "make use of" the old parts of a secondhand instrument by altering it to his novel design rather than construct a new one.[99] The correspondents evidently viewed the old instrument as "incomplete" in that it would serve readily to be altered into a new one, and Cavallo assumed Lind would be willing to pass on the instrument, for a fee, if he no longer wanted it. Cavallo assumed correctly and dispatched thirty-five pounds to Lind to pay for the quadrant, which Haas, presumably, collected.[100] Haas now needed to determine whether the relevant parts of the old instrument would serve for the new one. Those that did not were then put back into circulation rather than thrown away. As Cavallo wrote to Lind,

> On account of the alterations, many pieces belonging to the said quadrant have been rejected viz: the micrometrical screws, the adjusting collars, which held the telescope, &c. &c. These pieces are at my disposal, and therefore at yours; so that if you have any scheme in view wherein some of those pieces are likely to be useful, they are at your service.[101]

Parts could be altered, reused, and put into new "service" in many instruments over time, and Haas (who evidently was also put into service) altered the quadrant through a bricolage of old and new elements. Meanwhile, as noted above, Cavallo kept unused parts in a "rubbish box" whose contents he made available to Lind. "If you have not made yet my new electrical instrument, and if you want any article towards the construction of that or of any other instrument, remember that I have a great many odd pieces in my rubbish box, which are at your service."[102]

Such actions demanded both material and social negotiations which could reshape the relationships and identities of both actors and objects. Things could go wrong in these negotiations. In February 1793, Cavallo wrote to Lind

> to request you to find out, if you can, the eye piece with the diagonal mirror belonging to the telescope of your quondam quadrant; for now that the alterations of the same are completed the said eye piece can not be found, and they say that it was never sent to town. Be so good as to look for it immediately, and to let me know by the return of the post whether it is in being or not. It will be very troublesome to have another made.[103]

The missing piece changed the interaction between Cavallo and Lind from a favor between friends (Cavallo making use of Lind's old quadrant) to an awkward request that could prove "troublesome," since the altered quadrant might have to rely on new, expensive parts after all. In the event, the eyepiece could not be found, and Cavallo did have to order a new one, though the two men did not quarrel over it. Cavallo, not Lind, supplied the missing part, perhaps because he was considered the broker in the deal.

Substitution was another thrifty practice undertaken by the two men. After discussing the eyepiece Cavallo turned to chemical experiments proposed by Lind:

> I am glad that you are going to try a simpler still; for that of Argand (I think with you) is of a very complicated construction. I only saw the drawing of it. As for my improvement, I had sometime ago a head of a tin still made on purpose, which I tried and seemed to answer very well, but the body of the machine being spoiled I neglected to have another, and therefore the head remained by me, which I shall send to you by Mr. Cooper; for you can easily have a body made to it.[104]

Once again, Cavallo was insistent that an old, in this case literally "incomplete," instrument could be put to new service, as a substitute for Lind's Argand still. Both "Mr. Cooper" and the still-head would be put into service to achieve this, and Cavallo proposed the substitution on the grounds that his still was of a simpler design than Argand's. Three days later, Cavallo sent the still to Lind with the thrifty stipulation to make good use of it. "Herewith you will receive a head without the body.

Make what you can of it, and success may attend your experiments on distillation."[105] Within a month Lind wrote again to report the success of the new still.[106]

Conclusion

Despite the rising tide of manufactures in early modern Britain, buying new was not the only option for obtaining instruments and scientific goods. Thrift demanded a balance between buying new and making use of existing things, and while historical accounts of the consumption of new scientific goods are numerous, less has been written about the circulation of used and secondhand goods. These circulations were diverse. Households were closely interconnected with one another, and exchanges with friends and neighbors helped build networks of support, established trust, and cultivated relationships of mutual advantage. Besides the proverbial "cup of sugar," telescopes, medical recipes, seeds, quadrants, and books were among the material things exchanged. This secondhand circulation was productive of scientific knowledge. Used instruments were widely put into service to perform chymical experiments, to make navigational and astronomical measurements, or to disseminate science to public audiences. Present-giving, lending, and borrowing circulated knowledge, provided scholars with scientific goods they might otherwise be unable to access, and provided them with instruments that, in Franklin's case, led him to explain the nature of electricity. Exchanges were prompted by the desire to make the most of possessions and were often accompanied by thrifty practices of alteration, bricolage, and substitution. As Cavallo and Lind's case makes apparent, these acts did not exclude commercial considerations but coexisted with them.

SEVEN

Auctions and the Dismantling of Science

Another site where thrift and commerce mingled was the auction. The growth of the instrument trade and the changing circumstances of the "middling sort" in eighteenth-century England began to transform the nature of thrifty science. As chapter 1 indicated, definitions of oeconomy and thrift changed with the growth of a more commercial society. Thrift was increasingly associated with saving, while enlightened commentators on oeconomy cast the house as a site of consumption, contributing to the wealth of the nation by consuming the products of manufactures and trade. Jan de Vries has spoken of an "industrious revolution" in the period, when by devoting more time and resources to market activity, families earned enough to pay for a growing number of household and imported goods.[1] As Michael Kwass has put it, by the end of the eighteenth century, families

> filled their homes with beds, dressers, and pottery; expanded their wardrobes to include white linens and brilliant calicoes; sported new accessories such as fans, umbrellas, and pocket watches; sipped exotic beverages (tea, coffee, and chocolate) laden with sugar; and smoked, chewed, and snorted prodigious quantities of tobacco. Their ancestors would have been amazed.[2]

In the case of the sciences, this change was reflected in the growing consumption of scientific instruments, designed, made, and sold by commercial instrument makers. As in the case of Lind's quadrant discussed in the previous chapter, thrifty traditions of making use, alteration, substitution, etc., might, toward the end of the eighteenth century, be increasingly applied to dedicated instruments rather than to shifts and bricolages made with items found in the home. The next chapter will consider the long-term consequences of these changes for thrifty science. The present one examines an event that captured the possibilities and tensions between old and new conceptions of thrift and the transition between a domestic culture of "incomplete" household objects and a market economy of specialized instruments. This was the auction.

Auctions were another occasion for the circulation of secondhand goods in the seventeenth and eighteenth centuries. Historians have given scant attention to early modern auctions of scientific goods, with most of the focus being on the nineteenth century. J. M. Chalmers-Hunt's *Natural Historical Auctions, 1700–1972: A Register of Sales in the British Isles* (1976) recorded how insects, birds, shells, minerals, and fossils fell under the hammer, and noted how auctions contributed to knowledge, through the compilation of sale catalogs, for example.[3] In 1967 Gerard L'Estrange Turner printed details of the sale in 1793 of the earl of Bute's instruments, the largest collection of the time, and reckoned it demonstrated "the collecting fervour of the period."[4] Most commonly, auction catalogs have been used to reveal the values of instruments or to detail their ownership or provenance.[5]

Larry Stewart has proposed that eighteenth-century auctions "provided the space in which mathematics and experimental philosophy were promoted just as any other commodity of interest to the traders of London."[6] Auctions were not only places of buying and selling but highly charged public events that encouraged social and scholarly interactions. Auctions, Stewart argues, belonged to the "rise of public science," the efforts of natural philosophers to promote Newtonian and experimental philosophy to new audiences of polite and commercial society in the eighteenth century.[7] This chapter builds on Stewart's observations to explore the auction as a site where thrifty science met public science. Auctions emerge as sites where both secondhand goods circulated and people gathered to learn about natural and experimental philosophy. Critical to this story is an appreciation of the construction of auctioneers' credit. Auctions succeeded only once they had secured a reputation as reliable institutions, and in that regard they were similar to the experimental philosophy itself, which emerged consecutively with auctions in the second half of the seventeenth century. Auctioneers, like experimental philosophers, needed to establish their credit for the practice to succeed, and it is interesting to compare their strategies to those pursued in the sciences. Furthermore, as Chalmers-Hunt noted of the nineteenth century, auctions contributed to the production of knowledge through practices of cataloging and inventory. They also led, in the course of the eighteenth century, to a redistribution of scientific material culture from elite collectors to middling practitioners, what the historian Cynthia Wall has called a "dismantling" of goods.[8]

This chapter also highlights some of the limits of thrift. As the "buyer of bargains" discussed in chapter 1 had it, the problem with auctions was that while they could serve oeconomy by continuing the lives of goods through secondhand circulation, they were also designed to encourage spending and accumulation. The anthropologist Haidy Geismar has highlighted the ambiguities that auctions encourage, as places where different notions of value come together, blend, and blur. "The auction salesroom is a paradoxical space in which multiple perspectives may coexist and influence one another . . . a place in which dynamically opposed models of value and exchange are increasingly entangled."[9] It could be argued that this was already true in the eighteenth-century auction. In a fascinating mixing up of oeconomic, epistemic, and market values, what might appear to be thrift could end up as an unbalanced excess of expenditure. Scholarly inventories fused with commercial cataloging. Aristocratic collections transformed into everyday working tools. The auction, then, stands for the emergence of a complex of different values applied to thrifty science that will be explored further in the final chapter.

Establishing the Credit of Auctions

Selling to the highest bidder had ancient origins and was long used to sell slaves, prisoners, and the spoils of war. But in the seventeenth century, first in the Netherlands and then in England, auctions began to be held to sell the estates of deceased persons, ships' inventories, and then, from the 1670s, pictures and books.[10] In the eighteenth century, a new generation of enterprising auctioneers sold the libraries and collections of deceased physicians, antiquarians, and notables, taking a small commission on the sale and giving the rest to executors or widows.[11] By the close of the century, all manner of goods were put up for sale at auction, including furniture, paintings, curiosities, books, plants, shells, and specimens, either in specialized sales or more commonly as part of the inventory of household estates.

The economic definition of an auction is a process of buying and selling based on the offering of bids, with the sale going to the highest bidder. It comes from the Latin *augeo*, meaning "I augment, or increase." Sara Pennell records a variety of early modern bidding processes. Sales might go to the highest bidder at the fall of the hammer, or proceed by the flame of a specified length of candle, whereby the last bid given when the candle expired won the lot. In "Mineing," the auctioneer announced descending prices for a lot, with the item going to the first bidder to shout "mine" when the desired price arrived.[12] The auction, like experimental philosophy, emerged as a new practice in England in the second half of the seventeenth century. Its popularity increased rapidly after the first book auction was undertaken by the bookseller William Cooper in London in October 1676. By 1700, there had been more than one hundred book auctions, selling 350,000 works for a total of about £250,000, prompting sales to begin across the country.[13] The new method of selling by auction was itself considered an experiment. Moses Pitt, who organized one of the earliest book auctions in London in 1678, referred to recent "Experiments of . . . the Sale of Books by Auction" and explained how he had

> resolved to gratifie the Learned by exposing to Sale the Library of ye worthy and learned Person deceas'd, with a considerable Number of other choice Books of most Sciences, some of which have been bought out of the best Libraries abroad . . . and others, that for several years last past at great Expense, and out of the most Eminent Sects of Learning beyond the Seas have been Imported, which I hope will be Invitation sufficient to the Curious to embrace this occasion of Buying what hath not hitherto, and perhaps may not again be exposed in this way to Sale.[14]

The auction was a highly evocative site in the seventeenth and eighteenth centuries. It offered an opportunity to purchase goods cheaply but was much more than a place to find bargains. The significance of auctions to good oeconomy was fraught, because they mixed the stewardship encouraged in the household, which demanded making the most of things, with a profit motive that encouraged people attending to spend as much as possible. Like the wife who bought too many bargains discussed in chapter 1, attendees could ruin themselves through superfluous purchasing. This was the view, for example, of Grub Street hack and political writer James Ralph, whose 1728 book on public diversions criticized the prevailing "Notion of Oeconomy" that led people to attend auctions and buy "what they do not want."[15] Thrifty philosophers celebrated and berated auctions, a sign of their ambiguous nature. Benjamin Franklin's most thrifty of essays, "The Way to Wealth," which appeared in 1758 in his periodical *Poor Richard's Almanack*, began with a warning about auctions.

> Here you are all got together to this sale of fineries and nick-nacks. You call them goods; but, if you do not take care, they will prove evils to some of you. You expect they will be sold cheap . . . but, if you have no occasion for them, they must be dear to you. Remember what Poor Richard says, "Buy what thou hast no need of, and ere long thou shalt sell thy necessaries." . . . "Many have been ruined by buying good pennyworths." . . . and yet this folly is practiced every day at auctions, for want of minding the Almanack.[16]

Auctions were regularly attacked as fraudulent and wasteful. Book auctioneers were suspected of mixing "old Rubbish" and unwanted stock with genuinely valuable works, and commentators complained about what is still known today as the "bidder's curse," the tendency of bidders to end up paying a higher than retail price as they competed to win a perceived bargain.[17] Auctioneers encouraged a highly competitive atmosphere, fueled by tobacco and drink.[18] They used accomplices known as "sweetners" to push up bids, described worthless items as precious bargains to naïve bidders, and were associated with many other fraudulent tricks to increase prices.[19] Bidders also came in for criticism. Contemporaries caricatured them as fashionable buffoons, incapable of discerning the value of goods despite their pretense to connoisseurship. The *Oxford Magazine* of 1771 depicted distracted bidders and an auctioneer unable to tell that a painting was upside down (fig. 7.1).

Auctioneers and their customers responded to such criticisms by constructing an image of auctions as respectable, creditable, even fashionable institutions. In much the same manner as experimental philosophers, auctioneers did this by appealing to erudition, tradition, and gentility, and by using spectacle and sociability to enhance their practice's appeal. Auctioneers sought to present themselves as learned gentlemen of fine taste and tried to make the auction a site of public spectacle and polite education.

Appeals to erudition were made soon after the advent of auctions. Moses Pitt presented his book auction as a curiosity, staged to "gratifie the Learned."[20] The late seventeenth-century bookseller Edward "Ned" Millington appealed to the ancients for credit, claiming he had started auctions on the authority of Herodotus, "who," as he put it, "commends that way of sale for the disposal of the most exquisite and finest beauties to their amoroso's," a reference to the "bridal auctions" of the Babylonians described in the *Histories*.[21]

Figure 7.1 Anon., *The Auction, or Modern Connoisseurs*, etching, in *Oxford Magazine* 17 (1771). The painting is of Don Quixote tilting at windmills, an allusion to misinterpretation. British Museum, museum no. 1868,0808.9989. © The Trustees of the British Museum.

In the eighteenth century, auctions were presented as engaging events, midway between entertainment and commerce. Like the nascent pleasure gardens of the period, auctions were represented by their advocates as spaces of public leisure where the classes and genders might mix and observe one another. James Ralph claimed in 1728 that auctions "of late Years are become one of the principal Amusements of all Ranks, from the Duke and Dutchess to the Pick-pocket and Street-walker."[22] Several decades later Roger North, son of the author of *Observations and Advices Oeconomical*, described a prominent auction of paintings formerly belonging to Sir Peter Lely: "There is no play, spectacle, shew, or entertainment that ever I saw where people's souls were so engaged in expectation and surprise."[23] Auctions continued to be presented as educational, as places where the public might attend, not to buy anything, but because they "must needs show / Their veneration for Virtu."[24]

Auctioneers also appealed to gentility to gain credit, a successful strategy taken by auction houses such as Sotheby's, founded in 1744, Christie's, in 1766, and Bonham's, in 1793.[25] James Christie became one of the most successful auctioneers of the eighteenth century. He came from a poor background but created a new kind of auction business, exclusive, elite, and erudite. Christie associated with high society as much as with the usual traders and merchants, and had his portrait painted by his neighbor Thomas Gainsborough as a confident gentleman. His rostrum was made by Thomas Chippendale in mahogany. His auctions focused on painting, furniture, and wine, which had the highest cultural capital among bidders. He opened an auction room in Pall Mall in 1766 and allowed people to visit only by invitation.[26] Catalogs were thorough and accurate, and Christie held receptions and private views to cultivate his clientele. These soon included the likes of Horace Walpole, the duchess of Portland, and Sir Joseph Banks.[27] Christie and others succeeded in creating a positive reputation for auctions, which by the 1770s appeared among the occasions for fashionable society to attend. In *Town Eclogues* (1772) Charles Jenner wrote of a lady of fashion who "In one continual hurry rolled her days / At routs, assemblies, auctions, op'ras, plays."[28]

Undoubtedly criticism of auctions remained. The 1775 *Countryman's Guide to London*, a manual to help country folk avoid being ripped off in the capital, lambasted auctions as fraudulent:

> Having taken a petty shop, they bring together their . . . goods, which, for the most part, are faulty . . . Many novices and countrymen have thought they have bought bargains of this rabble; but time has soon convinced them of a palpable delusion.[29]

Nevertheless, Christie and others were on the way to securing a fashionable place for auctions in British commerce. The auctioneers thus resolved problems of credit similar to those facing experimental philosophers in this period. In the late seventeenth and eighteenth centuries, auctioneers and experimenters both appealed to gentility to secure themselves from critics, and both appealed to audiences by carefully balancing entertainment and education in their activities.[30] Simon Schaffer has described how, in the face of politicization surrounding the French Revolution, late eighteenth-century natural philosophers eschewed appeals to a broad public in favor of a more inward-looking and elite science.[31] Auctioneers made similar moves. It is perhaps not surprising, then, that the culture of experiment and the auction frequently overlapped in the seventeenth and eighteenth centuries.

Auctions and the Rise of Public Science

From the earliest book sales in London in the 1670s, experimental philosophers attended auctions, and the libraries and possessions of many ended up sold at auction after their deaths. If Franklin condemned auctions, Robert Hooke could not stay away. Hooke's diary evinces frequent visits to auctions, which provided him with important works of natural philosophy and opportunities for socializing. He was joined in auctions by aristocrats and artisans, suggesting the auction was a relatively classless occasion. On May 13, 1678, Hooke attended the auction of books formerly belonging to the divine Benjamin Worsley, sold by John Dunmore and Richard Chiswell at Paternoster Row.[32] He attended auctions again on May 21 and 22, where he bought eight works, "all too dear by half."[33] He was back again on May 27 and 28, June 3 and 4, and again in December, when he attended an auction run by Moses Pitt, winning Christian Melder's edition of Euclid for nineteen pence.[34] The next summer, on June 2, 1679, he attended William Cooper's auction of the books of Stephen Watkins and Thomas Shirley, again at Paternoster Row, where he bought a copy of Athanasius Kircher's *Mundus subterraneous*.[35] He was joined there by his friend the lawyer Sir John Hoskins. In April 1680 he attended another auction, this time of Sir Kenelm Digby's books at Paternoster Row, where he bought a German edition of Agricola's *De Re Metallica* for three shillings and two pence.[36]

Some of the auctions Hooke attended were held in the coffeehouses of Chancery Lane and Guildhall, prime locations of public science.[37] Stewart has documented the bustling culture of experimental trials and public lectures on Newtonian philosophy going on in coffeehouses across London in this period.[38] Auctions took place in the same venues, located between Covent Garden and the Royal Exchange, including Child's in St. Paul's Churchyard, Button's and the Bedford in Covent Garden, and Garraway's in Exchange Alley.[39] All manner of exotic animals, plants, and curiosities of nature and art could be had at the coffeehouse auctions. Property was bought and sold, including Jamaican plantations and slaves, making coffeehouse auctions serve colonialism.[40] Auctions were also connected to the sciences. Auctions and lectures on natural philosophy were often advertised adjacent to one another in the newspapers. The *Daily Courant* for February 3, 1710, advertised a variety of auctions "by the Candle" at the Temple-Change coffeehouse in Fleet Street, for wine, books, the cargo of a ship, and a selection of Italian drawings, and then described "A Course of Experiments" on gravitation, air, light, and electricity at Mr. Hauksbee's in Wine Office Court, Fleet Street. Pamphlets detailing the contents of the course and pamphlets cataloging auction lots were available ahead of time.[41] Experimental lecturers made use of their homes for auctions. In 1720, the Channel Row house of Royal Society demonstrator and lecturer Jean Theophilus Desaguliers provided the venue for an auction, of the collections of architect William Tallman, who had died two years earlier. On other occasions Desaguliers lectured there on experimental philosophy.[42]

Coffeehouses were open only to men, extending the more or less exclusively male spaces of experimentation established in the seventeenth century in sites such as university colleges and the Royal Society. However, an exception to this rule was the coffeehouse auction, when women were allowed entrance, though often in a gallery separated from the men. By the mid-eighteenth century, women and men attended auctions together.[43] Auctions beyond London offered opportunities for scholarly correspondence and exchange. Scholars offered advice to auctioneers on exotic naturalia in sales.[44] Physicians, antiquaries, collectors, and philosophers kept one another informed of impending auctions across Europe. Sir Hans Sloane's correspondents kept him abreast of continental auctions and indicated items they desired or might be able to procure.[45] In 1711 the botanist James Petiver traveled to Leiden to attend an auction of Paul Hermann's collection on Sloane's behalf, where he purchased, among other things, a bird of paradise.[46] In 1725, the physician Johann George Steigertahl attended a book auction in Hanover on Sloane's behalf.[47] Sloane's acquaintances also used books purchased at auction as gifts to Sloane in return for favors.[48]

In London, scholars' own goods were increasingly put up for auction.

From the 1670s, numerous scholars' libraries were auctioned following their deaths, including those of Robert Boyle and Robert Hooke.[49] As Alice Marples notes, by the 1720s, news of the death of a prominent collector could send a flurry of excitement across the Republic of Letters in anticipation of the opportunity to obtain rare and desirable items.[50] Scholars at auctions might appear shifty. In 1728, James Ralph noted how "Foes to our publick Auctions insinuate, that the Virtuosi go there to part with their old Curiosities at a dear rate, and pick up others more valuable for a Trifle."[51] As the making and circulation of specialized mathematical, optical, and philosophical instruments grew, so they increasingly appeared in estate sales. The estate auction of Yarmouth MP Hewer Edgeley Hewer in 1729 was composed of "curious Mathematical Instruments, and some Antique Curiosities" along with beds, curtains, clocks, paintings, and furniture.[52] The following year saw an auction exclusively for scientific instruments, organized by an unknown auctioneer to sell off the stock in trade of retiring instrument maker Richard Glynne of Fleet Street. The sale took place at the Bedford coffeehouse in Covent Garden after a catalog was made available at numerous coffeehouses across London. The lots included "Quadrants, Telescopes . . . Surveying Instruments . . . Dials, all sorts of Microscopes; some curious Spheres and Orrery's . . . also Globes and Air-Pumps."[53] The instruments were open for viewing three days before the sale.[54] The Bedford coffeehouse was well-known to the virtuosi, being the location of William Whiston's lectures on Newtonianism in 1719 and the residence of Desaguliers until 1744.

It was not self-evident that natural philosophical items might be appropriate goods for sale at auction. In *The Auction: A Poem* (1770), the Bristol poet Thomas Chatterton described the natural philosophers Benjamin Martin and James Ferguson vying to buy a portrait of Newton.

For Newton's head whose piercing eyes,
Explor'd the wonders of the skies;
Who could with certitude declare
The size and distance of each star.
Martin and Ferguson contended,
And how the contest would have ended
I know not, had not evening come,
And call'd them both to lecture home.
They gone, no bidders could I see,
So light was held philosophy.[55]

Nevertheless, the following decade saw auctions of scientific goods become more common, with at least ten in the 1770s, thirteen in the 1780s, and sixteen in the 1790s. Peter de Clercq has surveyed some of these, noting how they ranged from a small auction of 43 lots to a large one of 274.[56] The sales were prompted when instrument makers died or departed from the trade, selling off their stock.

Reflecting the tradition of making use of the home, instrument sales typically happened at the houses of makers or the deceased.[57] Auctioneers also used their own homes as showrooms. James Short's instruments were sold by Abraham Langford and Son of Covent Garden and shown at their house in the Grand Piazza, in agreement with Short's executors.[58] Other showrooms included the home of bookselling partners George Leigh and John Sotheby, where the effects of astronomer William Russell and mathematician Samuel Dunn were sold in the 1790s. Christie used a more dedicated space, the Great Room in Pall Mall, where he auctioned the instruments of surveyor William Roy and watchmaker Larcum Kendall.[59]

The attendees at instrument auctions were various, attracting a variety of scholars, artisans, merchants, and instrument makers. The Swiss astronomer Jean Bernoulli attended the auction of James Short's instruments, as did the secretary of the Levant Company William Russell, whose own collections were auctioned after his death in 1790. Bidders could spend a great deal at these sales. In 1789, the astronomer Alexander Aubert attended the auction by Christie of surveyor William Roy's collection of instruments and spent £135 on "a capital large equatorial instrument, which also serves for taking altitudes and azimuths."[60] The same instrument, made by Ramsden, was sold at another auction after Aubert died for £68 5s.[61] As de Clercq notes, bidders at these auctions typically did scientific work with the instruments they won, and lots also included parts and odds and ends of the kind one might have found in Tiberius Cavallo's "rubbish box"; indeed Cavallo was himself an attendee at auctions.[62]

Natural historical auctions also became popular in the middle decades of the eighteenth century. The contents of cabinets of curiosity were sold occasionally in the first half of the century. Fish, reptiles, insects, precious stones, and corals were among the items sold at an auction of the collections of Gresham College professor of physic John Woodward in 1728.[63] In 1766 Samuel Paterson sold the collection of "minerals, ores, fossils, earths, petrefactions, gems, and other stones," formerly belonging to the collector David Main, at his premises in Essex House on the Strand.[64] Paterson had begun his career as a London bookseller and auctioneer selling books, instruments, household goods, and anatomical preparations at the grandly titled Great Room in King Street, Covent Garden.[65] In 1781 he put up for sale a collection of "curiosities" from the Pacific islands belonging to an officer of Captain Cook's ship *Resolution*, and then in 1783 the shell cabinet of Cook's naturalist Daniel Solander.[66] There were also plant auctions, such as the sale of a "curious Collection of Hot-house and Greenhouse Plants" formerly belonging to Dr. John Fothergill, sold off at his gardens in Upton, Essex.[67]

By the end of the century, scholars anticipated that their goods could be sent for auction after their death and included instructions to this effect in their wills. In 1801 Royal Society fellow Humphrey Jackson requested in his will that

> two Great Globes and two Great Mirrors particularly with all my philosophical instruments whatsoever, Telescopes, Microscopes, Hydrostatic Balances of all sorts all which together cost me about one hundred and fifty pounds and . . . the utensils consisting of coppers and a variety of other utensils as experimental stills of curious constructions, my forge and bellows and other utensils in smithery used for my amusement crucibles Retorts and all other utensils in my Laboratory to be sold by auction and the produce to be laid out in the funds [bequeathed to his inheritors].[68]

Experimental goods thus became common features of bequests of the kind discussed in the previous chapter.

Auctions and Knowledge

By 1800 auctions had become a significant means to circulate natural philosophical goods from one owner to another. These goods no doubt helped to secure the credit of auctions by contributing to the sense that sales dealt in erudite subject matter. The production of catalogs provided a "literary technology" for representing auctions as enlightened and reputable sales, and fixed and identified, with varying degrees of detail and rigor, the contents of sales.[69] Unlike today's auction catalogs, those of the eighteenth century rarely amounted to more than a numbered list. The catalog for the sale of mathematical practitioner William Ludlam's effects, for example, consisted of four pages listing ninety-seven lots. The list indicated the number of items in a lot, in addition to the type, size, shape, and materials of each object. For instance,

> 44 Ferguson's Card-dial, and Luminarium, and Instrument to shew the position of Jupiter's Satellites

45 Brass Syringe and Turn-about Jet
46 Tuning Forks, and Tuning Horn. Two large and two small Flutes
47 Fragments, consisting of Vellum Tubes, Brass and Coco Cells, Brass Frames and Glasses for smoking, deep Convex eye Glass and Brass Tube, &c.
48 Plain Glass Barnacles, and ditto set in horn frame to keep the dust out of the eyes in working or travelling[70]

Such inventories mitigated against the thrifty sense that objects were incomplete by turning them into static entities with apparently fixed purposes, e.g., a "Card-dial . . . to shew the position of Jupiter's Satellites." Catalogs and sales themselves, with each item being brought before an audience of bidders one at a time, encouraged a vision of an object at a moment in time, so that its open-ended career of diverse services would have to be truncated into a more specific account. Elaborate description was unnecessary perhaps because lots could be viewed ahead of a sale and were held up and visible to bidders during the auction. Catalog writer's lack of expertise may also have curtailed descriptions. Indeed most catalogs classified instruments not as "optical," "philosophical," or "mathematical," the way instrument makers did, but in the same way they classified any other commodities. They might divide lots according to which day they were to be sold on, by the room from which the lots originated ("in the Garrets," "Back Room in the Yard"), or, less frequently, by the type of material being sold (mathematical instruments, shop fittings, stock-in-trade, working tools).[71] Catalogers occasionally identified the makers of instruments, but this was not common. Another advantage of brief lists was that it enabled bidders to record the prices realized for lots by writing them in the margin next to the lot details, a practice common in the eighteenth century, and one that has tended to lead historians to use catalogs as historical price guides.[72]

Not all catalogs of scientific sales were simple. Natural history sales offered grander fare than those for instruments and demonstrate that in the eighteenth century "secondhand" science was by no means restricted to cheap or second-rate commodities. The catalog of the sale of John Woodward's museum in 1728 was written in Latin and amounted to three hundred pages detailing, among other things, 4,666 books, complete with an index, which must have been a major undertaking to compile. One contemporary, Thomas Hearne, thought the volume to "exceed most of the Auction Catalogues I ever saw or heard of."[73] Experts contributed to the work. No less than Captain Cook's naturalist Daniel Solander was engaged to prepare the catalog for the 1786 sale in 344 lots of the museum of Margaret Cavendish Bentinck, the duchess of Portland, which he curated (fig. 7.2).[74]

Figure 7.2 Secondhand goods did not necessarily evoke poverty. Charles Grignion after E. F. Burney, frontispiece to *A Catalogue of the Portland Museum*, 1786. Wellcome Library, London.

Solander died before this could be finished, but the catalog was completed, most likely by the duchess's chaplain, John Lightfoot.[75]

Auction catalogs and the classificatory work of naturalists coincided in the Portland catalog. The preface insisted that "the celebrated Linnaeus, who had studied the Subject [of conchology], has not described One Fourth Part of the Objects contained in the Museum now offered to the Public."[76] Cataloging thus made a direct contribution to natural historical knowledge. The Portland catalog became the working catalog of the collection, in which Lightfoot identified many new species for the first time, placing an "S." beside the relevant species names. These included, for example, nine new species of *Conus*, of which five still remain valid classifications.[77]

Academic and commercial considerations needed to be balanced in catalogs. As the auctioneers Skinner & Dyke pointed out,

> Some Persons . . . may object to the Promiscuous Assemblage of the various Subjects here exhibited, and be ready to wish that they had been alloted . . . according to *Genus* and *Species* . . . But however desirable such an Attainment might have been to a few *Cognoscenti*, it is very certain that the Majority of the World are not *Methodists*. They love Variety more than Order, and would rather purchase Twenty different Species of Cones or Turbos in One Lot, than the same Number of *High Admirals* or *Wentletraps*. Yet to gratify every Palate, Care has been taken, as much as possible, to keep the *Grand Classes*, and often the *Genera* together; at least as far as Respect to *Size*, *Rarity*, or *Beauty*, would allow.[78]

Despite these concessions to commerce, the Portland catalog, a uniquely sumptuous production which included a frontispiece, provided detailed descriptions of lots. These included references to relevant literature, the location from which an item came, and a description of its rarity. The auction catalog was thus a hybrid of scientific and commercial order, whose balance of scholarly and marketable features was carefully managed.

Dismantling Science

The Portland sale of 1786 saw the duchess of Portland's natural history specimens sold to more lowly individuals. Buyers included the London jeweler and insect collector John Francillon, the surgeon and anatomist John Hunter, and the silversmith and insect collector Dru Drury.[79] Cynthia Wall has highlighted the role of the eighteenth-century auction in "narratives of dismantling," a reduction in the preeminence of the nobility through the dispersal of their material inheritance to lower social ranks. Cynthia Wall proposes that the popularity of auctions increased rapidly in the eighteenth century because auctions constituted a forum for social transformation.[80] Auctions "dismantled" the material lives of aristocratic or gentlefolk and made them available to other, aspiring social classes. Auctions made it possible to "buy into" a higher social order, and catalogs and salesmen did their best to fix images of social improvement through acquisition in the imaginations of their bidders. Auctions were sites where the identities of objects and persons were transformed through the process of their interaction.

Identifying the prestigious owners of goods served to give them credit and "pedigree." As chapter 6 noted in the case of recipe books and inherited instruments, some items gathered value with age, and in the case of an aristocratic collection, association with some prominent individual might serve a new owner's credit. Auctioneers played up pedigree, no doubt in part to secure the credibility of their sales. As purchasers of the Portland catalog read, possibly hinting at others' less scrupulous practices, "there is no one Article contained in it but was a Part of the Genuine Collection of . . . Margaret Cavendish . . . Nothing is foisted into it from the Cabinets of others."[81]

Instrument and natural history auctions also transferred the ownership of scientific goods from the aristocracy to persons practicing the study of nature. Certainly there was much horizontal transfer of ownership, as instrument makers and experimental philosophers passed away and left their belongings to be auctioned off. But "dismantling" was also common, as philosophers themselves noted. In the seventeenth century, John Evelyn lamented the dismantling of elite book collections. He told Samuel Pepys that the "Humour of Exposing books *sub hasta* [i.e., at auction]" had become "epidemical" and requested him to "secure (what with so much Cost, and Industrie you have Collected) from the sad dispersions many noble Libraries and Cabinets have suffer'd in these later times."[82] The process continued in the eighteenth century. Perhaps the best example of the redistribution of scientific goods from elites to working scholars was the sale in 1793 of the collections of John Stewart, the third earl of Bute, by Messrs. Skinner and Dyke, who had also sold the duchess of Portland's museum.[83] Bute served as a patron of science and organized a laboratory in his home at Luton House, Bedfordshire. Here he collected minerals, prints, books, maps, and various chymical, mathematical, optical, and philosophical instruments, which, after his death in March 1792, were sold over several auctions. These sales evidently broke up the collections of an aristocratic patron and redistributed them to aspiring philosopher nonaristocrats. An annotated catalog for the February 1793 sale indicates that some seventy natural philosophers were among those who helped to dismember the Bute collections.[84] Tiberius Cavallo attended, as did Francis John Hyde Wollaston, professor of natural and experimental philosophy at Cambridge, who purchased a philosophical table for making demonstrations in mechanics, probably made by George Adams, for £31. The astronomer John Pond purchased a solar microscope for sixteen guineas (£16 8s.). Other attendees included physician and chemist George Fordyce, minister and electrician Cadogan Morgan, and the instrument maker Edward Nairne. It seems reasonable to suppose that many of these practitioners, and certainly Cavallo, did not buy Bute pieces to put on display in a cupboard, but sought them out as working instruments. Thus the sale helped to transform the social order and access rights surrounding instrument use. Certainly Cavallo, who purchased five Dollond telescopes for forty guineas at the Bute sale, described the event as a process of redistribution, noting how "what had cost many years in collecting, arranging, &c. was dismembered and alienated in as many hours."[85]

Conclusion

Auctions contributed to a "dismantling" of science in the eighteenth century, as aristocratic collections were broken up and distributed to persons who would formerly have needed to behave as clients in order to obtain access to patrons' collections. This was one aspect of a complex cultural activity that concerned much more than simply buying and selling. As Geismar has written, auctions are "an exquisite ballet of negotiation, knowledge exchange, and sometimes corruption between collectors, dealers, auctioneers, academics, curators, and others."[86] This was as true of early modern auctions as those happening today. Auctions formed part of an emerging public sphere where thrifty, commercial, and public cultures intersected.

Cynthia Wall's argument suggests that bargain-hunting was not the principal motivation for purchasing things at auction in the eighteenth century. That different classes owned possessions on different scales, such that auctions passed these from one class to another, should not be taken to imply that different classes were more or less thrifty. Thrift was not about how much someone could afford, but about how they managed their wealth. Nevertheless auctions reveal a tension between different classes of scientific practitioners inasmuch as the accumulation of goods could be expensive to maintain, so that aristocratic or gentle collectors might be forced to sell them off in certain circumstances, to the benefit of less well-off experimenters and householders. But it should not be thought the latter bought at auction out of necessity.

Certainly some goods sold at auction were considered to be cheaper than normal. Cavallo reckoned the sale of Bute's mineral collection was "very cheap," but it is not clear if the prices paid at the instrument auction were lower than retail, where, as noted in chapter 6, new and secondhand goods were often sold interchangeably.[87] William and Samuel Jones, instrument makers of Holborn, sold new solar microscopes at this time for between £4 14s. and £19 19s., so John Pond's payment of £16 8s. at auction was within this range. One of the Jones brothers, whose firm sold secondhand instruments alongside new ones, purchased a ten-inch electrical machine at the Bute sale together with "an extensive variety of well chosen useful and entertaining apparatus" for £10 10s. (lot 208). In the same year, the catalog of instrument maker J. Bidstrup advertised ten-inch electrical machines for £10 10s., while the Jones brothers themselves sold new ten-inch electrical machines for £12 12s.[88]

This suggests that purchasers did not always obtain instruments at auction for "bargain" prices, so there may have been other motivations for bidding.[89] No doubt some people enjoyed the competition of bidding against others. The pedigree of objects likely allured others. But perhaps the key to understanding auctions is the same as applies to secondhand purchases more generally. In the eighteenth century there was little difference between buying new and used goods, since the distinction was not yet substantial. Being "secondhand" simply reflected a change in the state of ownership of an "incomplete object," with no sense of diminished utility. The mode of sale in auctions was controversial, and it is true that some people considered auctioned goods to be of poor quality, but this was not a complaint restricted to secondhand goods. Rather, the auction constituted one part of a broad network of what might be called "secondhand science," a busy circulation of goods in which householders bought, sold, lent, borrowed, and gifted goods not to make a profit but to benefit one another and make good use of things. The auction stood between the values of the marketplace and the values of thrifty oeconomy and encouraged a slippage between them. It was simultaneously a place to make a profit and a place to share. As Cavallo explained regarding the Bute sale,

I attended [the auction] each day, and purchased the following lots, viz: a lot of five telescopes by Dolland, viz: the large fine one of ten feet, another of 8 feet also with large aperture, and three small ones, all which were knocked down for 40 guineas, and I meant to keep only one large telescope for myself, but not being able to proportion the price to the different telescopes with a friend, who wished to have the rest, I gave them all up [to] thy friend.[90]

EIGHT

The Palatial Laboratory: Economy and Experiment

Auctions constituted an ambiguous space for thrifty science. They enabled the circulation of secondhand goods and the "dismantling" of collections, but as the "buyer of bargains" discussed in chapter 1 knew, auctions could also encourage unnecessary spending and wasteful accumulation. In this ambiguity the auction stands as a useful starting point for exploring the fate of thrifty science. Previous chapters have traced gradual transformations in notions of oeconomy, the evaluation and meaning of thrift, and competing visions of the relationship between experimenters and households. This chapter examines how these trends gave rise in the nineteenth century to a distinctive approach to materials that I shall call *economic science*. This refers to a form of experimental practice that occurred in spaces separate from the home, where "men of science" sought to assert independence and authority and develop a form of material and social management alternative to the traditional household oeconomy. Economic science, which had roots in the seventeenth-century move to create scientific academies separated from the home, promoted not homely balance and harmony, but national, especially industrial progress, profit, and growth, achieved through the articulation and rationalization of practice. Economic science also encouraged a different approach to material things, presenting them as complete rather than open-ended, as commodities to be consumed rather than continuously reworked. In the nineteenth century, proponents of a more domestic thrifty science criticized this approach, so that it was never uncontested. But economic science proliferated in the growing use of specialized instruments and dedicated spaces for experimentation that were represented as being quite apart from the family and domestic life. Thrifty science did not decline in this period, but it was marginalized as a trivial or immature set of practices located outside the realm of authoritative research. These developments laid the foundations for an unsustainable "Big Science" in the twentieth century.

The first section of this chapter traces debates over the growth of specialized instruments and experimental research spaces in the nineteenth century. Competing views of the home as a site of experiment that found their roots in seventeenth-century visions of domestic experiment and scientific academies found their analogue in nineteenth-century calls to extricate science from the home. These views were challenged by an ongoing use of the home as a site of thrifty inquiry and experimentation. The second section examines how changing ideas of thrift and oeconomy transformed attitudes to materials. Economic science presented a more utilitarian approach to materials, changing attitudes to reuse and adaptability. Exploring the growth of the scientific management of waste helps elucidate these changes and the meaning of "economic science." A third section considers the temporality of economic science, as thrifty practices were reimagined as only the beginning stage in the development of new science. A final section explores "pastoral asceticism" or a romantic enthusiasm for thrifty forms of science that flourished in the late nineteenth and early twentieth centuries in part as a reaction to the rise of economic science. This "string and sealing wax" approach to science would decline with the onset of the First World War.

Dedicated Spaces and Specialized Instruments

The idea that experimental science should happen outside the home had its roots in the seventeenth-century stipulation of figures like Bacon and Boyle that authoritative science needed to explain causes and be accredited in an academy. Against this view, many householders were satisfied that the home was self-sufficient as a space of "domestic experiment." A desire to take experiment outside the home endured in the eighteenth century in academies, coffeehouses, and oeconomic societies. But traditions of "domestic experiment" like those promoted by the Hartlib circle also continued to be valued in the period. Women excluded from academic sites external to the home organized salons in their homes to discuss natural philosophical topics and exchanged books and items for collections. They included the duchess of Portland, whose collections were discussed in the previous chapter, and Lady Anne Monson and the artist and botanist Mary Delany.[1] As noted earlier, the physician and secretary Charles Blagden operated much like a Hartlibian intelligencer, circulating philosophical news through a network of women and men in well-to-do households without much regard for publishing experiments.[2] His employer, Sir Joseph Banks, also emphasized a sociable exchange of knowledge over accreditation and publication, presiding over a variety of dining clubs and natural historical breakfasts at his home in Soho Square and collaborating with his sister Sarah Sophia Banks, another avid collector.[3] The home, in short, would remain a site for experimental inquiry throughout the eighteenth and nineteenth centuries.[4] Charles Darwin's residence at Down House in Kent was famously a critical resource for his work.[5] Darwin was a thrifty naturalist. Janet Browne notes his anxiety over household accounts, so that his "thrifty nature was a family joke."[6] Ultimately, his theory of evolution imagined nature as a "thrifty capitalist enterprise . . . 'continually trying to economise in every part of the organisation.'"[7] Many others would conduct scientific inquiries at home.

At the same time, the nineteenth century witnessed an unprecedented growth in nondomestic sites dedicated to scientific research. Alix Cooper has argued that in the nineteenth century men of science "increasingly came to work outside the home in institutional spaces . . . In the process, considerable ideological boundaries were erected between work and family, and between public and private realms, which have continued to shape modern thinking."[8] Typically spaces were complex and interesting hybrids of domestic and dedicated space, and there was much debate over the ideal nature of laboratories and instruments and their relationship to the home. These discussions make manifest how closely people continued to identify the home with experimental inquiry throughout the period. Of course, even the most "independent" institution still relied on domestic support and influence at a distance.

Sophie Forgan and Graeme Gooday have shown how autonomous spaces for experiment, separate from the home, proliferated in nineteenth-century Britain.[9] Just as mass production was beginning in manufactures, so the number of men requiring a standard training in science grew, prompting the growth of institutional laboratories to educate them. While these might initially use adapted space in a university or museum, as Gooday notes, "In many cases the laboratories became architecturally distinct and adjacent institutions."[10]

Early in the century "laboratory" was still a term closely associated with chemistry. Although the Royal Institution, founded in 1799, was first housed in a preexisting gentleman's residence on Albemarle Street in London, the building was substantially altered and included a dedicated laboratory in the basement made famous by the researches of Michael Faraday.[11] Other dedicated laboratories followed the model of Justus Liebig's teaching laboratory at the University of Giessen in Germany.[12] One was the laboratory of Birkbeck College in London, purpose-built by the architect Paul Hofmann in 1839. The building contained furnaces, fume cupboards, and rows of benches. Bottle racks (reagent shelves) were placed over sinks in the benches, and cupboards under them stored apparatus.[13]

Dedicated teaching laboratories for physics appeared in Britain from the 1840s, and included William Thomson's laboratory in Glasgow University (founded 1846), the Clarendon in Oxford (1872), and the Cavendish in Cambridge (1874). New laboratories might borrow from domestic design and layout in their architecture and spatial organization. They sometimes included domestic living quarters within them.[14] But by 1885 Thomson could claim that autonomous laboratories had become "universal" and essential for universities.[15]

Why did this division take place? No doubt the answer is complex, but several reasons might be suggested. First, the physical order and contents of homes made it harder to experiment inside them. Rooms became more differentiated and their functions more distinct. By 1800, the kitchen, bedroom, and parlor had quite dedicated uses. This distinction was also brought about as more commodities became available with which to decorate and so distinguish space in the home. The flexibility of domestic space that had been a hallmark of the seventeenth and eighteenth centuries was thus diminished, so that rooms were increasingly incapable of supporting scientific inquiry. At the end of the eighteenth century, William Herschel's sister Caroline recalled of their house in Bath:

> to my sorrow I saw almost every room turned into a workshop. A Cabinet maker making a Tube and stands of all descriptions in a handsome furnished drawing-room. Alex putting up a huge turning machine . . . in a Bedroom for turning patterns, grinding glasses and turning eye-pieces &c.[16]

For Herschel, the clash between the designated, handsomely furnished space of the drawing room or bedroom and scientific or mechanical labors provoked distress. An awkward mix of leisure and labor also led to anxiety. "Every leisure moment was eagerly snatched at for resuming some [astronomical] work which was in progress, without taking time for changing dress, and many a lace ruffle was torn or bespattered by molten pitch."[17] It was at this point that the laboratory might be said to have begun to fail as a domestic space.

Second, there was a growing tension between the rhetoric of science and the reality of practice. Enlightened philosophers championed the rational management of work, contrasting the minds who guided practice with the "hands" who carried it out. Philosophers were keen not to be seen as reliant on laborers, who, on the contrary, should appreciate the genius of those who guided their work.[18] The home, however, remained a space where male philosophers continued to depend on the embodied skills and experience of their families and servants to carry on.[19] Shifting to more exclusively male spaces, replacing domestic help with trained technicians, meant that the space of experiment might more closely resemble the rational ideal, even as men continued to rely on family support at home. In a sense, space caught up with rhetoric in the move outside the home. Natural philosophers had claimed an autonomous intellectual space since the seventeenth century. As the authority of natural philosophy increased in the eighteenth century, so its physical separation became more desirable and possible.

Nevertheless, to create a more autonomous male space for scientific research required a great deal of labor, and men of science undertook various strategies to distinguish these new sites from the home. First, they referred to themselves as "men of science" rather than "natural philosophers," now making their exclusive gender explicit in their self-designation.[20] Second, women were typically barred from participation in new scientific sites except as consumers of knowledge. Humphry Davy, working at the Royal Institution, announced in 1810 that it was "not our intention to invite them [women] to assist in the laboratories."[21] Women might hear Davy lecture on science, read about science and teach it to children, but their participation in spaces of research was to be discouraged.[22] Third, men of science needed to acquire new skills to replace the experience of the family, servants, and householders. Advocates of independent chemical laboratories thus proposed that chemists learn new methods or hire technicians to supply them: "every chemical experimenter will find a considerable advantage in so much mechanical talent as will enable him to make, or repair at least, the most common of his apparatus."[23] Finally, men of science now explicitly denounced the notion that the home could be a space for serious research. The English chemist and physicist William Crookes thus described Wilhelm Hofmann's Berlin laboratory as a "mansion of a noble kind—a palace" and insisted that "the idea is destroyed that a cellar or any hole may do for a laboratory."[24] What Iwan Morus has noted in respect of the politicization of experiment was also true of its domestic inheritance. In the nineteenth century, "the reordering of science and the establishment of new spaces of scientific activity was imperatively a way of disassociating themselves from past linkages that now appeared to have been disastrous."[25]

Another means by which men of science marked their distinction from the thrifty science of the home was through material culture, in the production and use of dedicated scientific instruments. These could not be improvised or adapted from domestic possessions and required training in the new autonomous laboratories to use. Specialized philosophical instruments had existed, as chapter 1 discussed, in the seventeenth and eighteenth centuries, but the nineteenth century witnessed a dramatic increase in their design and use. In 1809, William Nicholson's *British Encyclopedia* could still claim that although "complete sets of apparatus" were ideal for chemical inquiry, many great chemists had "made use of very simple, cheap, and small sized apparatus."[26] Ten years later, however, an article in Abraham Rees's *Cyclopaedia* banished any such idea. Now a laboratory was "a place . . . entirely devoted to the different operations of chemistry."[27] The author emphasized the point by denying that domestic utensils could serve chemistry:

> Although many of the most distinguished labourers in chemical science have been content with such apparatus as they have made themselves, or converted from the common domestic utensils; it must, nevertheless, be obvious, that they would have succeeded better with well contrived and appropriate apparatus, and their researches would, in all probability, have been much more extended.[28]

Nineteenth-century chemists thus explicitly acknowledged the thrifty nature of their predecessors' experimental practice and then condemned it. The article went on to insist that now one or two "distinct" rooms were required for a laboratory, containing a furnace, a stone sink, a bottle rack, cupboards, and shelves. These spaces should be filled with specialized instruments, including mortars, balances, Argand lamps, retorts, gasometers, and pneumatic troughs. The author contrasted these with older, more polychrest devices. Recommending Woulfe's distilling apparatus, the author lamented that "Before the discovery of this most useful apparatus by Mr. Woulfe . . . the common retort and receiver were used for all purposes."[29] The many uses of things now looked vague and unscientific.

Woulfe's distilling apparatus was just one of a plethora of dedicated, specialized instruments that served to distinguish nineteenth-century experiment from its thrifty, domestic past. These would be designed to perform a single function, while instruments serving "all purposes" were increasingly disdained. It was now that the term "shift," in its early modern sense, disappeared from scientific discourse. In contrast, *specificity* came to characterize instrument design in the period. The terminology of instruments changed, as optical, mathematical, and philosophical instruments of the eighteenth century were subsumed under the broader heading of "scientific apparatus," which also included the tools of biological, engineering, or physiological investigations.[30] These new "scientific" instruments were, as the physicist James Clerk Maxwell called them, "special." In 1876 Maxwell penned an essay, "General Considerations concerning Scientific Apparatus," to accompany the collection of instruments on display at the Great Exhibition in South Kensington.[31] In the essay, Maxwell defined a scientific instrument as "A piece of apparatus constructed specially for the performance of experiments."[32] He explained, "The fundamental principle [of instrument design] is, that the construction of the instrument should be adapted to the use that is to be made of it."[33] One instrument should have one use. Such a definition did not preclude the conversion of household possessions to scientific ends, but it did insist that only objects designed for experiment could count as instruments. In most cases, these would have to be specialized, designed to serve a single scientific purpose. As Maxwell wrote,

> Apparatus may be designed to produce and exhibit a particular phenomenon, to eliminate the effects of disturbing agents, to regulate the physical conditions of the phenomenon, or to measure the magnitude of the phenomenon itself. In many experiments, special apparatus is required for all these purposes.[34]

For Boerhaave in the eighteenth century, anything could *become* an instrument. But for Maxwell, only certain things could *be* scientific instruments. Maxwell's instruments were "complete." Certainly there was no sign of adapted household goods in Maxwell's essay or in the extensive catalog of instruments that it accompanied. Some thrifty considerations did remain in Maxwell's prescriptions, however. He insisted instruments be well made to avoid "warping, straining or wearing," which might distort their capacities, so durability remained a desirable property of apparatus.[35] But his remarks signaled a culture of instrumentation different from that of thrifty science, in which design and special function now predominated.

Such "special" apparatus would become a hallmark of nineteenth-century science and technology, and remains so today. This is not to say that thrifty instrumentation disappeared. Makeshifts continued to be constructed. In 1865, August Wilhelm Hofmann famously made what he called "a new mechanical dodge," the first physical model of molecules, using croquet balls standing for atoms connected with sticks to represent their valence. As Christoph Meinel has demonstrated, these prompted "one of the great revolutions in nineteenth-century chemistry" by showing atoms' arrangements in space.[36] But improvisations such as this were often looked down upon. Already in April 1851 the Armagh astronomer Thomas Romney Robinson noted how adaptable "polychrest" instruments were increasingly disdained, following objections that "polychrest machines seldom work well; that an object is best attained by undivided effort; and that the energy which, when confined in a single channel would be irresistible, is lost if you divide it into many streams."[37] Making instruments serve a single function caused the number of different instruments to proliferate. As a Philadelphia instrument dealer remarked in 1881, "The character and uses of scientific instruments are so varied and the stock so large as to require, in our establishment, their division into special departments."[38] By 1912, the instrument catalog of Baird and Tatlock Ltd. in Holborn, London, ran to 650 pages, listing many thousands of apparatus, most of which were dedicated to some specific task. Instruments might be diversified on the basis of different specialized functions, different disciplines, and different levels of expertise (fig. 8.1).

The sheer volume of specialized instruments makes it difficult to generalize about their character, but some features stand out. Besides being designed and constructed to serve a particular scientific purpose, specialized instruments offered, among other things, *precision*, in the production of exact measurements, which helped distinguish them from the shifty bricolages of household experiment.[39] Second, specialized instruments could increase in *scale*. While homely apparatus rarely expanded beyond a size convenient for a room or garden, dedicated laboratory apparatus could grow indefinitely. As John Heilbron notes, electrical apparatus in the later eighteenth century exhibited this tendency, with Leyden jar batteries and electrical machines expanding in size to gargantuan proportions.[40] In the twentieth century, this trend would continue as science became "Big." Third, the *cost* of instruments increased.

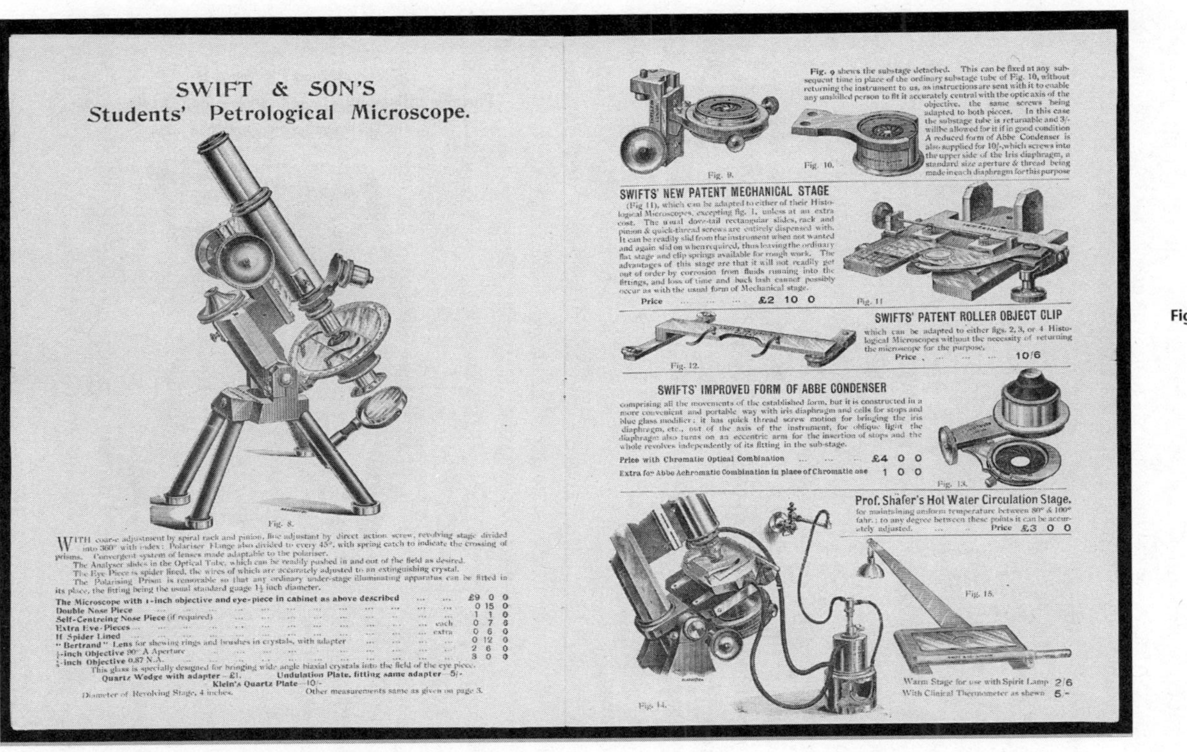

SWIFT & SON'S
Students' Petrological Microscope.

Fig. 8.

WITH coarse adjustment by spiral rack and pinion, fine adjustant by direct action screw, revolving stage divided into 360° with index: Polariser Flange also divided to every 45°, with spring catch to indicate the crossing of prisms. Convergent system of lenses made adaptable to the polariser.

The Analyser slides in the Optical Tube, which can be readily pushed in and out of the field as desired.

The Eye Piece is spider lined, the wires of which are accurately adjusted to an extinguishing crystal.

The Polarising Prism is removable so that any ordinary under-stage illuminating apparatus can be fitted in its place, the fitting being the usual standard guage 1½ inch diameter.

		£	s.	d.
The Microscope with 1-inch objective and eye-piece in cabinet as above described		£9	0	0
Double Nose Piece		0	15	0
Self-Centreing Nose Piece (if required)		1	1	0
Extra Eye-Pieces	each	0	7	6
If Spider Lined	extra	0	6	0
"Bertrand" Lens for shewing rings and brushes in crystals, with adapter		0	12	0
¼-inch Objective 90° A Aperture		2	6	0
⅛-inch Objective 0.87 N.A.		3	0	0

This glass is specially designed for bringing wide angle biaxial crystals into the field of the eye piece.

Quartz Wedge with adapter—£1. **Undulation Plate, fitting same adapter**—5/-

Klein's Quartz Plate—10/-

Diameter of Revolving Stage, 4 inches. Other measurements same as given on page 3.

Fig. 9. Fig. 10.

Fig. 9 shews the substage detached. This can be fixed at any subsequent time in place of the ordinary substage tube of Fig. 10, without returning the instrument to us, as instructions are sent with it to enable any unskilled person to fit it accurately central with the optic axis of the objective, the same screws being adapted to both pieces. In this case the substage tube is returnable and 3/- will be allowed for it if in good condition. A reduced form of Abbe Condenser is also supplied for 10/-, which screws into the upper side of the Iris diaphragm, a standard size aperture & thread being made in each diaphragm for this purpose

SWIFTS' NEW PATENT MECHANICAL STAGE

(Fig 11), which can be adapted to either of their Histological Microscopes, excepting fig. 1, unless at an extra cost. The usual dove-tail rectangular slides, rack and pinion & quick-thread screws are entirely dispensed with. It can be readily slid from the instrument when not wanted and again slid on when required, thus leaving the ordinary flat stage and clip springs available for rough work. The advantages of this stage are that it will not readily get out of order by corrosion from fluids running into the fittings, and loss of time and back lash cannot possibly occur as with the usual form of Mechanical stage.

Price £2 10 0

Fig. 11

SWIFTS' PATENT ROLLER OBJECT CLIP

which can be adapted to either figs. 2, 3, or 4 Histological Microscopes without the necessity of returning the microscope for the purpose.

Price 10/6

Fig. 12.

SWIFTS' IMPROVED FORM OF ABBE CONDENSER

comprising all the movements of the established form, but it is constructed in a more convenient and portable way with iris diaphragm and cells for stops and blue glass modifier; it has quick thread screw motion for bringing the iris diaphragm, etc., out of the axis of the instrument, for oblique light the diaphragm also turns on an eccentric arm for the insertion of stops and the whole revolves independently of its fitting in the sub-stage.

Price with Chromatic Optical Combination £4 0 0

Extra for Abbe Achromatic Combination in place of Chromatic one 1 0 0

Fig. 13.

Prof. Shafer's Hot Water Circulation Stage.

for maintaining uniform temperature between 80° & 100° fahr.; to any degree between these points it can be accurately adjusted. ... ... Price £3 0 0

Fig. 14. Fig. 15.

Warm Stage for use with Spirit Lamp 2/6
With Clinical Thermometer as shewn 5/-

Figure 8.1 This microscope was "specially designed" for petrological students to view crystals. Swift & Son, *Swift & Son's New Patent Microscope: Made in Various Forms for Histology, Physiology, Bacteriology & Petrology: With Suitable Apparatus Adapted for Prosecuting the Above Studies* (London: Swift & Son, 1890). Wellcome Library, London.

Domestic shifts were "free" in the sense that they were made with already purchased or made items, but specialized instruments needed to be bought from a maker or constructed using appropriate equipment. In the nineteenth century, the capacity to spend money on instruments became a sign of the credit of a laboratory or observatory, a form of scientific conspicuous consumption.[41] Far from being thrifty, men of science sought to obtain as much money as possible to expand the number, scale, and power of instruments at their disposal. Just as auctions shifted material possessions from the aristocracy to working scholars, so expanded funding enabled their scientific locations to become "palatial."

These changes did not go uncontested. The longstanding position that the home was sufficient for a new science to flourish found its analog in late eighteenth- and nineteenth-century critics who continued to champion domestic experiment. Thrifty apparatus might offer advantages that specialized equipment could not. William Nicholson's 1809 *British Encyclopedia* acknowledged that some chemical operations needed to occur in large manufactories, but insisted that small-scale experiments with homely apparatus still possessed "exclusive advantages." Heating a small sample with a blowpipe produced effects quickly, which were more easily visible than in a furnace. This was thrifty. "The saving of time is also an object of leading importance. The same considerations are likewise applicable to processes of fusion, or other applications of heat in a small vessel, such as a tobacco-pipe, placed in a common fire, urged by the bellows if necessary."[42] Nicholson indicated here that "common" items could yield effects that more elaborate spaces and instruments could not. As the scientific educator Richard Dennis Hoblyn complained in 1841, "It is not by the variety and costliness of his instruments that the philosopher is known, but by the knowledge and skill with which he uses them."[43]

Michael Faraday made a similar point in his *Chemical Manipulation* (1830). He noted that chemists differed greatly on the "essentials and requisites of a laboratory" and contrasted those who demanded special instruments and rooms with others who "will be satisfied with a small cupboard."[44] Faraday, no doubt appealing to a diversely resourced readership, did not take sides, and explained "Much of this variety of opinion depends upon the difference in the pursuits of the person."[45] He reckoned the home was still a good location for some researches, though his preferred laboratory was one "built for the purpose." Faraday's willingness to shift between different kinds of adapted and dedicated laboratories was reflected in his approach to apparatus. As Catherine Jackson has shown, in the 1820s and 1830s, Faraday joined Berzelius and Liebig in introducing cheap, home-blown glassware as an alternative to the expensive and highly specialized glass apparatus of French chemists. The resulting apparatus proved foundational for organic analysis, and Jackson has termed this a "Glassware Revolution" in chemistry.[46]

Evidently numerous experimenters still considered thrifty science to be full of possibilities for experiment. Attacks on domestic experiment such as those in Rees's *Cyclopaedia* simply sought to close down debate by flatly denying the home could serve as a space for natural inquiry. Advocates of "special" sites and instruments insisted that no more use should be made of domestic spaces or utensils, since only "complete" objects could serve science. In the long run, this was a very successful strategy. It made it seem that science could not progress if it stayed inside the home. But this was not inevitable.

Economic Science and Material Things

The proliferation of dedicated spaces for science and specialized scientific instruments makes sense if it is placed within a broader context of growing manufactures and changing notions of oeconomy and thrift. As chapter 1 showed, in the eighteenth century oeconomy was recast from moral advice on household living to a more general system of order, requiring rational management. Mandeville and Smith shared in a new definition of thrift as the saving of money, contrasting with the older notion of thrift as a balance or moderation that made the most of a material world provided by God. This view became dominant in the nineteenth century, and with it came a different, "economic" understanding of materials.

It might appear that the later eighteenth and early nineteenth centuries were a good time for the oeconomic management of materials. The decades around 1800 witnessed a new interest in diversifying the uses of things. One of the most famous instances of this was the Panopticon prison of Jeremy Bentham, first devised in the 1780s.[47] In recent times the Panopticon has been synonymous with the study of power, most famously by Foucault, but its thriftiness is also interesting.[48] One feature of the Panopticon was that it was not a design for a building but a design for the principle of a building, which could be applied in many different circumstances. Schools, hospitals, prisons, and workhouses could all follow the basic layout, which famously created an imbalance of power between inmates and their governors by making the latter periodically invisible. Jacques-Alain Miller and Richard Miller note how Bentham turned to Francis Bacon for a term to capture this idea. "Any Benthamic system can be dubbed with the term that, borrowing from Bacon, [Bentham] once used: it is a polychrest, a 'tool with multiple uses.' The Panopticon . . . embodies that definition because it is simultaneously a prison, a manufactory, a school, an asylum . . . Bentham has conceived a world without waste, a world in which anything left over is immediately used, a superusable world."[49]

One might see Bentham's Panopticon as the last word in polychrests and in thrifty science. But the notion of thrift Bentham employed equated it with saving only, following the views of his agrarian patriot friends. In a series of essays published in Arthur Young's *Annals of Agriculture* in 1798 under the title *Outline of a Work entitled Pauper Management Improved*, Bentham proposed the establishment of a company running "industry houses" to employ half a million of the poor of southern Britain.[50] Bentham gave the industry houses the logic and plan of the Panopticon design, with "circumferential screens occasionally interposed between the governing body in the centre of the building, and the governed classes all round."[51] He recommended *frugality* as key to their management. The clothing of poor inmates should be based, for example, on "*Frugality*—1. Materials, the cheapest, so as to afford sufficient warmth. 2. Form, excluding all useless parts—such as skirts to coats and waistcoats—brims to hats . . . *Necessity* and *use* the standards—not fashion . . . form being determined by frugality."[52] Form should follow function, a very modern notion and one echoed in James Clerk Maxwell's definition of scientific instruments. Frugality here meant *saving*, not a *balance* as in the old Aristotelian definition. For Bentham, frugality meant austerity, based on utility. Bentham sought a minimum of expenditure achieved through the stripping away of anything superfluous. If the Panopticon had many uses, it was not because it was an incomplete object but because it was cheap. Indeed, while early modern thrift was largely articulated through moral instructions, putting people and things into service, Bentham cast his discussion in terms of financial costs and benefits, a utilitarian accountancy of saving money. Thrift was an "Interest of the Purse," and "Frugality, economy" Bentham defined as "preservation of the quantity of wealth acquired."[53]

Bentham's position reflected changing ideas of oeconomy and thrift that had been developing for some time. In the early nineteenth century these coalesced into a distinctive approach to materials, which obscured the open-endedness of earlier approaches in favor of a utilitarian understanding that identified objects as having, ideally, a single use they were designed to fulfill. Now, when this use was "consumed," the object became "waste" and should then be discarded or "reused" in some manner. Eventually this reuse would be designated "recycling."

"Consumption" thus changed its meaning in the course of the eighteenth century. If it previously referred to the physical decay and wasting away of goods that thrifty practices were supposed to avoid, it now took on a more abstract, utilitarian meaning as the using up of some commodity designed for a particular function. As the political economist Jean-Baptiste Say put it, "by consumption is meant the destruction of utility, and not of substance, or matter."[54] Consumption was then represented as the province of the household, whose creativity and ongoing reworking of materials were now obscured, while "production" was identified with sites external to the home such as manufactories. Political economists continued to ally the order of the home to that of the state, but now did so by reckoning householders' contribution to the economy through consumption rather than by taking them as models of virtue for the realm. The first line of James Mill's *Elements of Political Economy* (1821) proclaimed, "Political economy is to the State, what domestic economy is to the family. The family consumes: and in order to consume it must be supplied by production."[55]

These changes were accompanied by a growing celebration of new goods, such that "novelty became an irresistible drug."[56] Celebrating novelty was nothing new, and the old was by no means dismissed in the decades around the turn of the nineteenth century. But it was increasingly distinguished from and compared with the new and found wanting. As commodities came to be identified with a single, predesignated use, so their newness grew in importance to "consumers" because a new product was an unused one. "Used goods," a term first appearing in the 1870s, thus emerged as a distinctive category of commodities, while the term "secondhand" shifted from being a designation of ownership to a characterization of how much a commodity had been used up. In utilitarian terms, secondhand or used goods were inferior because their utility was diminished by the first user. Hence shops exclusively devoted to secondhand books emerged in the early nineteenth century, since secondhand books would now appear inferior to new ones and could not be sold interchangeably with them. Historians have supposed that a decline in the price of new clothes in the 1820s and 1830s led to a growing disdain for secondhand garments, but this disdain equally resulted from a change in notions of what secondhand goods actually signified.

Like books, secondhand clothing now seemed inferior to the new, and as Alison Toplis notes, there was "a noticeable decline in the number of 'old clothes' sellers in London. Second-Hand clothing became increasingly associated with poverty and destitution."[57]

The new was by no means valued by everyone. Luddites and machine breakers violently opposed the modern machinery of mass production, while all sorts of communities reacted negatively to unfamiliar technologies such as gas lighting or railways.[58] Nevertheless, the value of the "new" continued to rise. A telling sign came in the 1850s, when the practice of "restoration" began, as a new class of expert craftsmen worked to restore objects in museums and private collections from a broken state to looking like new. The rivets and alterations to china and pottery that Hogarth had once made a feature in his family portraits now came to be considered ugly and were removed.[59]

Perhaps the most salient location for these new ways of discussing thrift and oeconomy was in the analysis of waste. Identifying goods with a specific use meant that once they ceased to serve that use, they should become "waste," a term, and what was arguably a new form of material, that proliferated across British cities in the later eighteenth and nineteenth centuries as never before. While a thrifty oeconomy identified specific leftover materials through their alternate uses in the home or workshop (rubbish, trash, junk, etc.), writers on economy explored a more general category of waste and did so through a calculus of utility. In the 1690s Sir George Mackenzie, an early advocate of equating thrift with saving money, had said that "Frugality is the true Mathematick of Christian Morality" because a financial view of thrift reduced it essentially to accountancy. In the early nineteenth century this notion of a quantitative frugality took on increasing importance. The fourth edition of the mathematician and inventor Charles Babbage's *Economy of Machines and Manufactures* (1835) makes this apparent.[60] For Babbage, as for Bentham, saving money was critical. Babbage argued that three advantages derived from industrial development for human society. The first and second of these concerned the value of machinery for enhancing human power, and the "economy they produce of human time."[61] Babbage's third advantage rested on "The conversion of substances apparently common and worthless into valuable products."[62] Thinking of objects as complete with a duration of specific use now enabled Babbage to ask what happened to them after this use was consumed. He observed of the iron trade,

The worn-out saucepans and tin ware of our kitchens, when beyond the reach of the tinker's art, are not utterly worthless. We sometimes meet carts loaded with old tin kettles and worn-out iron coal-scuttles traversing our streets. These have not yet completed their useful course.[63]

Although many people had practiced such circulations in previous centuries, no one had described this activity as a "useful course," suggesting that objects passed through some form of cycle that could be articulated and rationalized. This is just what Babbage did. In a remarkable appendix to the *Economy of Machines and Manufactures*, he recorded observations of the "profitable conversion of substances apparently of little value" at a horse-slaughtering yard in Montfaucon near Paris.[64] Based on Prussian studies of how "the discoveries of modern science" could make the best use of animal carcasses, Babbage described the process of converting a dead horse into a series of valuable commodities.[65] Early moderns used terms for old and broken things that designated how they might be put into service (rubble, junk, trash, etc.) but rarely referred to a general category of waste or use or discussed these in terms of monetary value. Babbage, however, presented a generalized idea, an abstraction, "substances apparently of little value," and quantified their worth. He went on to describe how the hair, skin, blood, hoofs, fat, flesh, tendons, and bones of horse carcasses were all converted into useful products such as animal food, manure, combs, lamp fuel, glue, fans, lathe strings, or used in the manufacture of sal ammoniac, Prussian blue dye, and ivory black. This could be subject to a precise calculation, so that a "dead horse . . . which can be purchased at from 8s. 6d. to 12s., produces from 2l. 9s. to 4l. 14s." in revenue.[66]

In Babbage's accountancy, things were complete, with a designated use, and once their use had been consumed, they were either "waste" or ready to be reused, an idea that would not have occurred previously, since things were understood as being made, unmade, and remade all the time. "Substances" moreover were now distinct from people, except insofar as they might be subjected to a common accountancy. It was now, of course, that the division of the objective and subjective was taking place, separating things from people. While early modern oeconomy dealt in "bodies," mixing the social, moral, and material, economic science separated these into passive "objects" and the human "subjects" who acted upon them. If humans and things were conflated, it was no longer in terms of the moral category of service but in terms of a quantifiable labor power.[67]

Bentham and Babbage thus marked the culmination of a gradual shift in thrifty practices away from household oeconomy. Their "economic science" proliferated in the nineteenth century. The example of waste management is indicative of its general trajectory. In the age of thrifty science, householders participated in a multitude of circulations of old materials, which were put into service in a variety of ways, either in the home, or by traders in old metals, rags, glass, etc. Scholars reckoned they were well placed to improve these circulations. Already in the seventeenth century, members of the Hartlib circle had promoted recycling of various materials as manures, to increase the prosperity of the nation.[68] In the eighteenth century, enlightened oeconomic societies explored and promoted similar schemes, setting out to optimize material circulations between town and countryside.[69] Scholarly plans and visions became more of a reality in the nineteenth century as men of science came to be seen as authorities on the proper management of "waste," which they considered on an industrial rather than a domestic scale. Many theorized the industrial use of "waste substances" for economic gain. The science of thermodynamics saw a replacement of notions of nature based on balance with a vision wedded to the new political economy of efficient industrial production. This industrial science is normally allied with the rise of steam engines and the effort to minimize heat waste. But it also emerged amidst attempts to transform material wastes.[70]

Principal among the Victorian scholars of waste was the journalist and Society of Arts protagonist Peter Lund Simmonds, who in 1862 published *Waste Products and Undeveloped Substances*, a compendium intended to survey possible uses of waste in manufactures.[71] Simmonds's book was squarely aimed at increasing the productivity and output of industry, being "full of the most practical business information," as one reviewer put it.[72] Tim Cooper has shown how Simmonds ingeniously claimed scientific authority over waste to serve economic and colonial interests. Simmonds argued that nature left alone amounted to "neglected utility," whereas experimentation could reveal the potential uses in all kinds of waste.[73] Simmonds thus employed the language of "making use" and "substitution" but now on utilitarian and commercial terms. Like Babbage, Simmonds sought to make waste practices visible, analyzable, and accountable in quantitative terms. He offered statistics on subjects such as the quantities of rags imported to Britain, the number of waste traders and secondhand dealers in London, and the scale of the Norwegian fishing industry. Surveys then followed on generating profitable industries of waste ranging from the use of coprolites for manure, to the American and German practice of using dog carcasses to manufacture gloves, fertilizers, and medicine. Counting 400,000 plus dogs in Britain, Simmonds reckoned, "there is scope enough for industry in this respect."[74] Simmonds also explored "undeveloped substances" that could operate as substitutes in manufactures. If substitution had been a part of the early modern Christian duty to explore divine gifts for their potential, it was now a means to grow profitable businesses. Simmonds assessed the value of potential substitutes for tea, paper, and fabric, and various plants and animals such as seaweed, lichens, and acorns. Dolphins, he argued, could be a profitable harvest, since their oil ("worth 6s. a gallon") could be used for lighting, while their skin (worth "from 6s. to 10s. a pound") served for making harnesses, mail bags, and shoes.[75] As Tim Cooper notes, Simmonds's analysis also extended to Britain's colonies, since he proposed that "uncivilized" nations were failing to make use of their natural resources at an appropriate level, warranting British intervention to ensure efficient exploitation. "Utilization is the great law of nature," Simmonds wrote in the 1873 edition of *Waste Products*. For "civilized man" there "must be no loss of anything once within his grasp. So much lost is so much power running to waste—it is the leak in the gas pipe, the hole in the water pipe."[76] Colonial subjugation would ensure this did not happen.

Simmonds was only one amongst a great number of entrepreneurs of waste in Victorian Britain.[77] As a review of his book noted, "At the present period, commerce is making such demands for increased supplies of various substances, that scientific men are carefully studying the residue of every manufacture, and the special qualities of each new product."[78] Economy was a process of maximizing profit through a quantified minimization of waste, achieved through experiment in new autonomous laboratories and applied in industry. This placed the scientific reuse of waste substances at the heart of capitalist growth. As the chemist and physicist William Crookes wrote in 1863, "The progress of our great chemical manufactures during the last ten years . . . appears chiefly to have been directed towards the utilization of waste substances."[79]

Perhaps the most exemplary case of this industrial management of waste was the use of coal tar derivatives to generate new commodities, to turn waste into profit. Around 1800, various individuals began using a waste product, coal gas, derived from the distillation of coal to make charcoal as an illuminant. After gas lighting took off in Britain in the first decades of the nineteenth century, increasing production generated another waste product, coal tar, which was itself analyzed and turned into various commodities in the middle decades of the century.[80] These included nitro-benzol, used as an almond scent in perfumes and confectionary, together with the famous aniline dyes, beginning with mauve, created by William Henry Perkin in 1856 as he was trying to synthesize quinine. Another coal tar byproduct was phenol, used variously to treat railway sleepers, to deodorize sewage in the streets, and then by Joseph Lister as the first antiseptic. Coal tar was quickly lionized as "the king of the waste products."[81]

The Victorians championed this economic science, but within it the home was rather lost, except as a site of consumption of new products. Men of science, not householders, families, or scavenging waste traders, were exclusively identified as those who were bringing about progress. The collective labor of domestic thrift was increasingly overlooked as male scientific accomplishments were given credit. An 1869 article in *Chambers's Journal of Popular Literature, Science and Arts* entitled "Waste Not!" explained, "One of the blessings of modern science presents itself in the form of economy, frugality, utilization. Things which were formerly thrown away as waste are now applied to man's purposes, to an extent far beyond our general supposition."[82] Husbandmen and housewives became metaphors for explaining how masculine science managed waste. The Edinburgh chemist Lyon Playfair proposed that science was bringing about "The Conquest of Waste" and exulted, "Chemistry, like a thrifty housewife, economizes every scrap."[83]

Playfair located the origins of the modern utilization of waste products in male forebears like Robert Boyle. In a lecture on the marvels of the coal-tar industry in 1862, Playfair exclaimed,

> Long ago, in the seventeenth century, Boyle wrote an Essay entitled "Man's Great Ignorance of the Uses of Natural Things . . ." The truth of the seventeenth century is still a truism in the nineteenth century, the whole progress of manufacture being merely an illustration of it. Substances which to-day are the most useless, to-morrow become embraced within the circle of industrial utilities.[84]

Economic science surely did find its roots in the work of figures like Boyle, but Playfair's account ignored the very different meanings of thrift that were entertained by early modern experimenters. In addition, the whole family might be involved in early modern experiments, whereas Playfair singled out an individual man as his key predecessor. Indeed, the rise of the man of science as a manager of waste displaced the networks and communities who traditionally dealt with old materials. Men and women who worked as scavengers to gather, transport, and sift the dust and debris ejected from homes were replaced by a new system of urban waste disposal. Already in 1829 a satirical cartoon captured the anxieties of traditional communities over rationalization (fig. 8.2). It told of the "Scavenger's Lamentation" when confronted with a mechanical road-sweeping device. "Ah," says the scavenger, "this is what

Figure 8.2 *The Scavenger's Lamentation; or, The Dreadful Consequences of Sweeping Streets by Machinery*. Science and Society Picture Library.

comes of Improvement—this is the happy effect of the March of Intellect—no employment for Scavengers now." Nothing became of regency road-sweeping machinery, but the general observation was prescient. Specialized mechanical apparatus in the form of incinerators and generators designed by men of science and engineers began to be used to burn rubbish in the last decades of the nineteenth century, wiping out the scavengers. By 1912, three hundred and thirty-eight dust incinerators in British cities had radically reduced the number of male and especially female scavengers and dust workers in employment.[85] The journalist Emily Hobhouse described these people as "forgotten by many and never known by more."[86] Not all schemes to enable scientists to manage waste were successful, but many were championed as the begetters of progress.[87] No less than the physicist William Thomson, Lord Kelvin, was guest of honor for the opening of an incinerator in Hoxton Square, Hackney, in 1897, designed with thermodynamic efficiency to generate electricity by burning rubbish.[88] By the eve of the First World War, the authority of the man of science over waste had come to appear self-evident. As an American journal put it in 1917, "Practically every big corporation is now employing large staffs of chemists and waste-saving geniuses to evolve methods whereby the dollar can be saved."[89]

Temporalities of Economic Science

The work of figures like Bentham and Babbage, Simmonds and Playfair constituted both a continuation of and a radical departure from the traditions of thrifty science. Thrift, frugality, and practices of "making use" continued to be evoked as core values of science in an industrial society but changed their meanings from earlier times. For many, though by no means all, thrift became equivalent to saving money and was served by efficiency and cost-cutting, while thriving amounted to the growth of manufactures and financial profit. Instead of placing value on open-ended bodies capable of being put into service in many ways, economic science celebrated objects designed to serve a single use as well as possible. When this was consumed, they became waste, and so capable of reuse in some ideally profitable enterprise.

Economic science differed from thrifty science in other ways. Although both gave an important role to frugality, thrift, and making use of things, they differed in the weighting of techniques. In economic science, substitution and the use of waste were given unprecedented importance because they best served the goal of profit, driving down the cost of materials if cheaper alternatives could be found. While dearth or scarcity remained motives for substitution, the search for profit replaced the desire for stewardship as the positive motivation to make substitutions. Using the same good many times or for many purposes, on the other hand, suffered, because profitability increased the more often consumers purchased the same product. Thrifty science sought out durability, but in economic science shortening the lifespan of commodities made sense, so that the nineteenth century witnessed a proliferation of goods intended to be used briefly, then thrown away. This began with the manufacture of cheap watches, newspapers, and paper collars in the United States.[90] By 1914, helped in part by concerns over the sanitary risk of germs, "disposable" goods had become abundant and included paper napkins, plates, cups, towels, tissues, and handkerchiefs. These are the roots of present-day unsustainability in science and medicine. Medical use of disposables is now extensive. It is estimated that "single-use disposable devices . . . currently comprise 85% . . . of medical equipment."[91]

Such transformations are apparent in the household literature of the nineteenth century. Writers continued to promote thrift in the home but now equated it with saving money and divided it from the economics of the state at large. In *Self-Help* (1859) and the sequel *Thrift* (1875), Samuel Smiles championed thrift and those who made good use of things in the home to study nature. Thrift, according to Smiles, was divided between "Private Economy," intended to promote the well-being of the family, and "Political Economy," doing the same for the nation. Like Bentham and Babbage, Smiles equated thrift with saving money and avoiding waste. His discussion also added a temporal dimension to thrift. Arguing that "Thrift began with civilization," Smiles assured readers that "savages" and "prehistoric" men were not thrifty because they did not save for tomorrow but only subsisted from day to day.[92] Taking a historical view, Smiles proposed that after men learned to labor, they achieved advantages that were inherited through the generations, accumulating as great art, literature, and science. Labor, industry, and hard work were therefore foundational to thrift and civilization. Capitalist accumulation, in the line of Adam Smith, was the basis of Smiles's notion of thrift: "The men who economize by means of labor become the owners of capital which sets other labor in motion. Capital accumulates in their hands, and they employ others to work for them. Thus trade and commerce begin."[93]

Smiles thus cast thrift as the initial stage on the path to accumulation and progress, an economic development that was reflected in culture. "The Parthenon began with a mud-hut; the Last Judgment with a few scratches on the sand."[94] In *Self-Help*, Smiles applied the same idea to thrift in science. He held up examples of significant discoveries achieved with little means to illustrate how simple beginnings led to great things. Like Playfair he attributed these to individual male innovators. He singled out Franklin's kite, Newton's prism, and Watt and the steam engine as exemplary instances of how thrift might begin some grand enterprise. "Gifford worked his first problems in mathematics, when a cobbler's apprentice, upon small scraps of leather, which he beat smooth for the purpose; whilst Rittenhouse, the astronomer, first calculated eclipses on his plough-handle."[95] Thrifty science was thus valuable, but only as the beginning of more mature research. It might take place in the home, but would need to be transferred to the laboratory to be properly explored. Thrifty science was domestic, accidental, romantic, but not normal science, in the Kuhnian sense.

These and other examples would be repeated in literature advocating simple experiments as the start of scientific investigations for many decades thereafter. In 1911 the Göttingen-trained chemist and president of Johns Hopkins University Ira Remsen invoked them in an essay, "The Simple Origin of Great Discoveries." Remsen took for granted that mature science involved highly specialized instrumentation and dedicated laboratories. "Within recent years the palatial laboratory has come into vogue, and everything is supplied. This is not objectionable; in fact, it is highly desirable."[96] Nevertheless, Remsen felt, like Smiles, that ingenious experiments with everyday items served to start the student on the path to such palatial heights. The student must learn to be like the "men of humble origin, with apparently everything against them," who had made important discoveries with "simple materials as were at hand."[97] Remsen romanticized thrift as a response to scarcity that innovators had managed to overcome.

Thrifty science was thus reimagined as a form of making do with limited resources, and as an immature first step in a process of scientific progress equated with specialization and autonomy from the home. This sense that thrifty science was an immature form of science was reflected in a strong association emerging in the late eighteenth and nineteenth centuries between children and domestic experimentation. In this period, many books and games appeared offering enticing lessons in science for children in the home.[98] In the 1830s the Liverpool lecturer John Smith offered "Easy and Amusing Experiments" to children, which would "require no apparatus that may not be generally found in every respectable dwelling-house."[99] Smith hoped that his book would "stimulate youth to inquiry and rational studies" in the future.[100] Charles Foote Gower's *The Scientific Phenomena of Domestic Life* (1847) promised to reveal the science behind children's experiences in the home, beginning with the bedroom window, whose frosty appearance in the morning he explained. Other chapters considered the phenomena of the breakfast parlor, the kitchen, and the study.[101] John Bower's 1894 *Simple Experiments for Science Teaching* was addressed to children and teachers who wanted to learn with little expense. Although Bower recognized that science now proceeded with "good apparatus and beautiful contrivances," he wished "to show that a good start can be made without them."[102] Consequently, "All the contrivances" in his book could "be made from the homeliest things, such as tumblers, saucers, basins, and pans employed for ordinary household purposes."[103] Many of the features of early modern thrifty science were thus reinscribed as childish practices at the start of a scientific career. This approach remains common today. As one website has it, "We can all agree that science is awesome. And you can bring that awesomeness into your very own home with these 20 safe DIY experiments you can do right now with ordinary household items."[104]

The temporalization of thrifty science as immature science also saw it associated with "uncivilized" peoples, meaning for the Victorians their own distant ancestors or "savage" races in foreign lands. Smiles supposed making use of things was always done out of expediency and so indicated a backward state. In the 1870s, the banker and man of science John Lubbock asserted a similar contrast when he described the flint "weapons" of prehistoric people. While modern science worked with specialized instruments, the "primitive savage" had to make do with "rude implements" put to a variety of uses.

It is useless to speculate upon the use of these rude yet venerable weapons. Almost as well might we ask, to what use could they *not* be applied? Numerous and specialised as are our modern instruments, who would care to describe the exact use of a knife? But the primitive savage had no such choice of weapons . . . with these implements, rude as they seem to us, he may have cut down trees, scooped them out into canoes, grubbed up roots, attacked his enemies, killed and cut up his food, made holes through the ice in winter, prepared firewood, &c.[105]

Polychrests were rude and uncivilized. Similar points were made about shifts. In his *Principles of Mechanics*, William Emerson derided early humans' attempts at mechanics, since they

would be content with very little theory . . . All their contrivances must be mere guessing, and they could but ill execute what they had so badly contrived; and must be continually mending their work by repeated trials, til they got it to such a form as to make a shift to serve for the use designed.[106]

"Primitive" peoples in the present had no need of specialization: "they will go about with any makeshift to cover their nakedness."[107] Shifts served to punctuate a historical and civilizational hierarchy. "[E]very contrivance for moving or raising water . . . from the most simple makeshift of the savage and the shepherd, to the most complicated modern machine . . . is the mark of an epoch . . . in the history of hydraulics."[108] Specialization thus marked the man of science as distinguished not only from the home but also from his inferior predecessors and contemporaries. This discourse did not go uncontested. *Punch* lampooned the man of science's self-superiority. In one satire a "simple savage" responded to the accusation by a "scientific sage" that his weapons were makeshifts or flukes by pointing out that newfangled European ships' guns were liable to burst: "I fail to see the use of your stultified scientific skill, or your uncertain arms of precision."[109] Nevertheless, Victorians took supposed makeshifts as a sign of inferiority in "primitive" technology and then, like Simmonds, accused their users of failing to make the most of natural resources. European intervention through colonization would resolve this.[110]

Pastoral Asceticism

The transition from the home to dedicated spaces for science was by no means inevitable or necessary, but it served a variety of commercial and colonial values. At the same time, it was not a ubiquitous transition and never went uncontested. Indeed in the second half of the nineteenth century, there emerged a trend for domestic scientific spaces that historians have examined closely. As Graeme Gooday, Simon Schaffer, and Donald Opitz have shown, the country house served as the location for a number of natural philosophical laboratories in the later decades of the nineteenth century. Since these examples are quite well-known, this final section will situate them in relation to the history of thrifty science.[111]

Gooday cites the example of James Edward Henry Gordon, a student of James Clerk Maxwell, whose wife, Alice, explained that in their youth they kept a laboratory at home.[112] In *Decorative Electricity* (1891), she wrote,

> Our early married life was spent in the country, where we owned a large laboratory and a small house attached. In those days I was bottle-washer and laboratory assistant to my husband, money for experimenting was scarce, and the makeshifts adopted in the laboratory were sometimes more ingenious than conducive to the comfort of the rest of the house. Well do I remember my dismay on finding one morning that all the kitchen crockery was full of acids and villainous compounds, and the cook informing me with an injured air "that we could have no more pastry till we released her rolling-pin," which was then covered with tin-foil and adjusted as one of the principal conductors in an experiment that I well knew would last many weeks.[113]

Alice recalled many practices evocative of thrifty science—family enterprise, cleaning, shifts, repurposing, and bricolage. Her anxiety, however, recalls Caroline Herschel's distress at the pursuit of astronomy in her home. Scientific materials sat awkwardly in a house with specialized rooms and furniture. Alice proposed that the house had its limits as a laboratory. "With the possible development of electric lighting [as a research topic] our country laboratory became inadequate."[114] The Gordons moved to London, and James continued his research in a factory.

The relative capacities of private country house and institutional laboratory were also a concern of John Strutt, Baron Rayleigh. Schaffer has examined how, in 1871, Rayleigh set up a laboratory in the billiard room at his country estate in Tofts, and two years later converted the conservatory and rooms in the stable block of his main home in Terling into "one of the largest private labs in Britain."[115] The house provided resources for significant physical researches, but when financial support and labor proved hard to find, Rayleigh accepted the position of head of the Cavendish Laboratory in Cambridge. Here Rayleigh's landed management skills helped organize the laboratory. As Schaffer remarks, "In experimental spaces like these, we see the pastoralization of the laboratory and the scientific transformation of the estate."[116] Like Alice Gordon, Rayleigh recognized the limits of the country house laboratory, but he did not think them insurmountable. On returning to Terling in 1885 he told a Cavendish colleague, "I hope to continue scientific work, although in a laboratory which though it can never compete with a public institution I hope to make effective."[117]

Such country house laboratories might be seen as late examples of thrifty science, mixing together the use of new and dedicated instruments and spaces with a regard for making use of existing possessions and the old and ready-to-hand. To begin with, this was a space of family collaboration. As Donald Opitz has shown, Evelyn Balfour, Lady Rayleigh, was an educated woman in her own right and contributed much to Rayleigh's researches, often joining him in the "book room" for scientific pursuits.[118] Then Rayleigh's scientific equipment was a peculiar mix of the new and the old. On the one hand, he used the latest specialized instruments, an auto-collimating spectroscope, a heliostat, and photographic equipment. Rooms in the house were given over to specialized functions: photography, spectroscopy, optical experiments, and chemistry. At the same time, Rayleigh saw himself, and was seen by others, as engaging in a form of thrifty science. He spoke favorably in public of the value of simple, domestic instruments. His speech of 1906 recalled claims that thrifty apparatus retained much potential, since

> it was just possible that nowadays scientific work was made too easy, or, at all events, too mechanical, for the full advantage of it to be reaped, and that the scientific spirit and method were, perhaps, better cultivated by the less perfect appliances of the past. Many of the original experimenters in science worked with exceedingly homely apparatus . . . As a rule, early experiments were made with whatever materials were at hand, and it was very often after the results sought for had been reached that more elaborate and suitable apparatus was contrived.[119]

Rayleigh's vision of thrifty science, like that of Remsen and Playfair, was inflected by late nineteenth-century assumptions: notably that thrifty science had been the practice of men alone. Rayleigh's portrait, by Philip Burne-Jones, presented him on his own working at a simple bench in a sparse room (fig. 8.3). Reliance on or collaboration with the family, as Thomas Wijck's painting portrayed, was nowhere to be seen. Nevertheless, the materiality of Rayleigh's work appeared thrifty. Observers claimed that he experimented with "Sealing wax, string, rough unplaned woodwork, and glass tubes joined together by bulbous and unsightly joints."[120] Rayleigh's country house laboratory, then, evoked the domestic experimental spaces of earlier times and an older notion of thrift as a mixture of adapted and dedicated things.

Rayleigh's thrift might be explained by his class. Simon Schaffer has shown that when the Cavendish laboratory was created at Cambridge University in the 1870s, it was met with skepticism, as smacking too much of the workshop and professionalism and implying that good scientists were made and not born. The country house's thrifty culture, imported into the Cavendish, helped secure it from such associations.[121] Economic science seemed to suggest that standardized training was what mattered in science, but for a certain class of observers this appeared vulgar, urban, pedestrian, and industrial. The thrifty science of the country house, what might be termed "pastoral asceticism," offered a more palatable image of cultivated ingenuity through private, homely experiment. Opitz notes Rayleigh's admiration of the physicist Henry Clifton Sorby, who, said Rayleigh,

> belonged to a class on whom England has special reason to congratulate herself—men who pursue science unprofessionally . . . It is to be feared that specialization and the increasing cost and complication of experimental appliances are having a prejudicial effect . . . On the other hand, the amateur is not without advantages . . . open to ingenuity unaided by elaborate appliances.[122]

Thrifty science thus appeared here as heroic gentlemanly amateurism, a romantic vision. This modern version of thrifty science proved enduring. Cavendish research in the late nineteenth and early twentieth centuries was famously identified as taking a "string and sealing wax" approach to science, exactly the pastoral asceticism of the country house. It may be that this image of the Cavendish, where Rayleigh was director from 1879 to 1884, was exaggerated. According to Jeff Hughes, such stories were "a caricature" because the laboratory contained highly specialized equipment such as particle accelerators, cloud chambers, and electromagnets.[123] J. A. Crowther recalled that the image had "little

Figure 8.3 Philip Burne-Jones, *John William Strutt (1842–1919), 3rd Baron Rayleigh*, oil on canvas, 1921. Wellcome Library, London.

foundation in fact."[124] But perhaps these readings mistake the presence of dedicated, specialized instruments as anathema to thrift, when in fact their presence was a part of it, a combination of the old and new, a balance between profligacy and avarice, a little Aristotle still left in the laboratories of Cambridge.

Conclusion

Economic science and pastoral asceticism stand for two forms of scientific endeavor that were indebted to the thrifty science of the seventeenth and eighteenth centuries. Home and family remained important for the sciences in the nineteenth century, either in their continuing role in experimental endeavors providing resources and expertise, or as entities against which autonomous, specialized, and dedicated practices and objects were defined. Thrifty science continued in the nineteenth century and had its advocates and champions. At the same time, trends that dated back to the seventeenth century culminated in the appearance of new spaces and identities that claimed to be autonomous from the home. Science came to be associated with specialized equipment and dedicated laboratories, with thrift reduced to an immature first step on the path to scientific progress. Thrifty science was now claimed to be done out of necessity in the face of scarcity, what came to be known as "making do," and was allied with primitivism, childhood, and tinkering.

With its celebration of efficient but increasing expenditure in science, and increasing scale, precision, and complexity in instrumentation, economic science might be said to have laid the foundations for "Big Science" in the twentieth century.[125] Even in the age of Big Science, however, thrifty science has not disappeared, though it has certainly been obscured. Scientists have continued to make use of things ready to hand, to improvise, substitute, and shift. But these activities now constitute a form of what the sociologist Harry Collins has called "tacit knowledge" or unarticulated labor, techniques crucial to investigations but having no presence in finished scientific papers or formal conceptions of scientific method.[126] After many decades of obscurity, it took outside observers like Collins to bring attention to these practices. Since the 1970s, laboratory studies and the sociology of science have made much of the way science is like a craft practice, not simply a logical progression of ideas. Karin Knorr-Cetina has compared the modern laboratory experimenter to a tinkerer, as someone who

> uses whatever he finds around him . . . to produce some kind of workable object. The tinkerer . . . manages with odds and ends . . . They are aware of the material opportunities they encounter at a given place, and they exploit them to achieve their projects . . . When we observe scientists at work in the laboratory, this sort of opportunism appears to be the hallmark of their mode of production.[127]

Of course it is no coincidence that the metaphor works.

Conclusion

In his 1911 essay on the simple origins of great discoveries, Ira Remsen wrote, "It may be asked why, if so many discoveries have been made with simple things in simple surroundings, should so much be spent on scientific work?" Remsen replied to himself:

> Times have changed. Many of the problems that in earlier times could be solved with simple things have been solved. The difficulties of scientific investigation are increasing. More and more refined apparatus is coming to be necessary . . . expensive apparatus is often required, and many profitable lines of investigation could not now be followed without large expenditures.[1]

Thrifty science, on this reading, was inadequate to the demands of modern, professional science, because its limits had already been reached and new knowledge demanded innovative, expensive apparatus. Since the end of the nineteenth century, many in the physical sciences have taken this view for granted and have moved in exactly this direction, toward more specialized instruments and more expensive "Big Science." Scientific spending and material culture have grown to unprecedented scales. Like the "buyer of bargains," we have now accumulated a gargantuan, and apparently unsustainable, quantity of scientific goods. So was Remsen correct to say that the era of thrifty science is now passed? Is there room for thrifty science in the future?

Remsen was wrong to suppose that thrifty science was only about homely apparatus, cheap instruments, and simple experiments. It was not. Although this book has focused

Figure 9.1 Glenn Brown, *The Loves of Shepherds (after 'Doublestar' by Tony Roberts)*, oil on canvas, 2000, 219.5 x 336 cm. © Glenn Brown. Courtesy of the artist and Gagosian.

on the domestic "making use of things," chapter 1 indicated that thrift was above all about achieving a balance between spending and buying new and saving and making use of things. The same approach would appear to be reemerging today and seems to make good sense given current anxieties over the environment and sustainability. While people in the late twentieth century supposed that the future of science would be mostly "big," as we enter the twenty-first century, it seems that the future will also be small. In today's derelict London there is little sign of interplanetary engineering, except perhaps in one of Glenn Brown's reworkings of a Tony Roberts space station, another piece of recycling (fig. 9.1). Clearly, there is still a need today for expensive, specialized, and dedicated science, just as there was for innovative air pumps and electrical machines back in the seventeenth and eighteenth centuries. But there is also room for a reevaluation of thrifty science, bringing it back from the diminished and marginal role it has been assumed to hold since the nineteenth century.

Thrifty Science

At the heart of this book has been an effort to unravel a quite different understanding of early modern experimental material culture than historians of the period normally assume, and an effort to explain why this approach is no longer evident. An important reason that thrifty science has been obscure, besides its marginalization among scientists themselves, is that historians of science have often tended to read a nineteenth-century "economic" notion of scientific material culture into the past, meaning that they focus on the emergence and circulation of dedicated and specialized instruments and sites, designed with the intention of pursuing natural inquiry. The "dedicated" history of science has been in the main a history of things like academies, observatories, cabinets of curiosity, universities and museums, microscopes, chronometers, and electrical machines. This book does not deny the importance of these instruments and sites for early modern science—clearly, they were essential. But a great number of early modern scientific sites and labors involved other things: creamware pottery, tobacco pipes, bedchambers, wineglasses, and so on. According to a thrifty notion of materials, people approached their goods and possessions as "incomplete," open to revision and change over a period of ideally long duration. Things should be "made use of" as well as possible and could be put into "service" in many different ways. Such an approach encouraged experimentation, as people *instrumented* their possessions to serve natural inquiry. Instrumenting is an awkward term, but that is useful. It reminds us that instruments were an active process as much as a static thing, a becoming of household possessions into a capacity for studying nature. Typically thrifty experiments involved a bricolage of adapted and specialized items, and while we have numerous histories of the latter, we have insufficient histories of the former. But to miss the common practices of "making use" and the more improvised culture that encouraged them is to overlook a significant element of experimental science in the seventeenth and eighteenth centuries.

Just to reiterate, this book has not argued that dedicated and specialized spaces and instruments were unimportant in the seventeenth and eighteenth centuries. But it has suggested that another side of scientific practice has been underestimated and warrants further consideration. The focus here has been on England and the home, but it might equally be asked what thrifty practices pertained in other places, among artisans, in the military, in churches and universities, and in the many other arenas that contributed to early modern science. No doubt different disciplines dealt with thrift in different ways in the course of the nineteenth and twentieth centuries, a distinction I have not attempted to unravel here. Evidently in the field sciences (botany, geology, entomology, etc.), thrifty practices have retained a more explicit and valued place than in physics or chemistry. Since field sciences continue to entail much work outside any dedicated scientific space, they continue to take the value of improvisation for granted.[2]

Science and Sustainability

Indeed, even in the case of the physical sciences, it may be that the era of "Big Science" has been only a brief hiatus in an otherwise long and thrifty history. Certainly in the twenty-first century there are plenty of signs that thrifty practices are regaining currency, from the recycling and refurbishment of scientific equipment to efforts to reduce reliance on disposable plastic goods in research.[3] Many of these practices are still discussed in the Victorian language of utility and saving, and it is hoped that this book may offer an alternative language that enables thrifty practices to be taken further. One place where an enthusiasm for thrifty practices is particularly evident is in hack- or makerspaces. These emerged in Germany and the United States in the mid-1990s and have proliferated in cities around the world. Participants, usually a mixture of artists, programmers, scientists, and craftspeople, make use of a variety of locations to learn craft and programming skills, to teach repair techniques, and to produce everything from computer code to a house made of ice for living on Mars.[4] Alongside spaces for making, some hack and makerspaces have been dedicated to "Do-It-Yourself Biology" (DIYBio) or "Garage Biology" using cheap or used equipment.[5] The London Hackspace, in Hackney, for example, includes the London Biohackspace.[6] Generally hack and makerspaces blur the boundaries of science, arts, and engineering, and work and home life.

Hack and makerspaces bring to the foreground a more open-ended and incomplete approach to things than do modern specialized laboratories. They make much of relocating research in local communities and encourage exchanges and the sharing of resources, an interest in encouraging repairs and recycling, and a concern to "hack" or improvise things (similar to the early modern notion of a "shift"). Many (though by no means all) participants pursue hacks and making in the name of sustainability. As one commentator has put it, in makerspaces,

> Found materials can be given new life through the vision of a crafts person, and the professional polish of a laser cutter, welder, or lathe. Having one central location for storage and fabrication can centralize materials and equipment and encourage resource sharing. With a smaller physical footprint in a living space, makers can reduce

personal energy consumption and find homes for unwanted, excess bits and pieces... "The Maker Movement and sustainability go hand in hand," says central California-based SLO MakerSpace co-founder Clint Slaughter. "The trends toward fixing instead of trashing, DIY instead of buying cheaply made goods from across the globe, and upcycling along with recycling have allowed millions to learn new skills, hack things to better suit their own needs, and even make a living just being a Maker."[7]

Old divisions of production and consumption give way to an explicitly thrifty making use of things in the hackspace, now envisioned in a consciously environmental framework.

Clearly these contemporary activities cannot be straightforwardly equated with the thrifty science of the past. But one goal of this book has been to offer a historical context for such practices and for ongoing changes today that are often assumed to be entirely novel. A historical appreciation of thrifty science shows that practices such as making use of old materials, sharing, and recycling are not entirely new, as they are sometimes represented, but emerge from a historical tradition. Their growing prominence in the twenty-first century might be seen as a reevaluation and return of existing practices rather than a radical innovation. The return of thrifty practices might be supported by a recognition that their eclipse by specialized spaces and equipment in the past was contingent, not inevitable. Thrift was eschewed in science not because the home placed some unavoidable limitations on the progress of knowledge but to serve the distinction of a new generation of men of science from their domestic (and often female) predecessors.

Thus, an appreciation of the history of thrifty science not only helps to make sense of current trends but also offers possibilities for taking them forward. If the decline of thrifty science was not a historical necessity, then there is room to pursue it once again, by revisiting and opening up for investigation, or at least reevaluating, smaller-scale, simpler phenomena of the kind that thrifty science was once well equipped to explore. As it turns out, this is just the sort of thing that philosophers of science are encouraging at present. The historian and philosopher of science Hasok Chang has argued for what he calls a "complementary science."[8] Chang proposes that many apparently straightforward scientific facts are more complicated than they may seem at first glance, so that textbook truisms such as "water boils at one hundred degrees" may in fact turn out to be complex and variable phenomena when subjected to experiment. The self-evidence of such claims is only an appearance, achieved after much debate and controversy was finally closed off to leave an apparently simple fact. In the case of boiling water, for example,

Chang shows how minimal apparatus—a kettle, a heat source, a thermometer—are needed to start unraveling the assumed fact. Examined alongside the history of past experiments on boiling water, complementary science may then, as Chang puts it, "recover past experimental knowledge that has been neglected by modern science, and extend the knowledge that has been recovered."[9] Contra Remsen, then, there is still a great deal of potential for doing science in "simple surroundings." Moreover, following the hackers and makers, such an approach might be seen as more sustainable, since it uses fewer resources than expensive, dedicated science. Thrifty science, then, might be seen as an environmental complement to Chang's "complementary science."

Is thrifty science scientifically productive? Remsen worried that there was nothing left to discover if science stayed at home. But he also shared with the Victorians an assumption that simple means had nevertheless furnished a tremendous store of knowledge in the past. The thrifty investigations of figures such as Newton, Watt, Priestley, and Franklin led to tremendous advances in natural science, giving rise to new theories of light and color, a new chemistry of airs, an appreciation of the electrical nature of lightning, and dramatic improvements in the steam engine. Remsen's was a romantic account failing to appreciate many features of the thrifty approach (not least the achievements of women experimenters). There is no reason to suppose that the thrifty science of hack and makerspaces or complementary science could not generate the same kind of revolutionary discoveries in the future.[10]

The history of thrifty science may offer fresh ways to think about sustainability itself. "Sustainability" is a contested term with diverse definitions.[11] This ambiguity might be regarded as valuable since it prompts debate and raises awareness about how humans should interact with their environments. Sustainability in that sense is valuable as a "polychrest," an open-ended idea whose incompleteness is its strength, not its weakness. Sustainability is often discussed in terms of large-scale indices, measures of energy expenditure and carbon dioxide emissions, an accountancy that sometimes reduces complex phenomena to a single parameter. This is of course one of the strengths of scientific definitions of sustainability, since they enable a global, comparative surveillance. However, such measures may also lose a sense of the complex interplay of moral, social, economic, political, and material interactions that shape and are shaped by the environment. A revival of the notion of *oeconomy*, a term that lay at the root of thrifty science, might serve to restore these to discussions of sustainability. An oeconomic reading of environmental problems and solutions would thus begin with the small scale of families, homes, and habitats and seek to manage them according to the values of thrift. Thrift means not minimizing expenditure (of money or energy or emissions) but rather achieving balance, between households and their environment, between the use of new and old things. Oeconomy asks for a more qualitative judgment than economy, based on balance, maintaining good order and harmony. The relatively simple goals of oeconomy might be practiced by households in a great variety of locations and contexts—indeed, globally speaking, most households outside the "developed" world already operate in just this way. And at the root of domestic oeconomy would be a renewed appreciation of the incompleteness of objects, valuing and encouraging the use of things capable of serving many ends. Thrifty science makes the home an experimental space, an oeconomic laboratory for working out solutions to the problems of sustainable living. A renewed domestic experimentation would also reconnect the public to science in ways that various efforts in the "Public Understanding of Science" have failed to do. Making the laboratory a room in the home again would allow science to reengage with a skill set belonging to householders that a more autonomous science has had to leave behind.[12] These activities would also be linked to broader communities through practices of sharing, exchange, and collaboration, moving up from local to community-scale practices.

The oeconomic prescription for sustainability might thus pose the following final question. If we were to create a twenty-first-century version of Thomas Wijck's painting of the alchemist's kitchen, what would it look like? Instead of alchemical vessels and hearths, hanging tortoises and flagstone floors, what would we see? The picture remains to be painted, but certain features might be discerned. The scene would have to be one in which the production of knowledge involved the whole family. It might see little difference between the utensils and furniture of the house and the laboratory, and would explore material things containing a wealth of possibilities, to serve not profit but a balanced relationship between householders and their environment. Beyond measures of energy consumption or economic indices this would be a simultaneous representation of the ideal relations between householders and their surroundings, between people and the objects and knowledge that they live with. Science would appear back in the sphere of everyday life and be thrifty. To some extent, through hackspaces and makerspaces and ingenious artistry, these changes are already happening. Perhaps they will continue in the future. Science, it seems, is coming home.

Acknowledgments

This book benefited from not one but two departmental homes. I began it in the Department of History at the University of Washington and finished it in the Department of Science and Technology Studies at University College London. Both departments supported my research, for which I am very grateful. I am indebted to the support and encouragement of my colleagues in Seattle, especially Bruce Hevly, Alison Wylie, Linda Nash, Tom Hankins, Karl Hufbauer, Elena Campbell, John Findlay, Lynn Thomas, Benjamin Schmidt, Joel Walker, Anthony Adler, the Simpson Center for the Humanities, and the members of the History Research Group (HRG) and Science Studies Network. At UCL, I am also grateful for support from Joe Cain, Jon Agar, Phyllis Illari, Leonie Hannan, Kate Smith, Margot Finn, Frank James, Mechthild Fend, Chiara Ambrosio, Hilary Powell, Mat Paskins, Hasok Chang, Haidy Geismar, Sarah Wilkes, Beth Munro, and the directors of the Institute of Making at UCL; also members of the Ad-Hoc History of Chemistry Reading Group, and many other inspiring friends and colleagues. My thanks also to Robert Anderson, Jenny Rampling, Anne Simmonds, Mike Jewess, and the Society for the History of Alchemy and Chemistry.

I learned a great deal about early modern recycling at a conference organized in Paris in June 2010 and I am grateful to the organizers, Ariane Fennetaux, Amélie Junqua, and Sophie Vasset. The conversations at this meeting really helped to make sense of this project and give it a trajectory.

Part of this project was completed during my time as a fellow of the Rachel Carson Center for Environment and

Society in 2011, and I am extremely grateful to colleagues, staff, and the directors Helmuth Trischler and Christof Mauch for their support and encouragement. The Carson Center offered a truly inspirational environment for research and conversation. I was also able to spend a term at the Norwegian University of Science and Technology in Trondheim, and my thanks go to the University of Washington's Scandinavian Studies Bergen Exchange Program, Christine Ingebritsen, Margrethe Aune, Terje Finstad, Thomas Berker, Kristine Ask, and Stine Helena Bang Svendsen. This book has been the result of many stimulating discussions, especially with the participants of the Situating Chemistry 1760–1840 network, and I am very grateful to the organizers John Perkins and Lissa Roberts. Lissa has been a terrific help in making sense of the history of science for me, and I am very grateful to her for continuing inspiration.

I acknowledge and am very grateful to the Beckman Center for the History of Chemistry in the Chemical Heritage Foundation for fellowship support to carry out research for this project. Particular thanks to Carin Berkowitz, Alison Roseberry-Polier, Michelle DiMeo, Elisabeth Berry Drago, Ann Elizabeth Wiener, Ronald Brashear, and James Voelkel for their help and advice during my stay. Thanks also to Gigi Naglak and Bernard Brown for being such kind hosts.

I am very grateful to Elaine Leong, Christine von Oertzen, Carla Bittel, and Lorraine Daston for inviting me to the Max Planck Institute for the History of Science in Berlin, to participate in the "Working with Paper" project which offered much to think about in relation to this book. My thanks for the support of the MPI and all the fellows of Department II, who offered excellent comments on parts of the manuscript.

It has been a great pleasure to work with my editor, Karen Merikangas Darling, at the University of Chicago Press, and I very much appreciate Karen's ongoing advice and direction. I am also indebted to two anonymous readers for the Press, who provided valuable critiques of the manuscript, and to Susan Tarcov for careful editing.

There are many other people I would like to thank for their contributions and support in this project. I could not have completed this book without the generous assistance of librarians and archivists at the British Library, the British Museum, the Science Museum, Hackney Archives, the Wellcome Library, and the National Maritime Museum in London; the Museum of the History of Science in Oxford; the Whipple Museum in Cambridge; and the Historical Society of Pennsylvania, the Chemical Heritage Foundation, the American Philosophical Society, and the Van Pelt Library, University of Pennsylvania, in Philadelphia.

My thanks also to Alexi Baker, Robert Bud, Catherine Jackson, Christian Vogel, Christina Wessely, Frédéric Graber, Graeme Gooday, Greg Radick, Gwen Ottinger, H. Otto Sibum, Jane Insley, Dominik Hünniger, Josie Kane, Ben Russell, John Tresch, Larry Stewart, Marie Thébaud-Sorger, Martin Willis, Neil Safier, Nicky Reeves, Rebekah Higgitt, Richard Dunn, Robert Brain, Simon Schaffer, Margarete Vöhringer, Sophie Forgan, Stephen Johnston, and Sophie Pitman.

"Oeconomy" tells us that knowledge is a product of the community, the household, and the family, and in addition to my academic community my family also made an enormous contribution to this book. So a big thank you to all the Werretts, Märkers, and Engelhards! Thank you above all to my dear and patient wife, Anna Märker, for your kindness and guidance in helping me to write this book.

Notes

INTRODUCTION

1. Hilary Powell, *Urban Alchemy* (London: by the author, 2015); Paul Talling, "Derelict London," accessed December 15, 2015, http://www.derelictlondon.com/; Celia Pym, "Home page," accessed August 17, 2017, http://celiapym.com.
2. Will Brooker, *Star Wars* (London: BFI, Palgrave Macmillan, 2009), 33–37.
3. The problem was highlighted in Vance Packard, *The Waste-Makers* (New York: D. Mackay, 1960); a more recent work is Giles Slade, *Made to Break: Technology and Obsolescence in America* (Cambridge: Harvard University Press, 2006).
4. Peter Galison and Bruce Hevly, eds. *Big Science: The Growth of Large-Scale Research* (Stanford: Stanford University Press, 1992); Jon Agar, *Science in the Twentieth Century and Beyond* (Cambridge: Polity Press, 2012), 330–53.
5. US Environmental Protection Agency/US Department of Energy, "Laboratories for the 21st Century: An Introduction to Low-Energy Design" (2008), PDF accessed May 20, 2016, www.i2sl.org/documents/toolkit/lowenergy_508.pdf, 1.
6. These include the Laboratory Research and Technical Staff (LabRATS) group at the University of California, Santa Barbara; LABS21, a US Environmental Protection Agency and Department of Energy-sponsored project focused on energy efficiency in laboratories; the National Institutes of Health Environmental Management System (NEMS) Sustainable Lab Practices Group; and private initiatives such as SBENCH (founded in 2007), a commercial venture based in New York encouraging the exchange and recycling of laboratory equipment. In the UK, the "S-Lab" program, modeled on the American LABS21 project, is seeking to create sustainable laboratories.

7. Ruth Oldenziel and Helmuth Trischler, introduction to *Cycling and Recycling: Histories of Sustainable Practices*, ed. Oldenziel and Trischler (New York: Berghahn Books, 2016), 2–3; E. F. Schumacher, *Small Is Beautiful: A Study of Economics As If People Mattered* (New York: Vintage, 1993).
8. Oldenziel and Trischler, introduction, 3–10.
9. E.g., David N. Livingstone, *Putting Science in Its Place: Geographies of Scientific Knowledge* (Chicago: University of Chicago Press, 2003); Kapil Raj, *Relocating Modern Science: Circulation and the Constitution of Scientific Knowledge, South Asia & Europe, 17th–18th Century* (London: Palgrave Macmillan, 2007).
10. Arjun Appadurai, ed., *The Social Life of Things: Commodities in Cultural Perspective* (Cambridge: Cambridge University Press, 1986); Karen Harvey, *History and Material Culture: A Student's Guide to Approaching Alternative Sources* (London: Routledge, 2009); Anne Gerritsen and Giorgio Riello, eds., *Writing Material Culture History* (London: Bloomsbury, 2014); on material culture in the history of science, see Lorraine Daston, ed., *Biographies of Scientific Objects* (Chicago: University of Chicago Press, 2000); Lorraine Daston, ed., *Things That Talk: Object Lessons from Art and Science* (New York: Zone Books, 2004); Ursula Klein and Emma C. Spary, eds., *Materials and Expertise in Early Modern Science* (Chicago: University of Chicago Press, 2010); Simon Werrett, "Matter and Facts: Material Culture in the History of Science," in *Material Evidence: Learning from Archaeological Practice*, ed. Robert Chapman and Alison Wylie (New York: Routledge, 2014), 339–52; Pamela H. Smith et al., "The Making and Knowing Project," accessed July 26, 2017, http://www.makingandknowing.org/.
11. Sven Dupré and Christoph Herbert Lüthy, eds., *Silent Messengers: The Circulation of Material Objects of Knowledge in the Early Modern Low Countries* (Berlin: LIT Verlag, 2011); Lissa Roberts, ed., *Centres and Cycles of Accumulation in and around the Netherlands during the Early Modern Period* (Berlin: LIT Verlag, 2011).
12. David Edgerton, *The Shock of the Old: Technology and Global History since 1900* (Oxford: Oxford University Press, 2007).
13. Simon Werrett, "Recycling in Early Modern Science," *British Journal for the History of Science* 46 (2013): 627–46.
14. Steven Shapin and Simon Schaffer, *Leviathan and the Air-Pump: Hobbes, Boyle and the Experimental Life* (Princeton, NJ: Princeton University Press, 1985); Larry Stewart, *The Rise of Public Science: Rhetoric, Technology, and Natural Philosophy in Newtonian Britain, 1660–1750* (Cambridge: Cambridge University Press, 1992).
15. E.g. Deborah E. Harkness, *The Jewell House: Elizabethan London and the Scientific Revolution* (New Haven: Yale University Press, 2007); Pamela H. Smith, *The Body of the Artisan: Art and Experience in the Scientific Revolution* (Chicago: University of Chicago Press, 2004); James Delbourgo, Kapil Raj, Lissa Roberts, and Simon Schaffer, eds., *The Brokered World: Go-Betweens and Global Intelligence, 1770–1820* (Sagamore Beach, MA: Science History Publications,

2009). Lawrence Principe uses the contemporary term "chymists" to designate a variety of practitioners in alchemy and chemistry in the seventeenth century, since the latter terms were seen as interchangeable. I will use the term "chymist" for the seventeenth and eighteenth centuries unless actors referred to "chemistry" or "alchemy" themselves. See Lawrence Principe, *The Aspiring Adept: Robert Boyle and His Alchemical Quest* (Princeton: Princeton University Press, 1998), 8–10.

16. Sujit Sivasundaram, ed., "Global Histories of Science," special section of *Isis* 101 (2010): 95–158; Daniela Bleichmar, *Visible Empire: Botanical Expeditions and Visual Culture in the Hispanic Enlightenment* (Chicago: University of Chicago Press, 2012); Dániel Margócsy, *Commercial Visions: Science, Trade, and Visual Culture in the Dutch Golden Age* (Chicago: University of Chicago Press, 2015).
17. Donald L. Opitz, Staffan Bergwik, and Brigitte Van Tiggelen, "Introduction: Domesticity and the Historiography of Science," in *Domesticity in the Making of Modern Science*, ed. Opitz, Bergwik, and Van Tiggelen (Basingstoke: Palgrave Macmillan, 2016), 2.
18. Deborah E. Harkness, "Managing an Experimental Household: The Dees of Mortlake and the Practice of Natural Philosophy," *Isis* 88 (1997): 247–62.
19. Steven Shapin, "The House of Experiment in 17th-Century England," *Isis* 79 (1988): 373–404. For more discussion of literature on science in the home, see the introduction to chapter 2 below.
20. Deborah R. Coen, *Vienna in the Age of Uncertainty: Science, Liberalism, and Private Life* (Chicago: University of Chicago Press, 2007); Deborah R. Coen, "A Lens of Many Facets: Science through a Family's Eyes," *Isis* 97 (2006): 395–419; see also Staffan Bergwik, "An Assemblage of Science and Home: The Gendered Lifestyle of Svante Arrhenius and Early Twentieth-Century Physical Chemistry," *Isis* 105 (2014): 265–91.
21. Linda A. Pollock, *With Faith and Physic: The Life of a Tudor Gentlewoman* (New York: St. Martin's Press, 1993); Alisha Rankin, "Medicine for the Uncommon Woman: Remedy Exchange among Noblewomen in Early Modern Germany" (PhD diss., Harvard University, 2005); Elaine Leong, "Making Medicines in the Early Modern Household," *Bulletin of the History of Medicine* 82 (2008): 145–68; Elaine Leong, "Collecting Knowledge for the Family: Recipes, Gender and Practical Knowledge in the Early Modern English Household," *Centaurus* 55 (2013): 81–103; Michelle DiMeo and Sara Pennell, eds., *Reading and Writing Recipe Books, 1550–1800* (Manchester: Manchester University Press, 2013); Andrew Wear, *Knowledge and Practice in English Medicine, 1550–1680* (Cambridge: Cambridge University Press, 2000), chap. 2.
22. Alisha Rankin, *Panaceia's Daughters: Noblewomen as Healers in Early Modern Germany* (Chicago: University of Chicago Press, 2013), esp. chap. 4.
23. Deborah Jean Warner, "What Is a Scientific Instrument, When Did It Become One, and Why?" *British Journal for the History of Science* 23 (1990): 83–93.

24. Shapin and Schaffer, *Leviathan and the Air-Pump*.
25. H. Otto Sibum, "Reworking the Mechanical Equivalent of Heat: Instruments of Precision and Gestures of Accuracy in Early Victorian England," *Studies in History and Philosophy of Science* 26 (1995): 73–106.
26. Richard Dunn and Rebekah Higgitt, *Ships, Clocks and Stars: The Quest for Longitude* (Glasgow: HarperCollins, 2014); on the telescope, see Mario Biagioli, *Galileo's Instruments of Credit: Telescopes, Images, Secrecy* (Chicago: Chicago University Press, 2006); Alison D. Morrison-Low, Sven Dupré, Stephen Johnston, and Giorgio Strano, eds., *From Earth-Bound to Satellite: Telescopes, Skills and Networks* (Leiden: Brill, 2011); on barometers, Jan Golinski, *British Weather and the Climate of Enlightenment* (Chicago: University of Chicago Press, 2007), chap. 4.
27. Livingstone, *Putting Science in Its Place*, chap. 2, provides an overview.
28. N. Gregson, A. Metcalfe, and L. Crewe, "Practices of Object Maintenance and Repair: How Consumers Attend to Consumer Objects within the Home," *Journal of Consumer Culture* 9 (2009): 250.
29. Karin Knorr-Cetina, "Objectual Practice," in *The Practice Turn in Contemporary Theory*, ed. Theodore R. Schatzki, Karin Knorr-Cetina, and Eike von Savigny (London: Routledge, 2001), 181–84.
30. Carolyn Merchant, *The Death of Nature: Women, Ecology and the Scientific Revolution* (San Francisco: Harper Collins, 1980); Merchant, "The Scientific Revolution and the Death of Nature," *Isis* 97 (2006): 513–33.
31. The same convention still applies among scholars today. When the sociologist of science Andrew Pickering first read Thomas Kuhn's *Structure of Scientific Revolutions*, he found it "very annoying," and on arriving at the Science Studies Unit in Edinburgh, "One of the first things I did . . . was give a little seminar to members of the group there on what I thought about Kuhn . . . at the end I held up *The Structure of Scientific Revolutions* and dropped it in the waste basket . . . I felt that I'd kind of disposed of it." My transcription from Canadian Broadcasting Company, "How to Think about Science: Episode 4—Ian Hacking & Andrew Pickering," accessed February 25, 2009, http://www.cbc.ca/ideas/features/science/index.html#episode2.
32. Donald Woodward, "Swords into Ploughshares: Recycling in Pre-industrial England," *Economic History Review* 38 (1985): 175–91; Tim Cooper, "Rags, Bones and Recycling Bins," *History Today* 56 (2006): 17–18.
33. Beverly Lemire, "Consumerism in Pre-industrial and Early Industrial England: The Trade in Second-Hand Clothes," *Journal of British Studies* 27 (1988): 1–24; Laurence Fontaine, ed., *Alternative Exchanges: Second-Hand Circulations from the Sixteenth Century to Today* (Oxford: Berghahn Books, 2008); Jon Stobart and Ilja Van Damme, eds., *Modernity and the Second-Hand Trade: European Consumption Cultures and Practices, 1700–1900* (Basingstoke, UK: Palgrave Macmillan, 2010); Ariane Fennetaux, Sophie Vasset and Amélie Junqua, eds., *The Afterlife of Used Things: Recycling in the Long Eighteenth Century* (New York: Routledge, 2014), 2.

34. Susan Strasser, *Waste and Want: A Social History of Trash* (New York: Metropolitan Books, 1999).
35. See the discussion in chapter 3.

CHAPTER ONE

1. Sri Mahaalakshmi Scientific Company, "Biotechnology Equipments," accessed June 30, 2016, http://www.mahaascience.com/biotechnology-equipments.html.
2. OED Online, accessed January 28, 2016, s.v. "Specialized."
3. Robert John Ackermann, *Data, Instruments, and Theory: A Dialectical Approach To Understanding Science* (Princeton, NJ: Princeton University Press, 1985), 87.
4. Edward Kellett, *Tricoenivm Christi in nocte proditionis suae, The Threefold Svpper of Christ in the Night that He vvas Betrayed* (London, 1641), 10; Thomas Urquhart, *The Trissotetras: or, a Most Exquisite Table for Resolving all Manner of Triangles*, 2nd ed. (London, 1645), 75.
5. See, e.g., Lawrence M. Principe, "Apparatus and Reproducibility in Alchemy," in *Instruments and Experimentation in the History of Chemistry*, ed. Frederic L. Holmes and Trevor H. Levere (Cambridge: MIT Press, 2000), 71.
6. Warner, "What Is a Scientific Instrument," 83.
7. Dotan Leshem, "Retrospectives: What Did the Ancient Greeks Mean by *Oikonomia*?," *Journal of Economic Perspectives* 30 (2016): 225–38.
8. Xenophon, *Xenophons Treatise of House Holde* (London, 1532); for a modern edition, Xenophon, *Memorabilia; Oeconomicus; Symposium; Apology*, trans. E. C. Marchant, O. J. Todd, Jeffrey Henderson (Cambridge: Harvard University Press, 2013); for commentary, see Michel Foucault, *The History of Sexuality*, vol. 2: *The Uses of Pleasure* (New York: Vintage, 1990), 152–65.
9. For an overview see Karen Harvey, *The Little Republic: Masculinity and Domestic Authority in Eighteenth-Century Britain* (Oxford: Oxford University Press, 2012); Harvey, "Oeconomy and the Eighteenth-Century 'House': a Cultural History of Social Practice," in *Domestic Practice in the Past: Historical Sources and Methods*, ed. Alison Blunt, special issue of the journal *Home Cultures* 11 (2014): 375–90; Natasha Korda, *Shakespeare's Domestic Economies: Gender and Property in Early Modern England* (Philadelphia: University of Pennsylvania Press, 2012), 15–51; Margaret Hunt, *The Middling Sort: Commerce, Gender and the Family in England, 1680–1780* (Berkeley: University of California Press, 1996).
10. References are to the second edition, anon., *England's Happiness Improved: or, an Infallible Way to get Riches, Encrease Plenty, and Promote Pleasure* (London, 1699).
11. Ibid., 157.
12. Ibid., 174.
13. Harvey, *Little Republic*, 34.

14. Michael McKeon, *The Secret History of Domesticity: Public, Private, and the Division of Knowledge* (Baltimore: Johns Hopkins University Press, 2009), chap. 3.
15. Harvey, *Little Republic*, 28–31.
16. See, e.g., Rob Iliffe and Frances Willmoth, "Astronomy and the Domestic Sphere: Margaret Flamsteed and Caroline Herschel as Assistant-Astronomers," in *Women, Science and Medicine 1500–1700: Mothers and Sisters of the Royal Society*, ed. Lynette Hunter and Sarah Hutton (Stroud, UK: Sutton, 1997), 235–65. Robert Hooke owned Gervase Markham's guide to housewifery, *Country Contentments or the English House-wife*. Items auct_BH_2263 (North); auct_BH-2010 (Markham), in the online database of Hooke's library, "Robert Hooke's Books," ed. Will Poole, Felicity Henderson, and Yelda Nasifoglu, accessed March 20, 2016, http://www.hookesbooks.com/.
17. Richard Drayton, *Nature's Government: Science, Imperial Britain, and the "Improvement" of the World* (New Haven: Yale University Press, 2000), 50–51.
18. Ibid., 51.
19. Harvey, *Little Republic*, 44–48.
20. Drayton, *Nature's Government*, chap. 3; for the French case, see Emma C. Spary, "'Peaches Which the Patriarchs Lacked': Natural History, Natural Resources, and the Natural Economy in France," *History of Political Economy* 35, special issue, *Oeconomies in the Age of Newton*, ed. Margaret Schabas and Neil De Marchi (2003): 14–41; for the Germanies, see Alix Cooper, " "The Possibilities of the Land": The Inventory of 'Natural Riches' in the Early Modern German Territories," in ibid., 129–53.
21. Gervase Markham, *Markhams Farwell to Husbandry or, The Inriching of all Sorts of Barren and Sterill Grounds in our Kingdome* (London, 1620), 2.
22. For discussion, see Margaret Schabas, *The Natural Origins of Economics* (Chicago: University of Chicago Press, 2005), 29–32; Donald Worster, *Nature's Economy: A History of Ecological Ideas*, 2nd ed. (Cambridge: Cambridge University Press, 1994), 33–36.
23. Lisbet Koerner, "Linnaeus' Floral Transplants," *Representations* 47 (1994): 144–169; Koerner, *Linnaeus: Nature and Nation* (Cambridge: Harvard University Press, 1999); Staffan Müller-Wille, "Nature as a Marketplace: The Political Economy of Linnaean Botany," *History of Political Economy* 35, special issue: *Oeconomies in the Age of Newton*, ed. Margaret Schabas and Neil De Marchi (2003): 154–72.
24. Drayton, *Nature's Government*, 98–102; C. A. Bayly, *Imperial Meridian: The British Empire and the World, 1780–1830* (London: Routledge, 2016), 121–25.
25. For Spanish and Dutch contexts, see Elena Serrano, "Making *Oeconomic* People: The Spanish *Magazine of Agriculture and Arts for Parish Rectors* (1797–1808)," *History and Technology* 30 (2014): 149–76; Joppe van Driel, "Ashes to Ashes: The Stewardship of Waste and Oeconomic Cycles of Agricultural and Industrial Improvement, 1750–1800," *History and Technology* 30 (2014):

177–206; van Driel, "The Filthy and the Fat: Oeconomy, Chemistry and Resource Management in the Age of Revolutions" (PhD diss., University of Twente, 2016).

26. Arthur Young, *Rural Oeconomy, or Essays on the Practical Parts of Husbandry: Designed to Explain Several Methods of Conducting Different Farms* (London, 1770), 2.
27. Koen Stapelbroek and Jani Marjanen, eds., *The Rise of Economic Societies in the Eighteenth Century: Patriotic Reform in Europe and North America* (Basingstoke, UK: Palgrave, 2012); Matthew Paskins, "Sentimental Industry: The Society of Arts and the Encouragement of Public Useful Knowledge, 1754–1848" (PhD diss., University College London, 2014); Elena Serrano, "Chemistry in the City: The Scientific Role of Female Societies in Late Eighteenth-Century Madrid," *Ambix* 60 (2013): 139–59.
28. Lissa Roberts, "Practicing Oeconomy during the Second Half of the Long Eighteenth Century: An Introduction," *History and Technology* 30 (2014): 135; Emma C. Spary, "Political, Natural, and Bodily Economies," in *Cultures of Natural History*, ed. Nicholas Jardine, James A. Secord, Emma C. Spary (Cambridge: Cambridge University Press, 1996), 178–96.
29. Roberts, "Practicing Oeconomy," 134.
30. Harvey, *Little Republic*, 57–61.
31. Sir James Steuart, *An Inquiry into the Principles of Political Economy. Being an Essay on the Science of Domestic Policy in Free Nations*, 2 vols. (London, 1767).
32. For more discussion of the relationship of "oeconomy" and "economy," see Lissa Roberts and Simon Werrett, "Introduction: A More Intimate Acquaintance," in *Compound Histories: Materials, Production, Governance, 1760–1840*, ed. Roberts and Werrett (Leiden: Brill, 2017), 6–8; Margaret Schabas and Neil De Marchi, introduction, *History of Political Economy* 35, special issue: *Oeconomies in the Age of Newton*; and chapter 8 below.
33. Anon., *The Good Hows-holder* (London, 1607), 1.
34. Anon., *England's Happiness Improved*, preface A3.
35. Richard Bradley, *The Country Housewife, and Lady's Director, for Every Month of the Year. Both in the Frugal Management of the House, and in the Delights and Profits of the Farm. Published for the Good of the Public*, 6th ed. (London, 1732).
36. For the history of meanings of thrift, see Peggy A. Knapp, "Thrift," in *Art and Context in Late Medieval English Narrative: Essays in Honour of Robert Worth Frank, Jr*, ed. Robert Edwards (Oxford: Boydell & Brewer, 1994), 193–205.
37. Ibid., 201–3.
38. William Shakespeare, *Richard II*, ed. Anthony B. Dawson and Paul Yachnin (Oxford: Oxford University Press, 2011), 269.
39. William Ames, *Marrow of Theology* (1629), quoted in James Calvin Davis and Charles Mathewes, "Saving Grace and Moral Striving: Thrift in Puritan

Theology," in *Thrift and Thriving in America: Capitalism and Moral Order from the Puritans to the Present*, ed. Joshua Yates and James Davison Hunter (Oxford: Oxford University Press, 2011), 104.

40. Walter de Henley, *The Booke of Thrift, containing a Perfite Order, and Right Methode to Profite Lands, and other Things belonging to Husbandry* (London, 1589), 12.
41. Joshua Yates and James Davison Hunter, introduction, in Yates and Hunter, *Thrift and Thriving in America*, 3–36.
42. David Evans, "Thrifty, Green or Frugal: Reflections on Sustainable Consumption in a Changing Economic Climate," *Geoforum* 42 (2011): 550–57.
43. Anon., *The Art of Thriving, or the Way To Get and Keep Money* (London, 1674), 1.
44. Samuel Johnson, *A Dictionary of the English Language: In which the Words are Deduced from their Originals, and Illustrated in their Different Significations*, 2 vols. (London, 1755–56), s.v. "Frugality."
45. Woodward, "Swords into Ploughshares"; Carlo M. Cipolla, *Before the Industrial Revolution: European Society and Economy, 1000–1700* (New York: Norton, 1976), 132–40.
46. Ayesha Mukherjee, *Penury into Plenty: Dearth and the Making of Knowledge in Early Modern England* (London: Routledge, 2015).
47. See, e.g., the advice of Henry Peacham, *The Worth of a Penny; or, A Caution to Keep Money. With the Causes of the Scarcity and Misery of the Want Thereof* (London, 1703).
48. Davis and Mathewes, "Saving Grace," 90–98; Yates and Hunter, *Thrift and Thriving in America*, 8–9; there have of course been long debates over the connection between the rise of modern science and the growth of Puritanism in England. Such a connection might support the claim that thrift was a prominent virtue among early modern English experimenters, but that claim will not be made here. I. B. Cohen, ed., *Puritanism and the Rise of Modern Science: The Merton Thesis* (New Brunswick, NJ: Rutgers University Press, 1990).
49. Lendol Calder, "Spending and Saving," in *The Oxford Handbook of the History of Consumption*, ed. Frank Trentmann (Oxford: Oxford University Press, 2012), 362–63.
50. On poverty in early modern England, see Steve Hindle, *On the Parish? The Micro-politics of Poor Relief in Rural England, c.1550–1750* (Oxford: Oxford University Press, 2004).
51. Annie Gray, "'A Practical Art': An Archaeological Perspective on the Use of Recipe Books," in DiMeo and Pennell, *Reading and Writing Recipe Books*, 51.
52. Michael Hunter, ed., *Robert Boyle: By Himself and His Friends: With a Fragment of William Wotton's 'Lost Life of Boyle'* (London: William Pickering, 1994), 28; thanks to Sadie Harrison for pointing this out.
53. Steven Shapin, "The Invisible Technician," *American Scientist* 77 (1989): 554–63; Shapin, *A Social History of Truth: Civility and Science in Seventeenth-Century England* (Chicago: University of Chicago Press, 1994).

54. Alisha Rankin, "Women in Science and Medicine, 1400–1800," in *The Ashgate Research Companion to Women and Gender in Early Modern Europe*, ed. Allyson M. Poska, Jane Couchman, and Katherine A. McIver (London: Routledge, 2016), 412–13.
55. Richard Baxter, *A Christian Directory* (London, 1673), 146; Ehrman notes of Martha Dodson, the wealthy wife of an eighteenth-century businessman, that "in spite of her comfortable circumstances, thrift and economy played a critical role in the management of her resources and household. In this way she was typical of her class and period." Edwina Ehrman, "Dressing Well in Old Age: The Clothing Accounts of Martha Dodson, 1746–1765," *Costume* 40 (2006): 35; see also Jane Whittle and Elizabeth Griffiths, *Consumption and Gender in the Seventeenth-Century Household: The World of Alice Le Strange* (Oxford: Oxford University Press, 2012), 242.
56. Dudley North (4th Baron North), *Observations and Advices Oeconomical* (London, 1669), 4.
57. Anon., *A Glasse for Housholders wherin thei maye se, bothe howe to rule Theim Selfes [and] Ordre their Housholde verye Godly and Fruytfull* (London, 1542), n.p.
58. According to Aristotle, "Virtue, then, is a state involving rational choice, consisting in a mean, relative to us, and determined by reason—the reason, that is, by reference to which the practically wise person would determine it. It is a mean between two vices, one of excess, the other of deficiency." *Nicomachean Ethics*, trans. and ed. Roger Crisp (Cambridge: Cambridge University Press, 2000), 31.
59. Anon., *Art of Thriving*, 1.
60. North, *Observations and Advices Oeconomical*, 99–100.
61. Roger North, "Essays," British Library, MS 32526, f. 59r.
62. Anon. [poss. Richard Burridge], "Oeconomie and Extravagance," *Gentleman's Magazine* 1, no. 11 (November 1731): 489; see also Simon Werrett, "Household Oeconomy and Chemical Inquiry," in *Compound Histories: Materials, Production, Governance, 1760–1840*, ed. Lissa Roberts and Simon Werrett (Leiden: Brill, 2017), 42.
63. On the issue of luxury, see, e.g., Maxine Berg, *Luxury and Pleasure in Eighteenth-Century Britain* (Oxford: Oxford University Press, 2005).
64. This is explored more fully in chapter 8.
65. Sir George Mackenzie, *The Moral History of Frugality, with its opposite Vices, Covetousness, Niggardliness, Prodigality, and Luxury* (London, 1711), in Sir George Mackenzie, *Essays Upon Several Moral Subjects* (London, 1713), 296.
66. Bernard Mandeville, *The Fable of the Bees: or, Private Vices, Publick Benefits* (London, 1732), 103–4; Nicholas Barbon, *A Discourse of Trade* (London, 1690), 11, 32; Istvan Hont, "The Early Enlightenment Debate on Commerce and Luxury," in *The Cambridge History of Eighteenth-Century Political Thought*, ed. Mark Goldie and Robert Wokler (Cambridge: Cambridge University Press, 2006), 387–95.

67. Mandeville, *Fable of the Bees*, 197.
68. Ibid., 105.
69. Ibid.
70. Adam Smith, *An Inquiry into the Nature and Causes of the Wealth of Nations*, 3 vols. (Dublin, 1776), 2: 99.
71. Ibid., 2: 105.
72. See, e.g., John Styles, "Product Innovation in Early Modern London," *Past & Present* 168 (2000): 124–69; Maxine Berg, "The British Product Revolution of the Eighteenth Century," in *Reconceptualizing the Industrial Revolution*, ed. Jeff Horn, Leonard N. Rosenband, and Merritt Roe Smith (Cambridge: MIT Press, 2010), 47–64.
73. North, *Observations and Advices Oeconomical*, 99.
74. Ibid., 64.
75. Anon., "Account of a Buyer of Bargains," *The American Museum, or, Repository of Ancient and Modern Fugitive Pieces, &c. Prose and Poetical* 1, no. 1 (April 1787): 307.
76. Ibid.
77. Ibid., 308.
78. Ibid.
79. Ibid.
80. Michel Foucault, *The Order of Things: An Archaeology of the Human Sciences* (London: Routledge, 1991), 17–45.
81. Thomas Tusser, *A Hundreth Good Pointes of Husbandry Lately Maried Vnto a Hundreth Good Poynts of Huswifery* (London, 1570), 28.
82. See, e.g., Hannah Woolley, *The Queen-like Closet, or Rich Cabinet stored with All Manner of Rare Receipts for Preserving, Candying and Cookery* (London, 1675).
83. Charles Carter, *The London and Country Cook, or, Accomplished Housewife*, 3rd ed. (London, 1749), v–vi.
84. Gray, "'A Practical Art,'" 52–53.
85. Noel Chomel, *Dictionaire Oeconomique: or, the Family Dictionary*, trans. Richard Bradley, 2 vols. (London, 1725), s.v. "Beech-Tree."
86. Henry Best, *Rural Economy in Yorkshire in 1641, being the Farming and Account Books of Henry Best of Elmswell, in the East Riding of the County of York* (Durham, 1857), 35.
87. Baxter, *Christian Directory*, 146.
88. Johnson, *Dictionary of the English Language*, s.v. "Nitre."
89. Ibid., s.v. "Dust."
90. Ibid., s.v. "Litter."
91. Ibid., s.v. "Refuse."
92. Ibid., s.v. "Rubbish."
93. Anon., *A New and Complete Dictionary of the Arts and Sciences*, 2nd ed., 5 vols. (London, 1763–64), 4: 2879.
94. Tusser, *Hundreth Good Pointes*, 34.

95. Joseph Priestley, *The History and Present State of Electricity*, 2nd ed. (London, 1769), x.
96. E.g., Lynette Hunter, "Women and Domestic Medicine: Lady Experimenters, 1570–1620," in Hunter and Hutton, *Women, Science and Medicine*, 89–107; Lynette Hunter, "Sisters of the Royal Society: The Circle of Katherine Jones, Lady Ranelagh," in ibid., 178–97; Wendy Wall, *Recipes for Thought: Knowledge and Taste in the Early Modern Kitchen* (Philadelphia: University of Pennsylvania Press, 2015); Elizabeth Spiller, ed., *Seventeenth-Century English Recipe Books: Cooking, Physic and Chirurgery in the Works of Elizabeth Talbot Grey and Aletheia Talbot Howard* (Aldershot, UK: Ashgate, 2008), x–xiii.
97. Gervase Markham, *Cheape and Good Husbandry for the VVell-ordering of all Beasts, and Fowles* (London, 1614), 79.
98. Leong, "Collecting Knowledge," 91–92, 95.
99. Stephen Gaukroger, *Francis Bacon and the Transformation of Early-Modern Philosophy* (Cambridge: Cambridge University Press, 2001).
100. Francis Bacon, "New Atlantis," *The Works of Francis Bacon*, ed. James Spedding, Robert Leslie Ellis, and Douglas Denon Heath, 14 vols. (London, 1858–74), 3: 119–66.
101. David Mallet, *The Life of Francis Bacon, Lord High Chancellor of England* (London, 1740), 56.
102. Francis Bacon, "Of Expense," *Works*, 6: 443.
103. Ibid., 6: 444.
104. Francis Bacon, "Maxims of the Law," *Works*, 7: 321.
105. Francis Bacon, "De Dignitate et Augmentis Scientiarum," *Works*, 1: 575 (book 3, chapter 5). My translation. I have used the original Latin here because Spedding's translation is vague.
106. Ibid. My translation.
107. Francis Bacon, "New Organon," *Works*, 4: 233.
108. Ibid., 4: 233, 242.
109. For discussion of the feast of the family, see Kate Aughterson, "'Strange Things So Probably Told': Gender, Sexual Difference and Knowledge in Bacon's *New Atlantis*," in *Francis Bacon's New Atlantis: New Interdisciplinary Essays*, ed. Glynn White (Manchester: Manchester University Press, 2002), 156–79.
110. Mark Greengrass, Michael Leslie, and Timothy Raylor, eds., *Samuel Hartlib and Universal Reformation: Studies in Intellectual Communication* (Cambridge: Cambridge University Press, 1994).
111. Samuel Hartlib, *Samuel Hartlib his Legacy: Or an Enlargement of the Discourse of Husbandry used in Brabant and Flandres* (London, 1651).
112. Samuel Hartlib [Gabriel Plattes], *A Description of the Famous Kingdome of Macaria, Shewing its Excellent Government wherein the Inhabitants Live in Great Prosperity, Health, and Happiness* (London, 1641), 5.
113. Hartlib, "An Introduction to the Legacy of Husbandry," in Hartlib, *Samuel Hartlib, his Legacy of Husbandry wherein are Bequeathed to the Common-wealth*

of England, not onely Braband and Flanders, but also Many More Outlandish and Domestick Experiments and Secrets (of Gabriel Plats and Others), 3rd ed. (London, 1655), n.p.

114. Ibid., 103.
115. Ibid., 33.
116. Ibid., 35.
117. William Wheeler to Benjamin Worsley, August 25, 1650, Hartlib papers online, 34/3/6B.
118. Ephemerides 1653 Part 1, Hartlib papers online, 28/2/51A; Ephemerides 1654 Part 1, Hartlib 29/4/3B; Ephemerides 1654 Part 2, Hartlib 29/4/20A; Ephemerides 1656 Part 2, Hartlib 29/5/74B; Ephemerides 1657 Part 2, Hartlib 29/6/18B.
119. Description of Henry Rivers's Wind-Powered Engine, Hartlib papers online, 63/6/1A and 1B.
120. John Evelyn, *Numismata, a Discourse of Medals, Ancient and Modern Together With Some Account of Heads and Effigies of Illustrious, and Famous Persons* (London, 1697), 164.
121. Drayton, *Nature's Government*, 52.
122. Carl Wennerlind, *Casualties of Credit: The English Financial Revolution, 1620–1720* (Cambridge: Harvard University Press, 2011), 61.
123. Leong, "Collecting Knowledge," 95; Leong, "Making Medicines"; Sara Pennell and Elaine Leong, "Recipe Collections and the Currency of Medical Knowledge in the Early Modern 'Medical Marketplace,'" in *Medicine and the Market in England and Its Colonies, c.1450–c.1850*, ed. Mark Jenner and Patrick Wallis (Basingstoke, UK: Palgrave Macmillan, 2007), 133–52.
124. See Harold Love, *Scribal Publishing in Seventeenth-Century England* (Oxford: Clarendon Press, 1993).
125. See, e.g., Francis Anthony, *The Apologie, or Defence of a Verity heretofore published concerning a Medicine called Aurum Potabile* (London, 1616), 5; anon., *The Best Way of Using the True Salt Polychrest of Messieurs Seignette of Rochel* (London, 1685).
126. Anon., "Medical recipe book, 17th to 18th century" (c. 1690–1710?), Wellcome Library MS 4054, f. 97r.
127. Alisha Rankin, "Empirics, Physicians, and Wonder Drugs in Early Modern Germany: The Case of the Panacea Amwaldina," *Early Science and Medicine* 14 (2009): 681; G. Papadopoulos, "The Admirable Effects of Panaceas: Ideas between Antiquity and Early Modern Times," *Wurzburg Medizinhistorische Mitteilungen* 30 (2011): 228–45.
128. Elizabeth Grey, Countess of Kent, *A Choice Manuall, or Rare and Select Secrets in Physick and Chyrurgery* (London, 1653), 175–76; for discussion, see Hunter, "Women and Domestic Medicine."
129. Michelle DiMeo and Joanna Warren, "The Countess of Kent's Powder: A Seventeenth-Century 'Cure-all,'" blog entry for "The Recipes Project," accessed September 22, 2015, http://recipes.hypotheses.org/tag/cure-all.

130. Grey, *Choice Manuall*, 175.
131. William Ellis, *The Country Housewife's Family Companion: or Profitable Directions for Whatever relates to the Management and Good Oeconomy of the Domestick Concerns of a Country Life* (London, 1750), i, ii.
132. Shapin and Schaffer, *Leviathan and the Air-Pump*.
133. J. V. Golinski, "A Noble Spectacle: Phosphorus and the Public Cultures of Science in the Early Royal Society," *Isis* 80 (1989): 11–39.
134. Abraham Cowley, "To the Royal Society," in Thomas Sprat, *The History of the Royal Society of London for the Improving of Natural Knowledge* (London, 1667), n.p., verse 1.
135. Robert Hooke, *Micrographia, or, Some Physiological Descriptions of Minute Bodies made by Magnifying Glasses with Observations and Inquiries thereupon* (London, 1665), preface.
136. Ibid.
137. Ibid., 165.
138. Michael Hunter, *Boyle: Between God and Science* (New Haven: Yale University Press, 2009); Shapin and Schaffer, *Leviathan and the Air-Pump*.
139. Michelle DiMeo, "Such a Sister became such a Brother. Lady Ranelagh's Influence on Robert Boyle." *Intellectual History Review* 25 (2015): 21–36; DiMeo, "Katherine Jones, Lady Ranelagh (1615–91): Science and Medicine in a Seventeenth-Century Englishwoman's Writing" (PhD diss., University of Warwick, 2009).
140. Boyle, quoted in DiMeo, "Such a Sister," 29.
141. John Aubrey, *Scientific Lives* (London: Hesperus Press, 2011), 12.
142. The essay forms the final part of "Some Considerations touching the Usefulness of Experimental Natural Philosophy, proposed in a familiar Discourse to a Friend," in Robert Boyle, *The Works of the Honourable Robert Boyle*, ed. Thomas Birch, 6 vols. (London, 1772; rpt. Hildesheim: Georg Olms, 1966), 3: 470–94.
143. Ibid., 3: 493.
144. Ibid., 3: 478.
145. On the gender division in seventeenth-century educational and scientific institutions, see Londa Schiebinger, *The Mind Has No Sex? Women in the Origins of Modern Science* (Cambridge: Harvard University Press, 1989), chap. 1.
146. John Securis, quoted in Jennifer K. Stine, "Opening Closets: The Discovery of Household Medicine in Early Modern England" (PhD diss., Stanford University, 1996), 82.
147. Jonathan Swift, *Travels into Several Remote Nations of the World . . . by Lemuel Gulliver* (Dublin, 1752), 169.
148. Mary Astell, *An Essay in Defence of the Female Sex in which are Inserted the Characters of a Pedant, a Squire, a Beau, a Vertuoso, a Poetaster, a City-critick, &c.* (London, 1696), 97–98.
149. Ibid., 110.
150. Aubrey, *Scientific Lives*, 46.

CHAPTER TWO

1. Shapin, "House of Experiment," 378.
2. Gadi Algazi, "Scholars in Households: Refiguring the Learned Habitus, 1480–1550," *Science in Context* 16 (2003): 32.
3. Harkness, "Managing an Experimental Household," 249.
4. Alix Cooper, "Homes and Households," in *The Cambridge History of Science*, vol. 3: *Early Modern Science*, ed. Katharine Park and Lorraine Daston (Cambridge: Cambridge University Press, 2006), 237.
5. E.g., Leong, "Collecting Knowledge."
6. See, e.g., Anita Guerrini, "The Ghastly Kitchen," *History of Science* 54 (2016): 71–97; Adi Ophir, "The Place of Knowledge Recreated: The Library of Michel de Montaigne," *Science in Context* 4 (1991): 163–89; Dora Thornton, *The Scholar in His Study: Ownership and Experience in Renaissance Italy* (New Haven: Yale University Press, 1998).
7. Derek Howse, *Greenwich Observatory: The Royal Observatory at Greenwich and Herstmonceux, 1675–1975*, vol. 3: *The Buildings and Instruments* (London: Taylor & Francis, 1975); Allan Chapman, "From Alchemy to Airpumps: The Foundations of Oxford Chemistry," in *Chemistry at Oxford: A History from 1600 to 2005*, ed. Robert J. P. Williams, Allan Chapman, and John S. Rowlinson (Cambridge: Royal Society of Chemistry, 2009), 40–46.
8. On the Society's locations, see Shapin, "House of Experiment," 381.
9. For discussion, see Owen Hannaway, "Laboratory Design and the Aim of Science: Andreas Libavius versus Tycho," *Isis* 77 (1986): 584–610.
10. See, e.g., Simon Werrett, "The Astronomical Capital of the World: Pulkovo Observatory in the Russia of Tsar Nicholas I," in *The Heavens on Earth: Observatories and Astronomy in Nineteenth-Century Science and Culture*, ed. Charlotte Bigg, Otto Sibum, and David Aubin (Durham, NC: Duke University Press, 2010), 33–57.
11. In the eighteenth century, however, the chemist Peter Shaw was quite insistent that chemistry could always be improved "without much Apparatus, or Expense, or without the Utensils and Instruments commonly employed in that Art: Which may, therefore, be practiced as well in a Study or Parlour, as in a Laboratory." Peter Shaw, *Chemical Lectures, Publickly Read at London in the Years 1731, and 1732, and at Scarborough, in 1733*, 2nd ed. (London, 1755), 463. On the history of dedicated chemical laboratories, see Peter Morris, *The Matter Factory: A History of the Chemical Laboratory* (London: Reaktion, 2015).
12. W. G. Hoskins, "The Rebuilding of Rural England, 1570–1640," *Past and Present* 4 (1953): 44–59; Matthew H. Johnson, "Rethinking the Great Rebuilding," *Oxford Journal of Archaeology* 12 (1993): 117–25.
13. Styles, "Product Innovation"; Berg, "British Product Revolution"; Peter Thornton, *Seventeenth-Century Interior Decoration in England, France and Holland* (New Haven: Yale University Press, 1978); Anthony Buxton, *Domestic*

Culture in Early Modern England (Woodbridge, UK: Boydell Press, 2015), 208–42.

14. Sara Pennell, "'Pots and Pans History': The Material Culture of the Kitchen in Early Modern England," *Journal of Design History* 11 (1998): 202–3; Frank E. Brown, "Continuity and Change in the Urban House: Developments in Domestic Space Organisation in Seventeenth-Century London," *Comparative Studies in Society and History* 28 (1986): 578–80, 586.
15. Knorr-Cetina, "Objectual Practice," 190; Knorr-Cetina draws on the idea of "epistemic things" in Hans-Jörg Rheinberger, *Towards a History of Epistemic Things: Synthesizing Proteins in the Test Tube* (Stanford: Stanford University Press, 1997).
16. The inventory (PROB 5/1324) is transcribed in Michael Hunter, "Hooke's Possessions at His Death: A Hitherto Unknown Inventory," in *Robert Hooke: New Studies*, ed. Michael Hunter and Simon Schaffer (Woodbridge, UK: Boydell Press, 1989), 292–94; of course, the inventory is not necessarily a complete or accurate reflection of Hooke's possessions.
17. Daniel Coxe to Robert Boyle, February 19, 1666, in Robert Boyle, *The Correspondence of Robert Boyle*, ed. Michael Hunter, Antonio Clericuzio, and Lawrence M. Principe, 6 vols. (London: Pickering & Chatto, 2001), 3: 70–71.
18. F. H. W. Sheppard, ed., *Survey of London*, vol. 29: *The Parish of St. James Westminster, Part One, South of Piccadilly* (London: Athlone Press, 1960), 367–68.
19. Evelyn, quoted in Adrian Tinniswood, *His Invention So Fertile: A Life of Christopher Wren* (Oxford: Oxford University Press, 2001), 43.
20. On Hauksbee, see Jim Bennett, "Shopping for Instruments in Paris and London," in *Merchants and Marvels: Commerce, Science, and Art in Early Modern Europe*, ed. Pamela H. Smith and Paul Findlen (New York: Routledge, 2002), 376; W. H. Quarrell and Margaret Ware, eds., *London in 1710: From the Travels of Zacharias Conrad von Uffenbach* (London, 1934), 77, 168.
21. John R. Milburn, *Adams of Fleet Street: Instrument-Makers to King George III* (Aldershot, UK: Ashgate, 2000), 24.
22. Adams, quoted in Milburn, *Adams of Fleet Street*, 38.
23. Amanda Flather, *Gender and Space in Early Modern England* (Woodbridge, UK: Boydell Press, 2007), 75–93.
24. Stine, "Opening Closets," 36–37.
25. John H. Appleby, "Erasmus King: Eighteenth-Century Experimental Philosopher," *Annals of Science* 47 (1990): 379.
26. Christa Jungnickel and Russell McCormmach, *Cavendish* (Philadelphia: American Philosophical Society, 1996), 242.
27. John Beale to Robert Boyle, May 7, 1666, in Boyle, *Correspondence*, 3: 163.
28. Isaac Newton to Hans Sloane, September 14, 1705, in Isaac Newton, "Original Letters of Sir Isaac Newton," *Monthly Magazine* 49 (1820): 235.
29. See, e.g., Thomas Birch, *History of the Royal Society*, 4 vols. (London, 1760), 2: 166, 168, 173; on Shortgrave, see M. B. Hall, *Promoting Experimental*

Learning: Experiment and the Royal Society, 1660–1727 (Cambridge: Cambridge University Press, 1991), 27.

30. Audrey T. Carpenter, *John Theophilus Desaguliers: A Natural Philosopher, Engineer and Freemason in Newtonian England* (London: Bloomsbury, 2011), 227–29, 239; Patricia Fara, *Pandora's Breeches: Women, Science and Power in the Enlightenment* (London: Pimlico, 2004), 19.
31. Samuel Hartlib noted the installation of a furnace in Alexander Akehurst's rooms in Trinity College, Cambridge, in 1653. See Mordechai Feingold, *Before Newton: The Life and Times of Isaac Barrow* (Cambridge: Cambridge University Press, 1990), 34.
32. Shapin, "House of Experiment," 380; Betty Jo Teeter Dobbs, *The Foundations of Newton's Alchemy, or, 'The Hunting of the Greene Lyon'* (Cambridge: Cambridge University Press, 1975), 98.
33. A. D. Atkinson, "William Derham, F.R.S. (1657–1735)," *Annals of Science* 8 (1952): 374–75.
34. Michael A. Hoskin, *The Herschel Partnership: As Viewed By Caroline* (Cambridge: Science History Publications, 2003), 25–56.
35. Stine, "Opening Closets," 37.
36. The inventory is reproduced in Douglas McKie, "Priestley's Laboratory and Library and Other of His Effects," *Notes and Records of the Royal Society* 12 (1956): 114–36.
37. M. J. Power, "East London Housing in the Seventeenth Century," in *Crisis and Order in English Towns 1500–1700*, ed. Peter Clark and Paul Slack (London: Routledge, 2007), 237–62.
38. Thornton, *Scholar in His Study*.
39. Andrew Cambers, *Godly Reading: Print, Manuscript and Puritanism in England, 1580–1720* (Cambridge: Cambridge University Press, 2011), 73.
40. Ibid., 73.
41. On closets, see ibid., 43–54; Stine, "Opening Closets," 86–87; Alan Stewart, "The Early Modern Closet Discovered," *Representations* 50 (1995): 76–100.
42. Kate Loveman, "Books and Sociability: The Case of Samuel Pepys's Library," *Review of English Studies* 61 (2010): 214–33.
43. Peter Thornton, *Authentic Decor: The Domestic Interior, 1620–1920* (London: Weidenfeld and Nicolson, 1984), 25.
44. Ibid., 77.
45. Rowan Watson, "Some Non-textual Uses of Books," in *A Companion to the History of the Book*, ed. Simon Eliot and Jonathan Rose (Oxford: Wylie-Blackwell, 2007), 480–92.
46. Shapin, "House of Experiment," 384; Steven Shapin, "'The Mind is its own Place': Science and Solitude in Seventeenth-Century England," *Science in Context* 4 (1991): 191–218.
47. Joannes Amos Comenius, *Orbis sensualium pictus*, cited in Cambers, *Godly Reading*, 71.
48. Shapin, "House of Experiment," 384–88.

49. John Dury to Samuel Hartlib, near Bremen, February 22, 1639. Hartlib papers online, 9/1/75B.
50. Jungnickel and McCormmach, *Cavendish*, 235.
51. Pennell, "'Pots and Pans History,'"; Sara Pennell, *The Birth of the English Kitchen, 1600–1850* (London: Bloomsbury, 2016); Doreen Yarwood, *The British Kitchen: Housewifery since Roman Times* (London: Batsford, 1981); Molly Harrison, *The Kitchen in History* (Reading, UK: Osprey, 1972).
52. Mukherjee, *Penury into Plenty*, 145–94; E. C. Spary, *Feeding France: New Sciences of Food, 1760–1815* (Cambridge: Cambridge University Press, 2014), 21–54.
53. Baxter, *Christian Directory*, 146.
54. Pennell, "'Pots and Pans History,'" 208.
55. Francis Bacon, "The History of Life and Death," *Works*, 5: 302; Werrett, "Recycling in Early Modern Science," 7.
56. Wall, *Recipes for Thought*, 212–18.
57. Hirst quoted in ibid., 241.
58. Michael Hunter and Alison Wiggins, eds., "The Workdiaries of Robert Boyle," accessed June 30, 2016, http://www.livesandletters.ac.uk/wd/index.html. For the recipes see workdiary no. 13, entries 6, 14, 35, 37, 40.
59. Amanda Vickery, *Behind Closed Doors: At Home in Georgian England* (New Haven: Yale University Press, 2009), 266; Karin Dannehl, "'To Families Furnishing Kitchens': Domestic Utensils and Their Use in the Eighteenth-Century Home," in *Buying for the Home: Shopping for the Domestic from the Seventeenth Century to the Present*, ed. David Hussey and Margaret Ponsonby (Aldershot, UK: Ashgate, 2008), 27–46; Rachel Field, *Irons in the Fire: A History of Cooking Equipment* (Marlborough, UK: Crowood Press, 1984).
60. Donald Woodward, "Straw, Bracken and the Wicklow Whale: The Exploitation of Natural Resources in England since 1500," *Past & Present* 159 (1998): 159.
61. Dannehl, "'To Families Furnishing Kitchens,'" 32.
62. Gray, "'Practical Art,'" 52.
63. Jean Theophilus Desaguliers, *A Dissertation Concerning Electricity* (London, 1742), 9; Priestley, *History and Present State of Electricity*, 65.
64. George Adams, *Lectures on Natural and Experimental Philosophy, considered in its Present State of Improvement*, 5 vols. (London, 1794), 2: 30.
65. Tiberius Cavallo, *A Complete Treatise on Electricity, in Theory and Practice, with Original Experiments* (London, 1777), 22–23.
66. Herman Boerhaave, *Elements of Chemistry*, 2 vols. (London, 1735), 1: 78.
67. Anon., "Medical recipe book, 17th to 18th century," (c. 1690–1710?) Wellcome Library MS 4054, f. 93r.
68. Stephen Hales, *Philosophical Experiments: Containing Useful, and Necessary Instructions for Such as Undertake Long Voyages at Sea* (London, 1739), 26.
69. Stephen Hales, *Statical Essays: containing Haemastatics; or, an Account of Some Hydraulic and Hydrostatical Experiments Made on the Blood* (London,

1740), 269; Tiberius Cavallo, *A Treatise on Magnetism, in Theory and Practice, with Original Experiments* (London, 1795), 221.

70. John Neale, *Directions for Gentlemen, who have Electrical Machines, how to Proceed in Making their Experiments* (London, 1747), 4.
71. Bridget Hill, *Servants: English Domestics in the Eighteenth Century* (Oxford: Clarendon Press, 1996); R. C. Richardson, *Household Servants in Early Modern England* (Manchester: Manchester University Press, 2010); Shapin, "Invisible Technician."
72. North, *Observations and Advices Oeconomical*, 43.
73. John Beale, "An Experiment to Examine, What Figure, and Celerity of Motion Begetteth, or Encreaseth Light and Flame," *Philosophical Transactions* 1 (1665–66): 227; John Beale to Robert Boyle, April 18, 1666, in Boyle, *Correspondence*, 3: 139.
74. John Beale to Robert Boyle, April 28, 1666, in Boyle, *Correspondence*, 3: 159.
75. Robert Boyle, "Some Observations about Shining Flesh, Made by the Honourable Robert Boyle; Febr. 15. 1671/72," *Philosophical Transactions* 7 (1672): 5108.
76. Amanda Vickery has called the kitchen "unequivocally a female domain." *Behind Closed Doors*, 266.
77. British Library, Add MSS 78337 (formerly MS 51). See Michael Hunter, *Science and the Shape of Orthodoxy: Intellectual Change in Late Seventeenth-Century Britain* (Woodbridge, UK: Boydell Press, 1995), 73–74; on Mary Evelyn, see Frances Harris, "Living in the Neighbourhood of Science: Mary Evelyn, Margaret Cavendish and the Greshamites," in Hunter and Hutton, *Women, Science, and Medicine, 1500–1700*, 198–217.
78. Elisabeth Berry Drago, "Thomas Wijck's Painted Alchemists at the Intersection of Art, Science, and Practice" (PhD diss., University of Delaware, 2016); McKeon, *Secret History of Domesticity*, 213–14.
79. Thomas Shadwell, *The Virtuoso*, ed. Marjorie Hope Nicholson and David Stuart Rodes (Lincoln: University of Nebraska Press, 1966), 111 (act 5, scene 2).
80. Guerrini, "Ghastly Kitchen."
81. Birch, *History of the Royal Society*, 2: 166, 168, 173, 176.
82. Edward Ravenscroft, *The Anatomist, or, The Sham-Doctor* (London, 1697), 13 (act 2, scene 1).
83. On early modern cellars, see Angela McShane and Nigel Jeffries, "I say 'shard,' you say 'sherd': Contrasting and Complementary Approaches to a Piece of Early Modern 'Venice Glass,'" in *The Routledge Handbook of Material Culture in Early Modern Europe*, ed. Catherine Richardson, Tara Hamling, and David Gaimster (London: Routledge, 2016), 452–65; George Sheeran, *Medieval Yorkshire Towns: Buildings, People, and Spaces* (Edinburgh: Edinburgh University Press, 1998), 148–49.
84. Hunter, "Hooke's Possessions at His Death," 293.
85. Werrett, "Recycling in Early Modern Science," 6.
86. Nehemiah Grew, *The Anatomy of Plants with an Idea of a Philosophical History of Plants* (London, 1682), 287.

87. Philiatros [Alethea Talbot], *Natura exenterata: or Nature unbowelled by the most exquisite anatomizers of her* (London, 1655), 155; see also 35, 378, 384.
88. Werrett, "Recycling in Early Modern Science," 5.
89. This was a postscript to Robert Boyle, "New Experiments and Observations touching Cold," *Works*, 2: 683–86.
90. Ibid., 684.
91. Robert Boyle, "Essays of the Strange Subtilty, Great Efficacy, Determinate Nature of Effluviums," *Works*, 3: 684.
92. Herman Boerhaave, *A New Method of Chemistry; Including the History, Theory, and Practice of the Art*, trans. Peter Shaw, 2 vols. (London, 1741), 2: 371–72.
93. Ibid., 2: 371–72.
94. Lucy Worsley, *If Walls Could Talk: An Intimate History of the Home* (London: Faber and Faber, 2011), 12.
95. John Evelyn to William Wotton, March 30, 1696, *The Diary and Correspondence of John Evelyn*, ed. William Bray, 4 vols. (London, 1857), 2: 350.
96. Carole Shammas, "The Domestic Environment in Early Modern England and America," *Journal of Social History* 14 (1980): 10; Woodward, "Straw, Bracken, and the Wicklow Whale," 65–67.
97. Joyce Rogers, *The Second Best Bed: Shakespeare's Will in a New Light* (Westport, CT: Greenwood Press, 1993).
98. See Clive Edwards, *Turning Houses into Homes: A History of the Retailing and Consumption of Domestic Furnishings* (Aldershot, UK: Ashgate, 2005), 19.
99. Philiatros [Alethea Talbot], *Natura exenterata*, 109.
100. Robert Boyle, "Observations made by Mr. Boyle about the Aerial Noctiluca contained in his Second Vial" (1680), *Works*, 4: 390; Boyle also used his four-poster bed to investigate the luminosity of diamonds. Robert Boyle, "Observations made this 27th of October, 1663, about Mr. Clayton's Diamond" (1664), *Works*, 1: 796.
101. Birch, *History of the Royal Society*, 2: 182.
102. Harvard University Collection of Historical Scientific Instruments, "Cylinder Electrical Machine, c. 1766, inventory number 0013," accessed February 15, 2016, http://waywiser.rc.fas.harvard.edu/view/objects/asitem/search@/1/displayDate-asc?t:state:flow=a96e45c1-f47b-469a-992b-0bec1581c403.
103. Sir Anthony Fitzherbert, quoted in Woodward, "Straw, Bracken, and the Wicklow Whale," 68.
104. Aubrey, *Scientific Lives*, 67.
105. Michael Tutton and Elizabeth Hirst, eds., *Windows: History, Repair and Conservation* (London: Routledge, 2015), 58–61, 258–62.
106. Dana E. Katz, "'Clamber not you up to the Casements': On Ghetto Views and Viewing," *Jewish History* 24 (2010): 127–53; Jill Colaco, "The Window Scenes in 'Romeo and Juliet' and Folk Songs of the Night Visit," *Studies in Philology* 83 (1986): 138–57.
107. Benjamin Franklin, "The Art of Procuring Pleasant Dreams," *Autobiography, Poor Richard, and Later Writings*, ed. Joseph A. Leo Lemay (New York: Library

of America, 1997), 378; Henry Oldenburg to Robert Boyle, March 17, 1666, in Boyle, *Correspondence*, 3: 115.

108. Ann Shackleford, *The Modern Art of Cookery Improved; Or, Elegant, Cheap, and Easy Methods, of Preparing Most of the Dishes Now in Vogue* (London, 1767), preface.
109. Ibid., 193.
110. Isaac Newton, *Opticks; Or, A Treatise of the Reflections, Refractions, Inflections and Colours of Light* (London, 1721), 21–53; Simon Schaffer, "Glass Works: Newton's Prisms and the Uses of Experiment," in *The Uses of Experiment: Studies in the Natural Sciences*, ed. David Gooding, Trevor Pinch, and Simon Schaffer (Cambridge: Cambridge University Press, 1989), 67–104.
111. Hooke, *Micrographia*, 168.
112. J. M. Neeson, *Commoners: Common Right, Enclosure and Social Change in England, 1700–1820* (Cambridge: Cambridge University Press, 1993).
113. Vittoria Di Palma, *Wasteland: A History* (New Haven: Yale University Press, 2014).
114. Chomel, *Dictionaire Oeconomique*.
115. See, e.g., Paula Findlen, "Sites of Anatomy, Botany, and Natural History," in Park and Daston, *Cambridge History of Science*, 3: 272–89; Clare Hickman, "The Garden as a Laboratory: The Role of Domestic Gardens as Places of Scientific Exploration," *Post-Medieval Archaeology* 48 (2014): 229–47; Golinski, *British Weather*.
116. See, e.g., Stine, "Opening Closets," 33.
117. Siena Louise Latham, "'Lady Alcumy': Elizabethan Gentlewomen and the Practice of Chymistry" (PhD diss., Victoria University, Wellington, 2010), 101–3.
118. Humphrey Newton, two letters to John Conduitt, Keynes MS 135, King's College, Cambridge, accessed December 17, 2015, http://www.newtonproject.sussex.ac.uk/view/texts/normalized/THEM00033.
119. Betty Jo Teeter Dobbs, "Newton's Alchemy and His Theory of Matter," *Isis* 73 (1982): 512–28; David Kubrin, "Newton and the Cyclical Cosmos: Providence and the Mechanical Philosophy," *Journal of the History of Ideas* 27 (1968): 325–46.
120. Henry Oldenburg to Robert Boyle, March 17, 1666, in Boyle, *Correspondence*, 3: 114.
121. Mary Terrall, *Catching Nature in the Act: Réaumur and the Practice of Natural History in the Eighteenth Century* (Chicago: University of Chicago Press, 2013), provides a detailed exploration of the study of garden insects and domestic fowl by Réaumur and his circle in eighteenth-century France.
122. E.g., Samuel Hartlib, *The Reformed Common-Wealth of Bees, Presented in Severall Letters and Observations to Sammuel Hartlib* (London, 1655); Lady Margaret Hoby, *The Private Life of an Elizabethan Lady: The Diary of Lady Margaret Hoby, 1599–1605*, ed. Joanna Moody (Stroud, UK: Sutton Publishing, 1998), 13.
123. Hooke, *Micrographia*, 165–69.

124. Mrs. [Anna Letitia] Barbauld, *Poems* (London, 1773), 37–40; for other uses of animals in early modern experiments, see Anita Guerrini, *Experimenting with Humans and Animals: From Galen to Animal Rights* (Baltimore: Johns Hopkins University Press, 2003).
125. Matthew Johnson, *English Houses, 1300–1800: Vernacular Architecture, Social Life* (London: Routledge, 2014), 37–39.
126. Jim Bennett, "Hooke's Instruments for Astronomy and Navigation," in Hunter and Schaffer, *Robert Hooke*, 21–32; Nicky Reeves, "Constructing an Instrument: Nevil Maskelyne and the Zenith Sector, 1760–1774" (PhD diss., University of Cambridge, 2008), 13–27.
127. Benjamin Franklin, *Experiments and Observations on Electricity, Made at Philadelphia in America*, 5th ed. (London, 1774), 120–22.
128. Robert Dossie, *The Handmaid to the Arts*, 2 vols. (London, 1758), 1: 364; thanks to Anna Maerker for this reference.
129. Joseph Banks, *The Endeavour Journal of Joseph Banks: The Australian Journey*, ed. Paul Brunton (Sydney: Angus & Robertson in association with State Library of New South Wales, 1998).
130. Francis Galton, *The Art of Travel, or, Shifts and Contrivances Available in Wild Countries* (London, 1855).
131. Richard Sorrenson, "The Ship as a Scientific Instrument in the Eighteenth Century," *Osiris* 11 (1996): 221–36.
132. As an eighteenth-century commentator put it, "my house has the appearance of a ship stored for a voyage across the Atlantic." Anon., "Account of a Buyer of Bargains," n.p.
133. Sprat, *History of the Royal Society*, 122–23.
134. Peter Borsay, *The English Urban Renaissance: Culture and Society in the Provincial Town, 1660–1770* (Oxford: Clarendon Press, 1989); James Ayres, *Building the Georgian City* (New Haven: Yale University Press, 1998).
135. H. J. Louw, "The Origins of the Sash-Window," *Architectural History* 26 (1983): 49–72, 144–50; Alison Stoesser, "Robert Hooke's Montagu House," in Hunter and Schaffer, *Robert Hooke*, 175.
136. Rachel Ramsey, "The Literary History of the Sash Window," *Eighteenth-Century Fiction* 22 (2009–10): 171–94.
137. See, e.g., Alice N. Walters, "Conversation Pieces: Science and Politeness in Eighteenth-Century England," *History of Science* 35 (1997): 121–54.

CHAPTER THREE

1. Robert Symmer, "New Experiments and Observations concerning Electricity," *Philosophical Transactions* 51 (1759–60): 340.
2. Ibid., 372; J. L. Heilbron, "Robert Symmer and the Two Electricities," *Isis* 67 (1976): 7–20.
3. Robert Symmer to John Mitchell, June 19, 1760, quoted in Heilbron, "Robert Symmer," 16.

4. Ehrman, "Dressing Well in Old Age"; Ehrman uses the accounts of the Berkshire widow Dodson to reconstruct her wardrobe and the diverse practices she used to purchase, maintain, and rework clothes.
5. See ibid., 35; the Museum of London has a pair of linen, silk, gold, and leather women's shoes made in 1720–50 from material dating to c. 1620. "Woman's shoes, 1720–50," ID no. A5999.
6. Pepys, quoted in Werrett, "Recycling in Early Modern Science," 7.
7. M. Bella Mirabella, ed., *Ornamentalism: The Art of Renaissance Accessories* (Ann Arbor: University of Michigan Press, 2011).
8. See, e.g., Coat and Breeches, England, c. 1705–15, Victoria and Albert Museum, London. Museum Number T.327 & A-1982. The coat was repaired quite extensively in the eighteenth century, while the breeches were altered in the nineteenth to turn them into theatrical costume.
9. Ariane Fennetaux, "Sentimental Economics: Textile Recycling in Eighteenth-Century Britain," in Fennetaux, Vasset, and Junqua, *Afterlife of Used Things*, 126; Woodward, "Swords into Ploughshares," 177–79.
10. George Washington to Robert Cary & Company, September 28, 1760, quoted in Fennetaux, "Sentimental Economics," 126.
11. Ehrman, "Dressing Well in Old Age," 36.
12. Karen Parker, "A Use for Broken Pipe Stems," *Society for Clay Pipe Research Newsletter* 17 (1988): 12–13; chymists used clay pipes as stirrers. See Robert Dossie, *The Elaboratory Laid Open: or the Secrets of Modern Chemistry and Pharmacy Revealed* (London, 1758), 273.
13. Sara Pennell, "Perfecting Practice? Women, Manuscript Recipes and Knowledge in Early Modern England," in *Early Modern Women's Manuscript Writing*, ed. Victoria E. Burke and Jonathan Gibson (Aldershot, UK: Ashgate, 2004), 237–58.
14. 12 Geo. 3 c. 48. See *The Statutes at Large From the Tenth Year of the Reign of King George the Third, Inclusive*, 11 vols. (London, 1774), 11: 529; *The Law-Dictionary: Explaining the Rise, Progress, and Present State of the English Law, in Theory and Practice*, ed. Thomas Edlyne Tomlins, 2 vols. (London, 1797), s.v. "Forgery."
15. C. Pajot des Charmes, *The Art of Bleaching Piece-Goods, Cottons, and Threads of Every Description* (London, 1799), 13.
16. Mary Cole, *The Lady's Complete Guide, or Cookery in all its Branches* (London, 1788), 533; no doubt this tradition shaped the close connection between hygiene and disposable goods that developed in the nineteenth century. Jane Celia Busch, "The Throwaway Ethic in America" (PhD diss., University of Pennsylvania, 1983).
17. J. A. Bennett, S. A. Johnston, and A. V. Simcock, *Solomon's House in Oxford: New Finds from the First Museum* (Oxford: Museum of the History of Science, 2000).
18. See, e.g., Robert E. Bergman and Thomas V. Moore, *Managing Interactive Video/Multimedia Projects* (Englewood Cliffs, NJ: Educational Technology Publications, 1990), 86; "repurpose, v.," OED Online, December 2015.

19. Lady Ayscough, "Receits of phisick and chirurgery," manuscript recipe book, Wellcome Library MS 1026, p. 29.
20. Boyle, "New Experiments and Observations touching Cold," 632.
21. Hannah Woolley, *The Cook's Guide: or, Rare Receipts for Cookery Published and Set Forth Particularly for Ladies and Gentlwomen* (London, 1664), 19.
22. [Thomas Henshaw], "Some Observations and Experiments on May-Dew," *Philosophical Transactions* 1 (1665–66): 33–36.
23. Stine, "Opening Closets," 36.
24. Maurice Crosland, "Priestley Memorial Lecture: A Practical Perspective on Joseph Priestley as a Pneumatic Chemist," *British Journal for the History of Science* 16 (1983): 233; Werrett, "Recycling in Early Modern Science," 8.
25. Joseph Priestley, *Experiments and Observations on Different Kinds of Air*, 3 vols. (Birmingham, 1790; Kraus Reprint 1970), 1: 16, 20, 51.
26. Elizabeth Fulhame, *An Essay On Combustion with a View to a New Art of Dying and Painting, wherein the Phlogistic and Antiphlogistic Hypotheses are Proved Erroneous* (London, 1794).
27. Ibid., x.
28. Ibid., x, 17.
29. Beale, "An Experiment to Examine," 227.
30. William Henly to John Canton, n.d., Canton Papers II, 86 (RS), quoted in John L. Heilbron, *Electricity in the 17th and 18th Centuries: A Study of Early Modern Physics* (Berkeley: University of California Press, 1979), 317.
31. Schaffer, "Glass Works," 78.
32. James Jurin noted how human blood separated out as it cooled in a porringer. James Jurin, "An Account of Some Experiments Relating to the Specifick Gravity of Human Blood," *Philosophical Transactions* 30 (1717–19): 1000.
33. Ronald F. Michaelis, "English Pewter Porringers," *Pewter Collectors Club of America Bulletin* 7 (February 1976): 115–21, and 7 (August 1976): 155–62.
34. Helen S. Foote, "Silver in the Service of Medicine," *Bulletin of the Medical Library Association* 32 (1944): 369–75.
35. Werrett, "Recycling in Early Modern Science," 8; Markus Krajewski, *Paper Machines: About Cards & Catalogs, 1548–1929* (Cambridge: MIT Press, 2011), 27–48.
36. Anna Marie Roos, *Web of Nature: Martin Lister (1639–1712), the First Arachnologist* (Leiden: Brill, 2011), 57.
37. Communication from Heather Wolfe, Folger Shakespeare Library.
38. Philip Luckombe and William Caslon, *A Concise History of the Origin and Progress of Printing: With Practical Instructions to the Trade in General* (London, 1770), 492.
39. P. J. Brownsey, "The Banks and Solander Collections—A Benchmark for Understanding the New Zealand Flora," *Journal of the Royal Society of New Zealand* 42 (2012): 132.

40. Banks, *Endeavour Journal of Joseph* Banks, 58 (July 2, 1770).
41. Anna-Marie Roos, "A Speculum of Chymical Practice: Isaac Newton, Martin Lister (1639–1712), and the Making of Telescopic Mirrors," *Notes and Records of the Royal Society* 64 (2010): 106.
42. Elizabeth Yale, "With Slips and Scraps: How Early Modern Naturalists Invented the Archive," *Book History* 12 (2009): 1–2.
43. For details of prices, see James Daybell, *The Material Letter in Early Modern England: Manuscript Letters and the Culture and Practices of Letter-Writing, 1512–1635* (Basingstoke, UK: Palgrave Macmillan, 2012), 34–36.
44. See, e.g., Edward Hatton, *An Intire System of Arithmetic: Or Arithmetic in all its Parts* (London, 1721), 19.
45. Bennett, Johnston, and Simcock, *Solomon's House in Oxford.*
46. Papers of John Pell, British Library, Add MS 4428, ff. 74r, 74v, 196r, 196v.
47. Aubrey, *Scientific Lives*, 63.
48. John Hood, *Tables of Difference of Latitude and Departure for Navigators, Land Surveyors, &c. with their Application to Plane Trigonometry* (Dublin, 1772), 11.
49. Harvard University Collection of Historical Scientific Instruments, "Octant Case, c. 1794, by Adrian Br. Hicks. Inventory No. 1996-1-0172b," accessed February 15, 2016, http://waywiser.rc.fas.harvard.edu/view/objects/asitem/search@/0?t:state:flow=68a4969a-f994-4632-a1f6-a2a28ec818c3.
50. Hindle, *On the Parish?*, chap. 1.
51. Thomas Tusser, *Fiue Hundreth Points of Good Husbandry vnited to as many of Good Huswiferie first deuised* (London, 1573), 58.
52. Thomas Sheridan, *A Complete Dictionary of the English Language*, second edition (London, 1789), s.v. "Expedient"; see also Mukherjee, *Penury into Plenty*, 11; Hindle, *On the Parish?*, 16.
53. Sheridan, *Complete Dictionary*, s.v. "Shift."
54. John Evelyn, *Silva, or a Discourse on Forest-Trees and the Propagation of Timber in His Majesties Dominions*, 2nd ed. (London, 1670), 189.
55. Daniel Defoe, *The Complete English Tradesman, in Familiar Letters* (London, 1725), 35.
56. John Ray, *Miscellaneous Discourses concerning the Dissolution and Changes of the World* (London, 1692), 238.
57. Henry Oldenburg, *The Correspondence of Henry Oldenburg*, ed. A. R. Hall and M. B. Hall, 13 vols. (Madison: University of Wisconsin Press; London: Mansel; London: Taylor & Francis, 1965–86), 4: 48–49.
58. Anon., "Medical recipe book, 17th to 18th century" (c. 1690–1710?), Wellcome Library MS 4054, f. 89r.
59. Hooke, *Micrographia*, 55.
60. William Barlow, *Magneticall Aduertisements: or Diuers Pertinent Obseruations, and Approued Experiments, concerning the Natures and Properties of the Loadstone* (London, 1618), 79–80.
61. John French, *The Art of Distillation, or, A Treatise of the Choicest Spagiricall Preparations performed by Way of Distillation* (London, 1653).

62. Fulhame, *Essay On Combustion*, v–vi.
63. Mukherjee, *Penury into Plenty*, 38–39.
64. Latham, "'Lady Alcumy,'" 52.
65. Alain Touwaide, "Quid pro Quo: Revisiting the Practice of Substitution in Ancient Pharmacy," in *Herbs and Healers from the Ancient Mediterranean through the Medieval West: Essays in Honor of John M. Riddle*, ed. Anne Van Arsdall and Timothy Graham (Ashgate, UK: Aldershot, 2011), 19–61.
66. Owen Gingerich, "Astronomical Paper Instruments with Moving Parts," in *Making Instruments Count: Essays on Historical Scientific Instruments presented to Gerard L'Estrange Turner*, ed. Robert Anderson, James A. Bennett, and W. F. Ryan (Aldershot, UK: Variorum, 1993), 63–74, esp. n. 14; Gingerich, "A Tusi Couple from Schöner's *De Revolutionibus*?" *Journal for the History of Astronomy* 15 (1984): 128–33.
67. Roos, "Speculum of Chymical Practice," 111.
68. Dossie, *Elaboratory Laid Open*, 351–75; on adulteration in this period, see William J. Ashworth, *Customs and Excise: Trade, Production, and Consumption in England, 1640–1845* (Oxford: Oxford University Press, 2003), 207–60.
69. Dossie, *Elaboratory Laid Open*, 369. See also Matthew Paskins, "One of These Things is Just Like the Others: Substitution as a Motivator in Eighteenth-Century Chemistry," in *Theory Choice in the History of Chemical Practices*, ed. Emma Tobin and Chiara Ambrosio (Dordrecht: Springer, 2016), 55–70.
70. John H. Appleby, "Humphrey Jackson, F.R.S., 1717–1801: A Pioneering Chemist," *Notes and Records of the Royal Society of London* 40 (1986): 147–68, esp. 163–164.
71. Silvio A. Bedini, *Thinkers and Tinkers: Early American Men of Science* (New York: Scribner, 1975).
72. A. D. Morrison-Low, *Making Scientific Instruments in the Industrial Revolution* (Aldershot, UK: Ashgate, 2007), 75, 122, 170, 204, 266–67.
73. William Marshall, *Experiments and Observations Concerning Agriculture and the Weather* (London, 1779), 161.
74. James Delbourgo, *A Most Amazing Scene of Wonders: Electricity and Enlightenment in Early America* (Cambridge: Harvard University Press, 2006), 227; on "onion" bottles, see Willy Van den Bossche, *Antique Glass Bottles: Their History and Evolution (1500–1850)* (Woodbridge, UK: Antique Collectors' Club, 2001).
75. James Woodhouse, "Dr Woodhouse's Economical Apparatus," appendix to James Parkinson, *The Chemical Pocket-Book, or Memoranda Chemica, arranged in a Compendium of Chemistry* (Philadelphia, 1802), 203.
76. Sara J. Schechner, "Instrumentation," in *A Companion to the History of American Science*, ed. Georgina M. Montgomery and Mark A. Largent (Oxford: Wiley, 2016), 414.
77. Bedini, *Thinkers and Tinkers*, 195.
78. Ibid., 203.
79. Ibid., 216, and see Bedini's figs. 47 and 48.

80. Ibid., 192–93.
81. Ibid., 195–97.
82. Ibid., 205.
83. Ibid., 206.
84. Ibid., 206.
85. Ibid., 209.
86. Matthew Jones, *Reckoning with Matter: Calculating Machines, Innovation, and Thinking about Thinking from Pascal to Babbage* (Chicago: University of Chicago Press, 2016), 126–56; Maxine Berg, "In Pursuit of Luxury: Global History and British Consumer Goods in the Eighteenth Century," *Past and Present* 182 (2004): 85–142; Berg, "From Imitation to Invention: Creating Commodities in Eighteenth-Century Britain," *Economic History Review* 55 (2002): 1–30; Patrick O'Brien, Trevor Griffiths, and Philip Hunt, "Political Components of the Industrial Revolution: Parliament and the English Cotton Textile Industry, 1660–1774," *Economic History Review* 44 (1991): 395–423; Prasannan Parthasarathi, *Why Europe Grew Rich and Asia Did Not: Global Economic Divergence, 1600–1850* (Cambridge: Cambridge University Press, 2011).
87. John Gascoigne, *Joseph Banks and the English Enlightenment: Useful Knowledge and Polite Culture* (Cambridge: Cambridge University Press, 1994), 205.
88. Stephen Hales, *A Description of Ventilators, whereby Great Quantities of Fresh Air May with Ease be Conveyed into Mines, Goals, Hospitals, Work-Houses and Ships in Exchange for their Noxious Air* (London, 1743), 61–63.
89. Ibid., 2.
90. Howse, *Greenwich Observatory*, 3: 116.
91. Simon Werrett, *Fireworks: Pyrotechnic Arts and Sciences in European History* (Chicago: University of Chicago Press, 2010), 57, 194, 200.
92. Benjamin Franklin to Peter Collinson, *Pennsylvania Gazette*, October 19, 1752; Priestley, *History and Present State of Electricity*, 170–72.
93. George Adams, *An Essay on Electricity, Explaining the Theory and Practice of that Useful Science*, 3rd ed. (London, 1787), supplement, li.
94. John Ash, *The New and Complete Dictionary of the English Language*, 2 vols. (London, 1775), s.v. "Alter" and "Alteration."
95. Richard Head and Francis Kirkman, *The English Rogue Described in the Life of Meriton Latroon*, 4 vols. (London, 1665–71), 2: 109.
96. Harvard University Collection of Historical Scientific Instruments, "Cylinder Electrical Machine," inventory number 0013.
97. Quoted in Harvard University Collection of Historical Scientific Instruments, "4-foot Gregorian Reflecting Telescope, c.1767, Inventory No. 0001a," accessed February 15, 2016, http://waywiser.rc.fas.harvard.edu/view/objects/asitem/search@/21/displayDate-asc?t:state:flow=0ece7490-6177-424e-ace7-831fa27eb82e; David P. Wheatland, *The Apparatus of Science at Harvard, 1765–1800* (Cambridge: Harvard University Press, 1968), 17–19.
98. Warner, "What Is a Scientific Instrument."

99. Boerhaave, *Elements of Chemistry*, 1: 500.
100. Ibid., 1: 78.
101. I avoid the term "instrumentalize" since this has many existing connotations. See, e.g., Andrew Feenburg, *Transforming Technology: A Critical Theory Revisited* (Oxford: Oxford University Press, 2002).
102. Cavallo to Lind, November 23, 1792, British Library, Add MSS 22898, ff. 24v–25r.

CHAPTER FOUR

1. Richard Allestree, *The Gentleman's Calling* (London, 1660), 46.
2. Tusser, *Hundreth Good Pointes*, 32.
3. Johnson, *Dictionary of the English Language*, s.v. "Durability."
4. Edgerton, *Shock of the Old*, 77; see also Andrew Russell and Lee Vinsel, "Hail the Maintainers," *Aeon* (April 7, 2016), accessed December 16, 2017, https://aeon.co/essays/innovation-is-overvalued-maintenance-often-matters-more.
5. Hermann Ludwig von Löwenstern, *The First Russian Voyage around the World: The Journal of Hermann Ludwig von Löwenstern, 1803–1806*, ed. Victoria Joan Moessner (Fairbanks: University of Alaska Press, 2003), 115.
6. Hooke, *Micrographia*, preface.
7. On approaches to scientific instrument design in this period, see Richard Sorrenson, *Perfect Mechanics: Instrument Makers at the Royal Society of London in the Eighteenth Century* (Boston, MA: Docent Press, 2013), chap. 4; W. D. Hackmann, "The Relationship between Concept and Instrument Design in Eighteenth-Century Experimental Science," *Annals of Science* 36 (1979): 205–24.
8. "Lake's New Invented Mathematical Lamps," *Daily Journal* 2469 (December 6, 1728), n.p.
9. *Public Advertiser*, May 21, 1753, quoted in Sara Pennell, "'For a Crack and a Flaw Despis'd': Thinking about Ceramic Semi-durability and the 'Everyday' in Early Modern England," in *Everyday Things: Medieval and Early Modern Material Culture*, ed. Tara Hamling and Catherine Richardson (Farnham, UK: Ashgate, 2010), 35.
10. Dossie, *Elaboratory Laid Open*, 3–4; Werrett, "Household Oeconomy and Chemical Inquiry," 44.
11. Robert Hooke, "Dr. Hook's Description of some Instruments for Sounding the great Depths of the Sea, and bringing Accounts of several Kinds from the Bottom of it. Being the Substance of some of his Lectures, in December, 1691," *Philosophical Experiments and Observations of the late Eminent Dr. Robert Hooke*, ed. William Derham (London, 1726), 235.
12. Mary Delany, *The Autobiography and Correspondence of Mrs. Delany*, ed. Sarah Chauncey Woolsey, 2 vols. (Boston: Roberts Brothers, 1879), 2: 411–12.
13. [Robert Boyle], "New Pneumatical Experiments about Respiration. Continued," *Philosophical Transactions* 5 (1670): 2047.

14. John Allen, *Specimina ichnographica: or, a Brief Narrative of several New Inventions, and Experiments* (London, 1730), 8.
15. Boerhaave, *Elements of Chemistry*, 1: 500–501.
16. Ibid., 1: 501.
17. Ibid., 1: 59.
18. Ibid.
19. William Smith [attrib.], Chemical Notebook, pre-1774. MS Coll. 155, Van Pelt Rare Books and Manuscripts, University of Pennsylvania, f. 9r.
20. Merry E. Wiesner, *Women and Gender in Early Modern Europe*, 2nd ed. (Cambridge: Cambridge University Press, 2000), 75.
21. Richard Lucas, *The Duty of Servants* (London, 1710), 101–2.
22. Fara, *Pandora's Breeches*, 16–17.
23. Shapin, "Invisible Technician."
24. R. T. Gunther, ed., *Early Science in Oxford*, vol. 10: *The Life and Work of Robert Hooke (Part IV)* (Oxford: for the author, 1935), 116 (April 24, 1689).
25. Alan Chapman, *England's Leonardo: Robert Hooke and the Seventeenth-Century Scientific Revolution* (Bristol: Institute of Physics Publishing, 2005), chap. 13.
26. Iliffe and Willmoth, "Astronomy and the Domestic Sphere"; Fara, *Pandora's Breeches*, chap. 8.
27. Anne Barker, *The Complete Servant Maid: Or Young Woman's Best Companion. Containing Full, Plain, and Easy Directions for Qualifying Them for Service* (London, c. 1770).
28. Ibid., 40.
29. Adams, *Essay on Electricity*, 32–33.
30. Such sensuous maintenance might be said to complement the "sensuous chemistry" discussed in Lissa Roberts, "The Death of the Sensuous Chemist: The 'New' Chemistry and the Transformation of Sensuous Technology," *Studies in History and Philosophy of Science* 26 (1995): 503–29.
31. Thomas S. Kuhn, *The Structure of Scientific Revolutions*, 4th ed. (Chicago: University of Chicago Press, 2012), 24.
32. Quoted in Amanda Vickery, *The Gentleman's Daughter: Women's Lives in Georgian England* (New Haven: Yale University Press, 1998), 149.
33. Robert Hooke recorded buying a mop and brooms in his diary. Gunther, *Early Science in Oxford*, 10: 116 (April 1689), 155 (October 10, 1689), 253 (June 24, 1693).
34. Robert Southwell, quoted in Keith Thomas, "Cleanliness and Godliness in Early Modern England," in *Religion, Culture and Society in Early Modern Britain: Essays in Honour of Patrick Collinson*, ed. Anthony Fletcher and Peter Roberts (Cambridge: Cambridge University Press, 1994), 65.
35. Ehrman, "Dressing Well in Old Age," 36.
36. Leonard Mascall, *A Profitable Booke, declaring Diuers Approoued Remedies, to take out Spots and Staines in Silkes, Veluets, Linnen and Woollen Clothes* (London, 1583).
37. Ibid., Aii.

38. See, e.g., Hannah Woolley, *The Compleat Servant-Maid; or, Young Maiden's Tutor* (London, 1677), 65–72; John White, *Arts Treasury: or, A Profitable and Pleasing Invitation to the Lovers of Ingenuity Contained in Many Extraordinary Experiments, Rareties, and Curious Inventions* (London, 1688), chaps. 3 and 4; Peregrine Montague, *The Family Pocket-book: or, Fountain of True and Useful Knowledge* (London, 1760), 55–59, 73–75; Barker, *Complete Servant Maid,* 40–43.
39. Hunter and Wiggins, "Workdiaries of Robert Boyle," workdiary no. 12, entry 69.
40. Anon., *The British Jewel; or Complete Housewife's Best Companion* (London, c. 1785), 80.
41. George Adams, *Essays on the Microscope; Containing a Practical Description of the Most Improved Microscopes* (London, 1784), 10.
42. Alexander Morrice, *A Treatise on Brewing,* 2nd ed. (London, 1802), 155–56.
43. Lady Ann Fanshawe, "Recipe Book, compiled from 1651," Wellcome Library, f. 263v.
44. Eliza Fowler Haywood, *A New Present for a Servant-Maid: Containing Rules for her Moral Conduct both with Respect to Herself and her Superiors* (London, 1771), 254.
45. Ibid., 256.
46. Ibid., 257.
47. See Jim A. Bennett, "The Art of Polishing: Practice and Prose in Eighteenth-Century Telescope Making," in Morrison-Low, Dupré, Johnston, and Strano, *From Earth-Bound to Satellite,* 103–21; Roos, "Speculum of Chymical Practice."
48. J. L. E. Dreyer, "A Short Account of Sir William Herschel's Life and Work, chiefly from unpublished sources," in William Herschel, *The Scientific Papers of William Herschel,* ed. John Louis Emil Dreyer, vol. 1 (Cambridge: Cambridge University Press, 1912), xxvii.
49. Anon., *An Address to Mothers, under the Following Heads: Maternal Authority. Domestic Attention. Diligence and Activity. Oeconomy. Simplicity* (Oxford, 1784), 118.
50. William Gilbert, *De magnete,* trans. P. Fleury Mottelay (New York: Dover, 1958), 74–97.
51. Adams, *Essay on Electricity,* 25.
52. Cavallo, *Complete Treatise on Electricity,* 172; Edward Nairne to John Fell, June 12, 1788, in Fell, "Miscellanea," Wellcome Library, MS 1175, p. 64; Neale, *Directions for Gentlemen,* 4.
53. Fell, "Miscellanea," Wellcome Library, MS 1175, unpaginated front matter.
54. Ibid.
55. Zoë Sofia, "Container Technologies," *Hypatia* 15 (2000): 181–201.
56. On such "restricted objects," see Gray, "'Practical Art,'" 51.
57. Charles Estienne and Jean Liébault, *Maison rustique; or, The Countrie Farme,* trans. Richard Surflet (London, 1600), 14.

58. Old Bailey Proceedings Online, version 7.2, "October 1781, trial of Edmund Arnold Samuel Wright (t17811017-54)," accessed 30 September 2015, www.oldbaileyonline.org.
59. Ibid. The miscreants were finally fined 1s. 6d. for their crime.
60. Vickery, *Behind Closed Doors*, 38–41.
61. On early modern English "storage furniture," see Buxton, *Domestic Culture*, 152–54, 198–203.
62. Tom Addyman, "Materia Chemica: Excavation of the Early Chemistry Stores at Old College, University of Edinburgh," typescript, 2012. My thanks to Robert Anderson for providing this essay; Werrett, "Household Oeconomy and Chemical Inquiry," 42–43.
63. Cavallo to Lind, April 25, 1792, British Library, Add MSS 22898, f. 7v.
64. Boyle, "Essays of the strange Subtilty," *Works*, 3: 725.
65. Broken box on the table and 'Epernay' crate, Watt's Workshop, Science Museum, London.
66. Minutes of a meeting of the Council of the Royal Society, November 1, 1722, Royal Society archive, CMO/2/305.
67. See, e.g., Robert Whytt, "Of the various Strength of different Lime-waters," *Essays and Observations, Physical and Literary: Read Before a Society in Edinburgh and Published by Them* 1 (1754): 380.
68. Jane Insley, "James Watt's Cookbook Chemistry," *Notes and Records of the Royal Society* 65 (2011): 301–8.
69. Francis Spilsbury, *Free Thoughts on Quacks and their Medicines, occasioned by the Death of Dr. Goldsmith and Mr. Scawen* (London, 1776), 48.
70. Francis Bacon, "Novum Organum," *Works*, 4: 233 (book 2, aphorism 50).
71. Lisa Jardine and Alan Steward, *Hostage to Fortune: The Troubled Life of Francis Bacon* (London: Victor Gollancz, 1998), 502–11.
72. Margaret Ponsonby, *Stories from Home: English Domestic Interiors, 1750–1850* (Aldershot: Ashgate, 2007), 80.
73. Amy Eyton (and others), "Collection of cookery receipts, with a few medical and household receipts" (c. 1691), Wellcome Library MS 2323, f. 124v.
74. On the materiality of early telescopes, see Marvin Bolt and Michael Korey, "The World's Oldest Surviving Telescopes," in *The Origins of the Telescope*, ed. Albert van Helden, Sven Dupré, Rob van Gent, and Huib J. Zuidervaart (Amsterdam: KNAW, 2010), 231–56; H. J. Zuidervaart, "The 'Invisible Technician' Made Visible: Telescope Making in the Seventeenth and Early Eighteenth-Century Dutch Republic," in Morrison-Low, Dupré, Johnston, and Strano, *From Earth-Bound to Satellite*, 41–102.
75. Isaac Newton, "An Extract of Another Letter of the Same to the Publisher, Dated March 30. 1672. by Way of Answer to Some Objection, Made by an Ingenious French Philosopher to the New Reflecting Telescope," *Philosophical Transactions* 7 (1672): 4034–35.
76. Ibid., 4035; a contemporary recipe for spirit of wine may be found in Lady Ayscough's recipe book, Wellcome Library, MS 1026, p. 37.

77. John Evelyn, "Treatise on Natural Philosophy and Chemistry," British Library, Add MSS 78345, f. 28r.
78. Hunter and Wiggins, "Workdiaries of Robert Boyle," workdiary no. 12, entry 74.
79. Alan Q. Morton and Jane A. Wess, *Public and Private Science: The King George III Collection* (Oxford: Oxford University Press/Science Museum, 1993).
80. Museum of the History of Science, Oxford, inventory number 10005.
81. Fell, "Miscellanea," Wellcome Library, MS 1175, 7.
82. [Edward Nairne], *Directions how to use the Compound Microscope, as Made and Sold by Edward Nairne, at the Golden Spectacles fronting the North-Gate of the Royal-Exchange, London* (n.d.). American Philosophical Society. Thomas Court Scientific Instruments Collection. MSS 509.078 M582.
83. Cavallo to Lind, May 6, 1786, British Library, Add MSS 22897, f. 51r.
84. Ibid.
85. See, e.g., Alexander Monro, "An Essay on the Method of Preparing and Preserving the Parts of Animal Bodies for Anatomical Uses," *The Works of Alexander Monro* (Edinburgh, 1781), 11–24.
86. For use as a temporary means to join vessels, see, e.g., Edward Lloyd [Lhwyd], "Chemical and other Processes, and Experiments on Metals," British Library, Add MSS 15071, f. 15r.
87. Latham, "'Lady Alchumy,'" 52. An eighteenth-century recipe book prescribed a lute for stills made of "clay & brine mingled." Anon., "Receipt-Book, 17th–18th century," manuscript recipe book, Wellcome Library, MS 4054, f. 65r.
88. Hunter and Wiggins, "Workdiaries of Robert Boyle," workdiary no. 12, entry 63.
89. Ibid.
90. George Wilson, *A Compleat Course of Chymistry* (London, 1703), 1–3.
91. Hannah Glasse, *The Art of Cookery, Made Plain and Easy* (London, 1796), 431.
92. Wilson, *Compleat Course of Chymistry*, 2.
93. Robert Boyle, "Certain Physiological Essays, and other Tracts written at distant Times, and on several Occasions," *Works*, 1: 360, 455.
94. Sir William Petty's Medical Studies, British Library, Add MSS 72891, f. 213r.
95. Anne Battam, *The Lady's Assistant in the Oeconomy of the Table* (London, 1759), 22.
96. Sir Kenelm Digby, *Two Treatises: in the One of which, the Nature of Bodies; In the Other the Nature of Mans Soule, is Looked Into, in Way of Discovery of the Immortality of Reasonable Soules* (London, 1645), 68–69.
97. Lloyd, "Chemical and other Processes," British Library, Add MSS 15071, f. 4r.
98. Peter Jacob Hjelm, "Experiments on Molybdaena, with a view to its reduction," *Crell's Chemical Journal*, 3 vols. (London, 1791–93), 3: 40–51, 166–88, 220–51.

99. Ibid., 44. See also 45, 170, 231.
100. Margaret Cavendish, Duchess of Newcastle, *Ground of Natural Philosophy Divided into Three Parts* (London, 1668), 150.
101. Joseph Moxon, *Mechanick Exercises, or, The Doctrine of Handy-works*, 2nd ed. (London, 1693), 81.
102. Smith, *Body of the Artisan.*
103. Anna Märker, "From Cookery to Chemistry: The Development of Preservation Techniques for Anatomical Research" (unpublished typescript, 2009); Harold J. Cook, "Time's Bodies: Crafting the Preparation and Preservation of Naturalia," in *Merchants and Marvels: Commerce, Science and Art in Early Modern Europe*, ed. Pamela H. Smith and Paula Findlen (London: Routledge, 2001), 223–47; Guerrini, "Ghastly Kitchen," 90; Wall, *Recipes for Thought*; C. Anne Wilson, ed., *Waste Not, Want Not: Food Preservation in Britain from Early Times to the Present Day* (Edinburgh: Edinburgh University Press, 1991).
104. Mary Bent, "Cookery book," c. 1664–1729, Wellcome Library MS 1127, f. 168v.
105. Anon., *England's Happiness Improved*, 21, 108, 127–28, 146, 164.
106. Ibid., 62–63.
107. Robin Pearson, "The Impact of Fire and Fire Insurance on Eighteenth-Century English Towns and Populations," in *Investing in the Early Modern Built Environment: Europeans, Asians, Settlers and Indigenous Societies*, ed. Carole Shammas (Leiden: Brill, 2012), 77.
108. Ken Albala, *Food in Early Modern Europe* (Westport, CT: Greenwood Press, 2003), 91.
109. Pearson, "Impact of Fire," 80.
110. North, *Observations and Advices Oeconomical*, 81.
111. Ibid., 133–34.
112. Frances Springatt (and others), "Collection of cookery and medical receipts," c. 1686–24, Wellcome Library MS 4683, f. 88r.
113. Mary Bent, "Cookery book," c. 1664–1729, Wellcome Library MS 1127, f. 175r.
114. Peter Shaw and Francis Hauksbee, *An Essay for Introducing a Portable Laboratory: By Means whereof all the Chemical Operations are Commodiously Perform'd* (London, 1731), 10.
115. Ibid., 13.
116. John Webster, *Academiarum examen, or, The Examination of Academies wherein is Discussed and Examined the Matter, Method and Customes of Academick and Scholastick Learning* (London, 1654), 71.
117. Fara, *Pandora's Breeches*, 14.
118. Shapin, "Invisible Technician"; Robert Boyle, "Additional Observations about the Aërial Noctiluca," *Works*, 4: 397.
119. Hunter and Wiggins, "Workdiaries of Robert Boyle," workdiary no. 13, entry 12. A muffler was a scarf or kerchief.

120. Anon., "Inventory of Dr. Priestley's Laboratory in 1791," in *Scientific Correspondence of Joseph Priestley: Ninety-seven Letters addressed to Josiah Wedgwood, Sir Joseph Banks . . . and Others*, ed. Henry Carrington Bolton (New York, 1892), 229.
121. Franklin, *Experiments and Observations on Electricity*, 436.
122. Benjamin Franklin, quoted in J. A. Leo Lemay, *The Life of Benjamin Franklin*, vol. 3: *Soldier, Scientist, and Politician, 1748–1757* (Philadelphia: University of Pennsylvania Press, 2009), 122.
123. Quoted in Thomas Hankins and Robert Silverman, *Instruments and the Imagination* (Princeton: Princeton University Press, 1995), 67.
124. Alexi Baker, " "Precision," "Perfection," and the Reality of British Scientific Instruments on the Move during the 18th Century," *Material Culture Review* 74–75 (Spring 2012): 14–29.
125. Dunn and Higgitt, *Ships, Clocks and Stars*.
126. J. A. Secord, "Knowledge in Transit," *Isis* 95 (2004): 654–72.
127. Cavallo to Lind, June 25, 1788, British Library, Add MSS 22897, f. 85v.
128. The crates have "I. Watt, Newhall" written on the side, suggesting this was their destination. James Watt's workshop, Science Museum, London.
129. Benjamin Vaughan to Ezra Stiles, September 5, 1789, in Richard Price, *The Correspondence of Richard Price*, vol. 3: *February 1786 to February 1791*, ed. W. Bernard Leach (Durham: Duke University Press, 1994), 255.
130. Dorian Gerhold, *Road Transport before the Railways: Russell's London Flying Waggons* (Cambridge: Cambridge University Press, 1993); Gerhold, *Carriers and Coachmasters: Trade and Travel before the Turnpikes* (Chichester, UK: Phillimore, 2005).
131. Fell, "Miscellanea," Wellcome Library, MS 1175, 11; carriage for electrical apparatus seems to have been 6–10 percent of the cost of the goods. See ibid., 13.
132. Ibid., 4.
133. Berg, *Luxury and Pleasure*, 123; see also Sally Newcomb, *The World in a Crucible: Laboratory Practice and Geological Theory at the Beginning of Geology* (Boulder, CO: Geological Society of America, 2009), 80.
134. Shaw and Hauksbee, *Essay for Introducing a Portable Laboratory*, 7–8.
135. Pitt and Pullein were booksellers. William Molyneux to John Flamsteed, May 13,1684, in John Flamsteed, *The Correspondence of John Flamsteed, The First Astronomer Royal*, ed. Eric Gray Forbes, Lesley Murdin, and Frances Wilmoth, 3 vols. (Bristol, UK: Institute of Physics Publishing, 1995–2002), 2: 168.
136. Cavallo to Lind, August 2, 1792, British Library, Add MSS 22898, f. 15r.
137. Cavallo to Lind, February 4, 1784, British Library, Add MSS 22897, ff. 15r–16v.
138. Franklin quoted in Bedini, *Thinkers and Tinkers*, 212.
139. See Joseph Priestley to John Vaughan, March 21, 1799, Joseph Priestley Papers, American Philosophical Society, B P931.

140. Shaw and Hauksbee, *Essay for Introducing a Portable Laboratory*, 1.
141. Ibid., 21.
142. John Berkenhout surveyed the development of portable furnaces. He also pointed out that if these were not available one could use "common garden-pots, of baked earth: they may be had of any size, and by means of a saw and file, may be easily adapted to any chemical purpose." John Berkenhout, *First Lines of the Theory and Practice of Philosophical Chemistry* (London, 1788), 244.
143. Anita McConnell, *Barometers*, 2nd ed. (Risborough, UK: Shire, 2003), 10–12.
144. Miles Partington to John Fell, June 2, 1786, in Fell, "Miscellanea," Wellcome Library, MS 1175, p. 144.
145. Fell, "Miscellanea," Wellcome Library, MS 1175, p. 5.
146. Robert K. Merton and Elinor Barber, *The Travels and Adventures of Serendipity: A Study in Sociological Semantics and the Sociology of Science* (Princeton: Princeton University Press, 2004); Steven Shapin, "The Accidental Scientist," *American Scientist* 92 (2004): 374–76.
147. Boyle, "Certain Physiological Essays," *Works*, 1: 455.
148. Ibid., 456–57; Leibniz later used Boyle's observations to prove that no bodies could be held to be in a state of absolute rest. G. W. Leibniz, *New Essays on Human Understanding*, trans. and ed. Peter Remnant and Jonathan Francis Bennett (Cambridge: Cambridge University Press, 1996), li.
149. Shaw and Hauksbee, *Essay for Introducing a Portable Laboratory*, 9.
150. Ibid., 10.

CHAPTER FIVE

1. Cavallo to Lind, September 22, 1794, British Library, Add MSS 22898, f. 51r.
2. Adams, *Essay on Electricity*, 148.
3. Pivotal works in this regard are Edgerton, *Shock of the Old*; Livingstone, *Putting Science in Its Place*; Trevor Pinch and Nelly Oudshoorn, eds., *How Users Matter: The Co-construction of Users and Technology* (Cambridge: MIT Press, 2003).
4. Pennell, "'For a Crack and a Flaw Despis'd.'"
5. Ibid., 39.
6. Edgerton, *Shock of the Old*, 80.
7. Simon Schaffer, "Easily Cracked: Scientific Instruments in States of Disrepair," *Isis* 102 (2011): 708.
8. Baker, "'Precision,' 'Perfection,'" 14.
9. Ibid., 27.
10. Ariane Fennetaux, Sophie Vasset and Amélie Junqua, introduction, in Fennetaux, Vasset, and Junqua, *Afterlife of Used Things*, 2.
11. Ibid., 3; Woodward, "Straw, Bracken, and the Wicklow Whale"; Woodward, "Swords into Ploughshares."

12. The Dutch case is examined in van Driel, "Ashes to Ashes."
13. Thomas Warren, *A True and Exact Particular and Inventory of All and Singular the Estate and Effects Whatsoever, of Thomas Warren* (London, 1732), 31–32.
14. Ponsonby, *Stories from Home*, 79–83.
15. Henley, *Booke of Thrift*, 2.
16. Matthew Bacon, *A Treatise on Leases and Terms for Years* (London, 1798), 318.
17. See, e.g., minutes of a meeting of the Council of the Royal Society, November 29, 1675, Royal Society Archives (hereafter RS), ref. CMO/1/228; July 13, 1709, RS ref. CMO/2/162, and July 3, 1729, RS ref. CMO/3/25.
18. Martha Ballard, "Diary of Martha Ballard," March 23, 1786 (breeches); June 28, 1786 (tea pot); November 3, 1786 (stockings); November 21, 1786 (trousers); July 6, 1789 (calico gown); Jan 4, 1790 (winter gown); April 22, 1796 (plow); October 3, 1796 (quilt), accessed December 10, 2015, http://dohistory.org/diary/.
19. Ibid., August 31, 1795.
20. Gunther, ed., *Early Science in Oxford*, 10: 138 (July 23, 1689).
21. Pennell, "'For a Crack and a Flaw Despis'd,'" 39.
22. Discussed in Lars Tharp, *Hogarth's China: Hogarth's Paintings and Eighteenth-Century Ceramics* (London, 1997), 32, 37.
23. Brian Dillon, *Ruin Lust* (London: Harry N. Abrams, 2014).
24. Nehemiah Grew, *Musaeum regalis societatis, or, A Catalogue and Description of the Natural and Artificial Rarities belonging to the Royal Society and preserved at Gresham Colledge* (London, 1685), 42, 60–65, quote on 165.
25. Ibid., 163.
26. Robert Boyle, "New Pneumatical Experiments about Respiration," *Philosophical Transactions* 5 (1670): 2016.
27. Ponsonby, *Stories from Home*, 81–85.
28. Harvey, *Little Republic*, 112–13.
29. Vickery, *Behind Closed Doors*, 271.
30. See, e.g., Henry Page, petition to the Middlesex Justices of the Peace, March 25, 1721, Middlesex Sessions, Sessions Papers, Justices' Working Documents SM/PS July 1721, London Lives, 1690–1800, accessed October 10, 2015, http://www.londonlives.org/browse.jsp?div=LMSMPS50195PS50195 0001&terms=mend#highlight.
31. Quoted in Joanna Martin, *Wives and Daughters: Women and Children in the Georgian Country House* (London: Hambledon and London, 2004), 157.
32. Old Bailey Proceedings Online, version 7.2, Ordinary of Newgate's Account, May 1717 (OA17170520), accessed February 11, 2016, www.oldbai leyonline.org.
33. Diary of Martha Ballard, February 15, 1794, accessed December 10, 2015, http://dohistory.org/diary/.
34. John Oliver's trade card, British Museum, Sarah Banks trade cards collection, Banks 66.35.

35. Edmund Morris's trade card, British Museum Heal Trade Cards collection, Heal 37.37; on mending china, see Isabelle Garachon, "Old Repairs of China and Glass," *Rijksmuseum Bulletin* 58 (2010): 35–54.
36. Morrison-Low, *Making Scientific Instruments*, 135–74; Alexi Baker, "Symbiosis and Style: The Production, Sale and Purchase of Instruments in the Luxury Markets of Eighteenth-Century London," in *How Scientific Instruments Have Changed Hands*, ed. Sara J. Schechner, Alison D. Morrison-Low, and Paolo Brenni (Leiden: Brill, 2016), 1–20.
37. Richard Dunn, "Touching and Cleaning: The Routine Work of an East London Instrument Supplier," *Bulletin of the Scientific Instrument Society* 89 (2006): 22.
38. Robert Hooke, *The Diary of Robert Hooke, MA, MD, FRS, 1672–1680*, ed. Henry W. Robinson and Walter Adams (London: Wykeham, 1968), 144 ("Nell took clothes to mend"), 164 ("Left Deans watch with Tompion who mended balance"), 407.
39. John Evelyn, "Trades. Secrets & Receipts Mechanical, as they came casualy to hand." British Library, Add MSS 78339, f. 112r.
40. Baker, "'Precision,' 'Perfection,'" 18.
41. Jonathan Thornton, "A Brief History and Review of the Early Practice and Materials of Gap-Filling in the West," *Journal of the American Institute of Conservation* 37 (1998): 5.
42. Ibid.
43. See the examples in the Museum of London, "Glass vessels repaired with lead & copper braces," ID nos. 16945, 13322, 86.240/2.
44. For an example of a plate with a lead patch, see the "eighteenth-century porcelain plate repaired with iron rivets" in the Museum of London, ID no. A9228; see also "Lead plugs used to repair ceramic vessels," ID nos. BWB83 & SWA81.
45. Thornton, "Brief History and Review," 6–7.
46. Dunn, "Touching and Cleaning," 22–23; London's National Maritime Museum possesses several lodestones from c. 1600 that were probably used for touching compass needles. See, e.g., ref. nos. NAV0707, NAV0713.
47. John Wilkes, *Encyclopaedia Londinensis, or, Universal Dictionary of Arts, Sciences, and Literature*, 24 vols. (London, 1801–28), s.v. "Materia Chemica."
48. Johnson, quoted in Pennell, "'For a Crack and a Flaw Despis'd,'" 34.
49. Priestley, *History and Present State of Electricity*, 496.
50. John Ewing, Astronomy Notebook, MS Coll. 135, Van Pelt Rare Books and Manuscripts, University of Pennsylvania, entry for July 20, 1784.
51. Moxon, *Mechanick Exercises*, 151. Moxon also explained that "Such Stuff as is crackt either with the heat of the Sun, or the droughth of the wind, is called Shaken Stuff" (169).
52. Schaffer, "Easily Cracked," 711.
53. Priestley, *History and Present State of Electricity*, 498.
54. Schaffer, "Easily Cracked."

55. Cavallo to Lind, February 13, 1789, British Library, Add MSS 22897, f. 92r.
56. Silvio A. Bedini, "Christina of Sweden and the Sciences," in Anderson, Bennett, and Ryan, *Making Instruments Count*, 113–16.
57. John Flamsteed to Henry Oldenburg, April 1, 1672, in Flamsteed, *Correspondence*, 1: 143.
58. Flamsteed to Oldenburg, April 1, 1672, ibid., 143.
59. Morrison-Low, *Making Scientific Instruments*, 75, 122, 170, 204, 266–67.
60. Bedini, *Thinkers and Tinkers*, 185, 203.
61. The card is pasted into the case of an octant made by Hicks. See Harvard University Collection of Historical Scientific Instruments, "Octant Case."
62. See, e.g., the comments on globes of Sir Thomas Bodley, quoted in Alexander Marr, "Learned Benefaction: Science, Civility and Donations of Books and Instruments to the Bodleian Library before 1605," in *Documenting the Early Modern Book World: Inventories and Catalogues in Manuscript and Print*, ed. Malcolm Walsby and Natasha Constantinidou (Leiden: Brill, 2013), 43.
63. David Rittenhouse's notes of observations used to determine the longitude of Wilmington, 1784. David Rittenhouse Papers, Historical Society of Pennsylvania, collection 552, vol. 3, f. 3.
64. Jonathan Belcher to Benjamin Franklin, December 18, 1751, *The Papers of Benjamin Franklin*, ed. Leonard W. Labaree, 41 vols. (New Haven: Yale University Press, 1959–), 4: 216.
65. See James Raven, *London Booksellers and American Customers: Transatlantic Literary Community and the Charleston Library Society, 1748–1811* (Columbia: University of South Carolina Press, 2002), 180–81.
66. For a general discussion see, e.g., Stephen Graham and Nigel Thrift, "Out of Order: Understanding Repair and Maintenance," *Theory, Culture and Society* 24 (2007): 1–25.
67. For the details, see Eric Robinson and Douglas McKie, eds., *Partners in Science: Letters of James Watt and Joseph Black* (London: Constable, 1970), 418, 434.
68. Schaffer, "Easily Cracked," 709.
69. Hugh S. Torrens, "The Geological Work of Gregory Watt, His Travels with William Maclure in Italy 1801–02 and Watt's 'Proto-Geological' Map of Italy of 1804," in *The Origins of Geology in Italy*, ed. Gian Battista Vai and W. Glen E. Caldwell, *Geological Society of America Special Papers* 411 (Boulder, CO: Geological Society of America, 2006): 179–97.
70. James Watt, quoted in Samuel Smiles, *Lives of Boulton and Watt* (London, 1865), 466.
71. James Watt's notebook of 1764–66 in Robinson and McKie, *Partners in Science,* 435.
72. Thanks to Jane Insley for pointing this out. For the syringe, see item 2020 in "Inventory, James Watt's Workshop, Objects presented by Maj. J. M. Gibson Watt 1924, R.P.Sc.M.1814," Science Museum, London.
73. Priestley, *Experiments and Observations*, 1: 21.

74. Ibid., 24.
75. Sara Pennell, "Material Culture in Seventeenth-Century 'Britain': The Matter of Domestic Consumption," in *The Oxford Handbook of the History of Consumption*, ed. Frank Trentmann (Oxford: Oxford University Press, 2012), 81.
76. Recipe book of Sir William Lowther and family, Wellcome Library MS 3341, f. 46v; see also the recipe book MS 7822, f. 49r.
77. Dossie, *Handmaid to the Arts,* 2: 21–31.
78. Hooke, *Micrographia*, preface, n.p.; Jim Bennett and Scott Mandelbrote, *The Garden, the Ark, the Tower, the Temple: Biblical Metaphors of Knowledge in Early Modern Europe* (Oxford: Museum of the History of Science, 1998), 69–70.
79. Quoted in Hunter, *Boyle: Between God and Science*, 71.
80. Samuel Clarke, *A Collection of Papers, Which passed between the late Learned Mr. Leibnitz, and Dr. Clarke, In the Years 1715 and 1716* (London: 1717), 112–13.
81. Florin-Stefan Morar, "Reinventing Machines: The Transmission History of the Leibniz Calculator," *British Journal for the History of Science* 48 (2015): 123–46.
82. Kubrin, "Newton and the Cyclical Cosmos."
83. John Beale to Robert Boyle, July 13, 1666, in Boyle, *Correspondence*, 3: 192.
84. Recycling is a term that originated in distilling processes in the oil industry in the twentieth century. See, e.g., T. G. Delbridge, "The Cracking of Petroleum," *Journal of the Franklin Institute* 202, no. 5 (November 1926): 579–81.
85. On early modern brooms, see Woodward, "Straw, Bracken, and the Wicklow Whale," 69.
86. Mary Douglas, *Purity and Danger: An Analysis of Concepts of Pollution and Taboo* (London: Routledge Classics, 2002), 44–50; William Viney, *Waste: A Philosophy of Things* (London: Bloomsbury, 2014).
87. Sarah Tarlow, *The Archaeology of Improvement in Britain, 1750–1850* (Cambridge: Cambridge University Press, 2007), 184–88.
88. A dung pot had a trapdoor to release filth onto the ground.
89. Miles Ogborn, *Spaces of Modernity: London's Geographies, 1680–1780* (New York: Guilford, 1998), 91–103, and for a list of relevant legislation, 306–7; Emily Cockayne, *Hubbub: Filth, Noise and Stench in England, 1600–1770* (New Haven: Yale University Press, 2007), 183–91; Peter Hounsell, *London's Rubbish: Two Centuries of Dirt, Dust and Disease in the Metropolis* (Stroud, UK: Amberley, 2013), chap. 1.
90. Woodward, "Straw, Bracken, and the Wicklow Whale," 57, 59.
91. Ibid., 70–71.
92. Tim Hitchcock, *Down and Out in Eighteenth-Century London* (London: Hambledon and London, 2004), 42.

93. The vamp is the "upper Leather of a Shoe. To vamp or new-vamp, to mend or furbush up." John Kersey, *A New English Dictionary: Or, Compleat Collection of the Most Proper and Significant Words, Commonly Used in the Language* (London, 1702), s.v. "Vamp."
94. Woodward, "Straw, Bracken, and the Wicklow Whale," 72–73.
95. Mukherjee, *Penury into Plenty*, 114.
96. Gabriel Plattes, *The Profitable Intelligencer, Communicating his Knowledge for the Generall Good of the Common-wealth and All Posterity* (London, 1644). Hartlib reprinted this as "Mercurius Laetisicans" in *Samuel Hartlib, his Legacy of Husbandry*, 175–82. I have used Hartlib's text, as it includes page references.
97. Ibid., 180.
98. Ibid., 181.
99. Ibid.
100. Woodward, "Swords into Ploughshares," 179–82; Allan Potofsky, "Recycling the City: Paris 1760–1800," in Fennetaux, Vasset, and Junqua, *Afterlife of Used Things*, 71–88.
101. E. Walter Maunder, *The Royal Observatory Greenwich: A Glance at Its History and Work* (Cambridge: Cambridge University Press, 1900), 42.
102. Michael Cooper, *A More Beautiful City: Robert Hooke and the Rebuilding of London after the Great Fire* (Stroud, UK: Sutton Publishing, 2003), 180–83.
103. Philiatros [Alethea Talbot], *Natura exenterata*, 91, 136, 157, 188, quotation on 359.
104. Ibid., 368.
105. L. C., *Fundamenta Chymica; or, A Sure Guide into the High and Rare Mysteries of Alchemy* (London, 1658), 153.
106. Daniel Coxe to Robert Boyle, c. June 1666, in Boyle, *Correspondence*, 3: 179.
107. Hunter and Wiggins, "Workdiaries of Robert Boyle," workdiary no. 9, entry 39.
108. Ibid., entry 40.
109. Priestley, *Experiments and Observations*, 22.
110. Woodward, "Swords into Ploughshares," 183–86; Werrett, "Recycling in Early Modern Science," 7.
111. Richard S. Westfall, *Never at Rest: A Biography of Isaac Newton* (Cambridge: Cambridge University Press, 1980), 551–67; Ming-Hsun Li, *The Great Recoinage of 1696 to 1699* (London: Weidenfeld and Nicolson, 1963).
112. Woodward, "Swords into Ploughshares," 186.
113. Hunter, "Hooke's Possessions at His Death," 293.
114. Anita McConnell, "Instruments and Instrument-Makers, 1700–1850," in *The Oxford Handbook of the History of Physics*, ed. Jed Z. Buchwald and Robert Fox (Oxford: Oxford University Press, 2013), 330–31.
115. John Harris, *Lexicon Technicum: Or, An Universal English Dictionary of Arts and Sciences*, 2nd ed., 2 vols. (London, 1723), s.v. "Brass."
116. Robert Smith, *A Compleat System of Opticks*, 2 vols. (London, 1738), 2: 304.

117. William Arderon and Henry Baker, "Abstract of a Letter from Mr. William Arderon, F.R.S. to Mr. Henry Baker, F.R.S. on the Giving Magnetism and Polarity to Brass," *Philosophical Transactions* 50 (1757–58): on 774–75.
118. D. W. Crossley, "The English Glassmaker and His Search for Materials in the 16th and 17th Centuries," in *The Prehistory and History of Glassmaking Technology*, ed. Patrick McCray (Westerville, OH: American Ceramic Society, 1998), 167–79; D. Dungworth, "Vauxhall, London: The Scientific Examination of Glass and Glassworking Materials from the Late Seventeenth Century Glasshouse, Research Department Report 83/2006 (English Heritage, 2006)," accessed December 10, 2015, http://research.historicengland.org.uk/Report.aspx?i=14503&ru=%2FResults.aspx%3Fp%3D619, 12, 17, 29; D. Dungworth and C. Mortimer, "Examination of Glassworking Materials from Cheese Lane, Bristol. Centre for Archaeology Report 6/2005 (English Heritage, 2005)," accessed December 10, 2015, services.english-heritage.org.uk/ResearchReportsPdfs/006-2005.pdf, 3, 7.
119. "Glass-house Clerk," *The Plate-Glass Book, consisting of the following Tables* (London, 1757), xvii.
120. Hooke, *Micrographia*, preface.
121. Francis Hauksbee, *Physico-Mechanical Experiments on Various Subjects. Containing an Account of several Surprizing Phenomena touching Light and Electricity* (London, 1709), 140.
122. Addyman, "Materia Chemica," n.p.
123. David Baker, "Optical Connections," *Optician*, February 10, 2012: 29–30; Mary Webster, *Johan Zoffany, 1733–1810* (New Haven: Yale University Press, 2011), 56, 71.
124. Thomas Skeete, *Experiments and Observations on Quilled and Red Peruvian Bark* (London, 1786), 36; William Nicholson, *An Introduction to Natural Philosophy*, 2 vols. (London, 1807), 2: 76; "Chemical and other Processes," British Library, Add MSS 15071, f. 27r.
125. John Evelyn, "Treatise on natural philosophy and chemistry." British Library, Add MSS 78345, f. 38r.
126. Philiatros [Alethea Talbot], *Natura exenterata*, 183.
127. Birch, *History of the Royal* Society, 1: 349.
128. Joseph Black, "An Analysis of the Waters of some Hot Springs in Iceland," *Transactions of the Royal Society of Edinburgh* 3 (1794): 106.
129. Ibid.
130. Woodward, "Swords into Ploughshares," 187.
131. See, e.g., Leonard N. Rosenband, *Papermaking in Eighteenth-Century France: Management, Labor, and Revolution at the Montgolfier Mill, 1761–1805* (Baltimore: Johns Hopkins University Press, 2000).
132. Evelyn, quoted in Amélie Junqua, "Unstable Shades of Grey: Cloth and Paper in Addison's Periodicals," in Fennetaux, Vasset, and Junqua, *Afterlife of Used Things*, 187.

133. Chloe Wigston Smith, *Women, Work, and Clothes in the Eighteenth-Century Novel* (Cambridge: Cambridge University Press, 2013), 49.
134. Ibid., 53–55.

CHAPTER SIX

1. Leong, "Collecting Knowledge"; Leong, "Making Medicines"; Pennell and Leong, "Recipe Collections"; DiMeo and Pennell, *Reading and Writing Recipe Books*.
2. Anon., *A Catalogue of the Portland Museum, lately the Property of The Duchess Dowager of Portland, Deceased: Which will be Sold by Auction by Mr. Skinner and Co. on Monday the 24th April, 1786* (London, 1786), iii.
3. Mario Biagioli, *Galileo, Courtier: The Practice of Science in the Age of Absolutism* (Chicago: University of Chicago Press, 1993), 36–53.
4. Ilana Krausman Ben-Amos, *The Culture of Giving: Informal Support and Gift-Exchange in Early Modern England* (Cambridge: Cambridge University Press, 2008); see also Felicity Heal, *The Power of Gifts: Gift Exchange in Early Modern England* (Oxford: Oxford University Press, 2015).
5. See Elaine Leong, *Recipes and Everyday Knowledge: Medicine, Science, and the Household in Early Modern England* (Chicago: University of Chicago Press, 2018).
6. Ezreel Tonge to Robert Boyle, March 17, 1666, in Boyle, *Correspondence*, 3: 116; William Eamon has made a similar point about books of secrets serving as early experiments, in *Science and the Secrets of Nature* (Princeton: Princeton University Press, 1994), 7.
7. Banks, quoted in Hickman, "Garden as a Laboratory," 234; Arlene C. Leis, "Sarah Sophia Banks: Femininity, Sociability and the Practice of Collecting in Late Georgian England" (PhD diss., University of York, 2013), 19, 24–25.
8. Ben-Amos, *Culture of Giving*, 66; see also Craig Muldrew, *The Economy of Obligation: The Culture of Credit and Social Relations in Early Modern England* (London: Palgrave Macmillan, 1998).
9. Ben-Amos, *Culture of Giving*, 67.
10. Elizabeth Yale, *Sociable Knowledge: Natural History and the Nation in Early Modern Britain* (Philadelphia: University of Pennsylvania Press, 2016).
11. Diary of Charles Blagden, Royal Society, 8 vols., Ref: CB/3., vol. 3, f. 40v. Transcription by Hannah Wills. The quotation refers to James Rennell, the surveyor of India, William Wilberforce the abolitionist, and John Pratt, 1st Marquess Camden.
12. Sloane's letters to Richardson are reproduced in John Nichols, ed., *Illustrations of the Literary History of the Eighteenth Century*, vol. 1 (London, 1817), 269–89.
13. Delany, *Autobiography*, 2: 232.

14. Franklin to Peter Collinson, March 28, 1747, *Papers of Benjamin Franklin*, 3: 118.
15. Cavallo to Lind, May 28, 1787, British Library, Add MSS 22897, f. 74r.
16. Cavallo to Lind, May 31, 1789, British Library, Add MSS 22987, f. 93v.
17. Johann Hevelius to Henry Oldenburg, August 8, 1674, Oldenburg, *Correspondence*, 11: 62.
18. Ibid.
19. On *Ruinenlust*, see chap. 5.
20. Samuel Clarke, *A Paraphrase of the Four Evangelists* (London, 1738), 260.
21. Anon., *Observations on Some Papers in that Very Useful Collection: Intitled, Museum Rusticum* (London, 1766), 14.
22. Jungnickel and McCormmach, *Cavendish*, 235.
23. Gian Domenico Cassini was much vexed by the loss of a large glass concave mirror to Queen Christina of Sweden as he was never able to make another mirror of the same quality. Bedini, "Christina of Sweden and the Sciences," 109–10.
24. Gunther, *Early Science in Oxford*, 10: 69 (November 2, 1688), 174 (December 23, 1689), 223 (March 21, 1692/3).
25. D. R. Woolf, *Reading History in Early Modern England* (Cambridge: Cambridge University Press, 2000), 168–202.
26. Gunther, ed., *Early Science in Oxford*, 10: 200 (December 24, 1692).
27. Lady Ranelagh to Lady Broghill, undated letter quoted in DiMeo, "Katherine Jones," 269–70.
28. Jennifer Summit, *Memory's Library: Medieval Books in Early Modern England* (Chicago: University of Chicago Press, 2008).
29. Georges Louis Leclerc, comte de Buffon, *Buffon's Natural History: Containing a Theory of the Earth*, 10 vols. (London, 1797), 3: 82–83.
30. William Derham, "Experiments about the Motion of Pendulums in Vacuo," *Philosophical Transactions* 24 (1704–5): 1785.
31. Ibid.
32. Adams, *Essay on Electricity*, vi.
33. Priestley to John Vaughan, May 30, 1796, Joseph Priestley Papers, American Philosophical Society, B P931.
34. William Derham to Hans Sloane, January 10, 1706/7, British Library, Sloane MS 4040, accessed October 16, 2015, https://drc.usask.ca/projects/sloane letters/doku.php?id=letter&letterid=1309, f. 288.
35. Jonathan Belcher to Benjamin Franklin, January 20, 1752, *Papers of Benjamin Franklin*, 4: 255.
36. J. H. Stallard, "On the Economical Construction of Workmen's Dwellings and Especially in Reference to Improving the Health and Habits of the Class," *Journal of the Society of Arts* 19 (1871): 468.
37. B. Lisa Gröger, "Of Men and Machines: Co-operation among French Family Farmers," *Ethnology* 20 (1981): 167.
38. Derham to Hans Sloane, January 10, 1706/7, f. 288.

39. Anon., *A Word in Season: or, an Essay to promote Good-Husbandry in Hard and Difficult Times: Being, in part, Advice from a Gentleman, to his Son a Tradesman in London* (London, 1697), 6.
40. Leong, "Collecting Knowledge," 85.
41. Ibid., 81–82.
42. Ibid., 85–88.
43. Robert Boyle, "An Exact Copy of the Last Will and Testament of the Honourable Robert Boyle," *Works*, 1: clx.
44. Delany, *Autobiography*, 2: 478.
45. Benjamin Franklin, "Last Will and Testament," in *The Life of Benjamin Franklin Written by Himself*, ed. John Bigelow, 3 vols. (Cambridge: Cambridge University Press, 2011), 3: 488.
46. James Beattie, "Copy of the last Will and Testament of James Beattie, LL.D. written by his own Hand, and dated 20th July, 1799," in Sir William Forbes, *An Account of the Life and Writings of James Beattie* (New York, 1807), 512.
47. Ibid.
48. Boyle, "Exact Copy of the Last Will," *Works*, 1: clxi.
49. Schiebinger, *Mind Has No Sex?*, 79–82.
50. Simon Werrett, "From the Grand Whim to the Gasworks: Philosophical Fireworks in Georgian England," in *The Mindful Hand: Inquiry and Invention from the Late Renaissance to Early Industrialisation*, ed. Lissa Roberts, Simon Schaffer, and Peter Dear (Amsterdam: Edita; Chicago: University of Chicago Press, 2007), 325–48.
51. Will of Charles Diller Godenne, otherwise Charles Diller, National Archives, Prob 11/1175.
52. Ibid.
53. See the classified advertisements in *Felix Farley's Bristol Journal*, no. 2112 (April 11, 1789); John Langford, *A Century of Birmingham Life*, 2 vols. (Birmingham, 1868), 1: 399.
54. See "Copy of the Will of Lady Sadleir," September 25, 1701, Royal Society, MM/13/23.
55. Alexander Marr, "Learned Benefaction: Science, Civility and Donations of Books and Instruments to the Bodleian Library before 1605," in *Documenting the Early Modern Book World: Inventories and Catalogues in Manuscript and Print*, ed. Malcolm Walsby and Natasha Constantinidou (Leiden: Brill, 2013), 27–50.
56. Ibid., 31.
57. Ibid., 45.
58. James Manning to Samuel Stennert, November 8, 1783, reproduced in Reuben Aldridge Guild, *The Life, Times, and Correspondence of James Manning and the Early History of Brown University* (Boston, 1864), 314.
59. James Watt, "James Watt's Recollections of his Friend Dr. J. Robison," in Robinson and McKie, *Partners in Science*, 410.

60. Robert P. Multhauf, *A Catalogue of Instruments and Models in the Possession of the American Philosophical Society* (Philadelphia: American Philosophical Society, 1961), 17.
61. Jungnickel and McCormmach, *Cavendish*, 307–8; the telescope is still in the Royal Society, item no. MO/1/1/2 and MO/1/1/6; see also the correspondence with the Society from William Derham on the telescope, William Derham to Hans Sloane, February 23, 1712 (Royal Society EL/D1/58) and William Derham to John Chamberlayne, August 22, 1713 (Royal Society EL/D1/59).
62. Alexander Gordon, "John Horsley (1685–1732)," *Oxford Dictionary of National Biography*, 60 vols. (Oxford: Oxford University Press, 2013), 9: 1276–77; Robert Schofield, *The Enlightenment of Joseph Priestley: A Study of His Life and Work from 1733 to 1773* (University Park: Pennsylvania State University Press, 1997), 142; Werrett, "Recycling in Early Modern Science," 10.
63. A. D. Morrison-Low, "Early Navigational Instruments in Scotland," in Anderson, Bennett, and Ryan, *Making Instruments Count*, 221.
64. John Longfield and Neville Maskelyne, "Extracts of Three Letters from John Longfield, M.D. at Corke in Ireland, to the Astronomer Royal, Containing Some Astronomical Observations; Together with the Longitude of Corke, Deduced from the Said Observations, by the Astronomer Royal," *Philosophical Transactions* 69 (1779): 173.
65. DiMeo and Pennell, introduction to *Reading and Writing Recipe Books*, 14.
66. National Maritime Museum, London, item no. 064 NAV1298, in W. F. J. Mörzer Bruyns and Richard Dunn, *Sextants at Greenwich: A Catalogue of the Mariner's Quadrants, Mariner's Astrolabes, Cross-staffs, Backstaffs, Octants, Sextants, Quintants, Reflecting Circles and Artificial Horizons in the National Maritime Museum, Greenwich* (Oxford: Oxford University Press, 2009), n.p.
67. National Maritime Museum, London, item no. NAV1140. Thanks to Richard Dunn for pointing this instrument out to me.
68. On pedigree as a basis for calibration, see Simon Werrett, "Russian Responses to the Voyages of Captain Cook," in *Captain Cook: Explorations and Reassessments*, ed. Glyn Williams (New York: Boydell & Brewer, 2004), 194.
69. Sir H. C. Englefield, "Description of a New Transit Instrument," *Philosophical Magazine and Journal* 43, no. 189 (January 1814): 1.
70. Johnson, *Dictionary of the English Language*, s.v. "Second-Hand."
71. Elizabeth C. Sanderson, "Nearly New: The Second-Hand Clothing Trade in Eighteenth-Century Edinburgh," *Costume* 31 (1997): 38–48; Miles Lambert, "'Cast-off Wearing Apparell': The Consumption and Distribution of Second-Hand Clothing in Northern England during the Long Eighteenth Century," *Textile History* 35 (2004): 1–26, 38–48; Clive Edwards and Margaret Ponsonby, "Desirable Commodity or Practical Necessity? The Sale and Consumption of Second-Hand Furniture, 1750–1900," in *Buying for the Home: Shopping for the Domestic from the Seventeenth Century to the Present*, ed. David Hussey and Margaret Ponsonby (Aldershot, UK: Ashgate, 2008),

117–38; Beverly Lemire, *The Business of Everyday Life: Gender, Practice and Social Politics in England, c.1600–1900* (Manchester: Manchester University Press, 2005); Lemire, "Plebeian Commercial Circuits and Everyday Material Exchange in England, c. 1600–1900," in *Buyers & Sellers: Retail Circuits and Practices in Mediaeval and Early Modern Europe*, ed. Bruno Blondé, Peter Stabel, Jon Stobart, and Ilja Van Damme (Turnhout: Brepols, 2006), 245–66; Fontaine, *Alternative Exchanges*; Ponsonby, *Stories from Home*, 85–87; Ilja Van Damme, "Second-Hand Trade and Respectability: Mediating Consumer Trust in Old Textiles and Used Clothing (Low Countries, Seventeenth–Eighteenth Centuries)," in *Selling Textiles in the Long Eighteenth Century: Comparative Perspectives from Western Europe*, ed. Jon Stobart and Bruno Blondé (London: Palgrave Macmillan, 2014), 193–209.

72. Quoted in Sara Pennell, "'All but the Kitchen Sink': Household Sales and the Circulation of Second-Hand Goods in Early Modern England," in Stobart and Van Damme, *Modernity and the Second-Hand Trade*, 37.
73. Jon Stobart and Ilja Van Damme, introduction to Stobart and Van Damme, *Modernity and the Second-Hand Trade*, 4.
74. Jon Stobart, "Clothes, Cabinets and Carriages: Second-Hand Dealing in Eighteenth-Century England," in Blondé et al., *Buyers & Sellers*, 225–44.
75. "Ordinary of Newgate's Account, August 1740 (OA17400806)," Old Bailey Proceedings Online, version 7.2, accessed October 20, 2015, www.oldbailey online.org.
76. On Phillips Garden, see Gillian Wilson, "The Kedleston Fountain: Its Development from a Seventeenth-Century Vase," *J. Paul Getty Museum Journal* 11 (1983): 1, 5.
77. *London Courant Or the New Advertiser*, no. 490 (January 23, 1747; *London Evening Post*, no. 2980 (December 9, 1746; *London Evening Post*, no. 3029 (April 2, 1747); "Now selling cheap," *Public Ledger*, no. 2041 (July 19, 1766).
78. See, e.g., "To be Sold at the Coach and Horses," *Daily Courant*, no. 6591 (December 5, 1722) (chariots); "To be sold at Second-Hand," *Daily Courant*, no. 4406 (December 7, 1715) (shop fittings).
79. "Advertisements," *Weekly Journal or Saturday's Post*, no. 72 (Saturday, April 16, 1720), 431.
80. Ian Mitchell, *Tradition and Innovation in English Retailing, 1700 to 1850: Narratives of Consumption* (Farnham, UK: Ashgate, 2014), 71–72.
81. Ibid., 73.
82. "To prevent the Decoys and Impositions," *London Evening Post*, no. 1432 (January 18, 1737).
83. "Stafford and John Briscoe," *London Evening Post*, no. 2440 (June 28, 1743).
84. "Preston's Second-Hand Musical-Instrument Warehouse," *Morning Post and Daily Advertiser*, no. 4728 (May 13, 1788).
85. See, e.g., Patrick Colquhoun, *A Treatise on the Police of the Metropolis*, 6th ed. (London, 1800), 74–80, 548–49.

86. Ian Mitchell, "'Old Books—New Bound'? Selling Second-Hand Books in England, c. 1680–1850," in Stobart and Van Damme, *Modernity and the Second-Hand Trade,* 139–57.
87. John Feather, "The Merchants of Culture: Bookselling in Early Industrial England," *Studies on Voltaire and the Eighteenth Century* 217 (1983): 13.
88. Ibid., 12–13.
89. Cavallo to Lind, August 23, 1791, British Library, Add MSS 22897, f. 114r.
90. "This day is published," *Public Advertiser,* no. 6862 (October 16, 1756); on Heath and Wing, see Gloria Clifton, *Directory of British Scientific Instrument Makers, 1550–1851* (London: Zwemmer, National Maritime Museum, 1995), 131, 301; on their trade cards, British Museum Heal, 105.49 and Banks, 105.23.
91. "To be Sold cheap," *Public Advertiser* no. 8510 (February 11, 1762), n.p.
92. Further announcements of Heath and Wing's second-hand sales may be found at e.g. "To be Sold cheap," *Public Advertiser,* no. 7964 (May 17, 1760); "To be Sold cheap," *Public Advertiser,* no. 8609 (June 7, 1762); "To be Sold cheap," *Public Ledger,* no. 1798 (October 9, 1765).
93. Peter de Clercq, "Second-Hand Instruments in a W. & S. Jones Catalogue of 1795," *Bulletin of the Scientific Instrument Society* 85 (2005): 38–39.
94. "James Hull, Teacher of Geography," *General Advertiser,* no. 3736 (October 16, 1746).
95. Henry Greatorex to John Flamsteed, April 22, 1706 (letter 1086), in Flamsteed, *Correspondence,* 3: 307.
96. Richard Price to Ezra Stiles, July 31, 1789, in Price, *Correspondence,* 3: 244.
97. Paola Bertucci, "Medical and Animal Electricity in the Work of Tiberius Cavallo, 1780–1795," in *Luigi Galvani International Workshop Proceedings,* ed. Marco Bresadola and Giuliano Pancaldi (Bologna: Universita di Bologna Dipartimento di Filosofia Centro Internazionale per la Storia delle Universita e della Scienza, 1999), 147–66.
98. Cavallo to Lind, November 23, 1792, British Library, Add MSS 22898, f. 24r.
99. On Haas, see E. Lefebvre and J. G. de Bruijn, eds., *Martinus Van Marum: Life and Work,* 6 vols. (Haarlem and Leyden: Noordhoff International and H. D. Tjeenk Willink for Hollandsche Maatschappij der Wetenschappen, 1969–76), 6: 141.
100. Cavallo to Lind, December 8, 1792, British Library, Add MSS 22898, f. 27r.
101. Cavallo to Lind, February 26, 1793, British Library, Add MSS 22898, f. 31r.
102. Cavallo to Lind, April 13, 1795, British Library, Add MSS 22898, f. 57r.
103. Cavallo to Lind, February 26, 1793, British Library, Add MSS 22898, f. 31r.
104. Cavallo to Lind, March 4, 1793, British Library, Add MSS 22898, f. 33v.
105. Cavallo to Lind, March 7, 1793, British Library, Add MSS 22898, f. 35r; presumably this was also a private joke about King Louis XVI, who had been executed six weeks earlier in January 1793.
106. Cavallo to Lind, April 10, 1793, British Library, Add MSS 22898, f. 39r.

CHAPTER SEVEN

1. Jan de Vries, *The Industrious Revolution: Consumer Behavior and the Household Economy, 1650 to the Present* (Cambridge: Cambridge University Press, 2008).
2. Michael Kwass, review of *The Industrious Revolution: Consumer Behavior and the Household Economy, 1650 to the Present,* by Jan de Vries, *American Historical Review* 114, no. 3 (2009): 706.
3. J. M. Chalmers-Hunt, ed., *Natural Historical Auctions, 1700–1972: A Register of Sales in the British Isles* (London: Sotheby Parke Bernet, 1976).
4. Gerard L'Estrange Turner, "The Auction Sales of the Earl of Bute's Instruments, 1793," *Annals of Science* 23 (1967): 213–42.
5. Peter de Clercq, "Private Instrument Collections Sold at Auction in London in the Late 18th Century," *Bulletin of the Scientific Instrument Society* 95 (2007): 28–36; de Clercq, "Private Instrument Collections Sold at Auction in London in the Late 18th Century," *Bulletin of the Scientific Instrument Society* 100 (2009): 27–35.
6. Larry Stewart, "Other Centres of Calculation, or, Where the Royal Society Didn't Count: Commerce, Coffee-Houses and Natural Philosophy in Early Modern London," *British Journal for the History of Science* 32 (1999): 150.
7. Stewart, *Rise of Public Science*; see also Jan Golinski, *Science as Public Culture: Chemistry and Enlightenment in Britain, 1760–1820* (Cambridge: Cambridge University Press, 1992).
8. Cynthia Wall, "The English Auction: Narratives of Dismantlings," *Eighteenth-Century Studies* 31 (1997): 1–25.
9. Haidy Geismar, "Alternative Market Values? Interventions into Auctions in Aotearoa/New Zealand," *Contemporary Pacific* 20 (2008): 300, 315.
10. On the history of auctions, see Brian Learmont, *A History of the Auction* (Iver, UK: Barnard and Learmont, 1985); Jeremy Cooper, *Under the Hammer: The Auctions and Auctioneers of London* (London: Constable, 1979).
11. Giles Mandelbrote, "The Organization of Book Auctions in Late Seventeenth-Century London," in *Under the Hammer: Book Auctions since the Seventeenth Century,* ed. Robin Myers, Michael Harris, and Giles Mandelbrote (New Castle, DE: Oak Knoll Press; London: British Library, 2001), 15–36.
12. Pennell, "'All but the Kitchen Sink,'" 37–56; Rosie MacArthur and Jon Stobart, "Going for a Song? Country House Sales in Georgian England," in Stobart and Van Damme, *Modernity and the Second-Hand Trade,* 175–95.
13. John Lawler, *Book Auctions in England in the Seventeenth Century (1676–1700): With a Chronological List of the Book Auctions of the Period* (London: Elliot Stock, 1898), xvii.
14. [Moses Pitt], *Catalogus variorum librorum instructissimae Bibliothecae praestantissimi Doctissimiq. Viri in Angliâ defuncti . . . clarissimi Gisberti Voetii:*

quorum auctio habebitur Londini . . . Novembris 25, 1678 (London, 1678), quoted in "Catalogue raisonnée of the collection of Catalogues of Libraries, prints, paintings, coins, curiosities, etc. sold by auction from 1676. to 1824 inclusive, and of some other works relating to bibliography, belonging to William Upcott; in his own handwriting." British Library, Add MSS 15918, f. 5r (hereafter "Catalogue raisonnée").

15. James Ralph, *The Touch-Stone; or, Historical, Critical, Political, Philosophical, and Theological Essays on the Reigning Diversions of the Town* (London, 1728), 232.
16. Benjamin Franklin, "The Way to Wealth," in anon., *Miscellanies in Prose and Verse: Selected from Pope, Swift, Addison, Goldsmith, Sterne, Hume, Smollet, Gay, Shenstone, Prior, Murphy, and Brooke* (Leominster, c. 1770), 67.
17. Mandelbrote, "Organization of Book Auctions," 15; Ulrike Malmendier, and Young Han Lee, "The Bidder's Curse," *American Economic Review* 101, no. 2 (2011): 749–87.
18. Pennell, "'All but the Kitchen Sink,'" 49.
19. See, e.g., anon., "On Sham Auctions," *Yearly Chronicle . . . Essays . . . which appeared in the St. James's Chronicle*, no. 96 (October 22, 1761), 331–33.
20. [Pitt], *Catalogus variorum librorum*, f. 5r.
21. Mandelbrote, "Organization of Book Auctions," 31–32; John Nichols, *Literary Anecdotes of the Eighteenth Century*, 6 vols. (London, 1812), 4: 29.
22. Ralph, *Touch-Stone*, 231–32.
23. Quoted in Mandelbrote, "Organization of Book Auctions," 27.
24. Thomas Chatterton, *The Auction: A Poem* (London, 1770), 5.
25. Robert Lacey, *Sotheby's: Bidding for Class* (London: Warner, 1998), 19–33; on Bonham's, see Cooper, *Under the Hammer*, 135.
26. Lacey, *Sotheby's*, 24–31.
27. Frank Herrmann, "Christie, James (1730–1803)," *Oxford Dictionary of National Biography* (Oxford: Oxford University Press, 2004) , accessed October 20 2015, http://www.oxforddnb.com/view/article/5362; William Roberts, *Memorials of Christie's: A Record of Art Sales from 1766 to 1896*, 2 vols. (London: George Bell and Sons, 1897), 1: 1–11, 24, 25; Christie's early career is the focus of Lynn Frances Schibeci, "The London Auction Market and the Commodification of English Taste, 1766–1823" (PhD diss., Northwestern University, 1999), esp. 89–159.
28. Charles Jenner, *Town Eclogues* (London, 1772), 17. "You ought never to allow them to take up too much of your time," a mother warned her daughter in a book of instructions for polite ladies; "it is the practice, I'm afraid, of too many young ladies. What with the auctions in the forenoon, visiting in the afternoon, and plays, operas, Ranelagh, Vauxhall, &c. in the evening, I doubt they find but little time for more useful employments." Charles Allen, *The Polite Lady; or, a Course of Female Education: in a Series of Letters, from a Mother to her Daughter* (London, 1788), 115.
29. Anon., *The Countryman's Guide to London. Or, Villainy Detected. Being a Clear*

Discovery of all the Various Tricks and Frauds that are Daily Practiced in that Great City (London, 1775), 29–30.

30. Shapin, *Social History of Truth*; Golinski, "Noble Spectacle."
31. Simon Schaffer, "Natural Philosophy and Public Spectacle in the Eighteenth Century," *History of Science* 21 (1983): 1–43.
32. Hooke, *The Diary of Robert Hooke*, 358 (May 13, 1678); see also Leona Rostenberg, *The Library of Robert Hooke: The Scientific Book Trade of Restoration England* (Santa Monica, CA: Modoc Press, 1989), 66–81.
33. Hooke, *Diary of Robert Hooke*, 359 (May 21 and 22, 1678).
34. Ibid., 387–88 (December 9 and 10, 1678).
35. Ibid., 414 (June 2, 3, and 4, 1679).
36. Ibid., 443 (April 17 and 19, 1680).
37. Stephen Inwood, *The Man Who Knew Too Much: The Strange and Inventive Life of Robert Hooke, 1635–1703* (London: Macmillan, 2002), 153.
38. Larry Stewart, "Other Centres of Calculation"; Rob Iliffe, "Material Doubts: Hooke, Artisan Culture and the Exchange of Information in 1670s London," *British Journal for the History of Science* 28 (1995): 285–318.
39. Garraway's was the location of wholesale auctions of drugs to druggists and apothecaries from the mid-eighteenth century until 1866. Anna Simmons, "Sites, Production and Networks: Wholesale Pharmaceutical Manufacturing in London, c. 1760–c. 1840," in Roberts and Werrett, *Compound Histories*, 295.
40. *Public Advertiser*, no 11870 (April 22, 1773): n.p.
41. *Daily Courant* no 2584 (February 3, 1710): n.p.
42. See Carpenter, *John Theophilus Desaguliers*, 229.
43. Brian Cowan, *The Social Life of Coffee: The Emergence of the British Coffee House* (New Haven: Yale University Press, 2005), 249.
44. James Reilly to Sir Hans Sloane, December 14, 1719, British Library, Sloane MS 4045, ff. 275–76.
45. Sir Robert Sibbald to Sir Hans Sloane, December 25, 1712, British Library, Sloane MS 4043, ff. 116–17; Arthur Charlett to Sir Hans Sloane, October 26, 1697, British Library, Sloane MS 4036, f. 364; William Sherard to Sir Hans Sloane, September 20, 1698, British Library, Sloane MS 4037, ff. 123–24; Philip Henry Zollman to Sir Hans Sloane, August 25, 1725, British Library, Sloane MS 4048, ff. 49–50; see also George Bere to Sir Hans Sloane, June 1, 1717, British Library, Sloane MS 4045, ff. 5–6.
46. James Petiver to Sir Hans Sloane, June 18, 1711, British Library, Sloane MS 4042, f. 305.
47. Johann George Steigertahl to Sir Hans Sloane, November 17, 1725, British Library, Sloane MS 4048, ff. 96–97.
48. Humfrey Wanley to Sir Hans Sloane, August 9, 1698, British Library, Sloane MS 4037, ff. 106–7.
49. Iordan Avramov, Michael Hunter, and Hideyuki Yoshimoto, "Boyle's Books: The Evidence of His Citations," *Robert Boyle Project Occasional Papers* 4 (2010): x; Hunter, "Hooke's Possessions at His Death."

50. Alice Marples, "Collecting and Correspondence in the Papers of Sir Hans Sloane (1660–1753)" (PhD diss., Kings College London, 2016), 245.
51. Ralph, *Touch-Stone*, 232.
52. "To be Sold by Auction, This Day at 11 . . . ," *Daily Journal*, no. 2604 (May 13, 1729).
53. *Daily Journal*, no. 2949 (June 19 1730).
54. "Mathematical Instruments," *Daily Journal*, no. 2949 (June 19, 1730).
55. Chatterton, *Auction*, 11.
56. de Clercq, "Private Instrument Collections" (2007), 35.
57. E.g., anon., *A Catalogue of all the Large and Valuable Stock in Trade, of Mr. John Urings, Deceased; At his Late House, No. 174, in Fenchurch-Street . . . Consisting of a Great Variety of Mathematical, Philosophical and Optical Instruments . . . Which will be Sold by Auction by T. Aldersey, and S. Martin* (London, 1773).
58. Anon., *A Catalogue of the Large and Valuable Stock of the Well-known and Ingenious Mr. James Short, F.R.S., Late of Surry-Street, in the Strand . . . Which will be Sold by Auction, by Mr. Langford and Son, at their House in the Great Piazza, Covent-Garden, (by Order of the Executors), on Wednesday and Thursday the 5th and 6th of April 1769* (London, 1769).
59. de Clercq, "Private Instrument Collections" (2007), 29.
60. Quoted in ibid., 31.
61. Anon., *A Catalogue of . . . Astronomical and Mathematical Instruments of the Late Alexander Aubert, Esq. of Highbury, Islington . . . Sold at Auction by Leigh and Sotheby* (London, 1806), lot 335; de Clercq, "Private Instrument Collections" (2007), 36n26.
62. The sale of Samuel Dunn's navigation instruments included "A Parcel of Old Brass and Mahogany Sextants, &c." Quoted in ibid., 35; on Cavallo, see below.
63. Christopher Bateman and John Cooper, *A Catalogue of the Library, Antiquities, &c. of the Late Learned Dr. Woodward . . . Which will begin to be Sold by Auction, at Mr Cooper's in the Great Piazza, Covent Garden, on Monday the 11th Day of November, 1728* (London, 1728).
64. Samuel Paterson, *A Catalogue of the Entire and Valuable Collection of Minerals, Ores, Fossils, Earths, Petrefactions, Gems, and other Stones, made in Great Britain and Ireland by Mr. David Main, Deceased* (London, 1766); see also "Catalogue of a Select Cabinet of Foreign Ores, Minerals and Fossils chiefly from the Continent," March 6, 1780, entry in the "Catalogue raisonnée," f. 47r.
65. William Noblett, "Samuel Paterson and the London Auction Market for Second-Hand Books, 1755–1802," *Papers of the Bibliographical Society of America* 108 (2014): 139–90; Samuel Paterson, *A Catalogue of the Genuine Household Furniture, Philosophical and Mathematical Instruments, Various Improvements in Mechanics . . . Plate, Watches, Books . . . of Henry Mill . . . Which Will be Sold by Auction, by Samuel Paterson . . . on Thursday, April the 18th, 1771* (London, 1771); Samuel Paterson, *Museum Falconarianum. A Catalogue*

of the Entire and Capital Museum of Anatomical Preparations, . . . Instruments; Medicaments, . . . And Other Effects; of the Late Mr. Magnus Falconar, Surgeon (London, 1778).

66. "Catalogue of a Choice Collection of Natural & Artificial Curiosities made in the Pacific ocean by an Officer in the Ship *Resolution*, during the Voyage," February 6–8, 1781, entry in "Catalogue raisonnée," f. 47r; "Catalogue of [Daniel Solander's] Cabinet of Shells, and other subjects of Natural History, Antiquities and other effects," February 13–19, 1783, entry in "Catalogue raisonnée," f. 47r.
67. "Catalogue of [John Fothergill's] curious Collection of Hot-house and Green-house Plants—sold on the premises at Upton, near Stratford in Essex," August 20–22, 1781, entry in "Catalogue raisonnée," f. 47r.
68. Will of Humphrey Jackson, March 8, 1801, National Archives, Kew, Prob. 11/1359.
69. Steven Shapin, "Pump and Circumstance: Robert Boyle's Literary Technology," *Social Studies of Science* 14 (1984): 481–520.
70. Peter de Clercq, "Transcription of the Ludlam sale catalogue, in an untitled PDF transcription of fourteen sale catalogues in the University of Oxford Museum of the History of Science," accessed October 21, 2015, http://www.mhs.ox.ac.uk/collections/library/ephemera/auction-catalogues/auction03/.
71. Quotations are from ibid.
72. See, e.g., the annotated catalog in the Oxford Museum of the History of Science, Anon., *A Catalogue of the Very Large, Capital, and Genuine Stock in Trade of Mr. Benjamin Martin, Optical, Mathematical, and Philosophical Instrument-Maker, No. 171, facing Serjeant's Inn, Fleet-Street, a Bankrupt, deceased, Which will be Sold by Auction, By Mr. Herring, On Friday, March 8, 1782, and the Four Following Days* (London, 1782).
73. Quoted in Joseph M. Levine, *Dr. Woodward's Shield: History, Science, and Satire in Augustan England* (Ithaca: Cornell University Press, 1977), 128.
74. Anon., *Catalogue of the Portland Museum*; on the collection, see Beth Fowkes Tobin, *The Duchess's Shells: Natural History Collecting in the Age of Cook's Voyages* (New Haven: Yale University Press, 2014); Stacey Sloboda, "Display-ing Materials: Porcelain and Natural History in the Duchess of Portland's Museum," *Eighteenth-Century Studies* 43 (2010): 455–72.
75. S. P. Dance, "The Authorship of the Portland Catalogue (1786)," *Journal of the Society for the Bibliography of Natural History* 4 (1962): 30–34; E. Alison Kay, "The Reverend John Lightfoot, Daniel Solander, and the Portland Catalogue," *Nautilus* 79 (1965): 10–19.
76. Anon., *Catalogue of the Portland Museum*, iii.
77. Alan J. Kohn, *Conus of the Southeastern United States and Caribbean* (Princeton: Princeton University Press, 2014), 168.
78. Anon., *Catalogue of the Portland Museum*, iv.
79. Chalmers-Hunt, *Natural Historical Auctions*, 4; on the buyers, see T. D. A. Cockerell, "Dru Drury, an Eighteenth Century Entomologist," *Scientific*

Monthly 14 (1922): 67–82; Simon Chaplin, "Anatomy and the 'Museum Oeconomy': William and John Hunter as Collectors," in *William Hunter's World: The Art and Science of Eighteenth-Century Collecting,* ed. E. Geoffrey Hancock, Nick Pearce, and Mungo Campbell (Farnham, UK: Ashgate, 2015), 29–44.

80. Wall, "The English Auction."
81. Anon., *Catalogue of the Portland Museum,* iii.
82. John Evelyn, *The Letterbooks of John Evelyn,* ed. Douglas D. C. Chambers and David Galbraith (Toronto: University of Toronto Press, 2014), xlii, 907.
83. Turner, "Auction Sales of the Earl of Bute's Instruments."
84. Reprinted in ibid., 237–40.
85. Cavallo to Lind, February 12, 1793, British Library, Add MSS 22898, f. 29v.
86. Geismar, "Alternative Market Values?," 294.
87. Cavallo reported, "The articles in general sold very well, excepting a few." Ibid.
88. J. Bidstrup, *A Catalogue of Optical, Mathematical & Philosophical Instruments, Made and Sold by J. Bidstrup, (no. 36.) St. Martin's Street, Leicester Square* (London, c. 1793), 7; William Jones and Samuel Jones, *A Catalogue of Optical, Mathematical, and Philosophical Instruments, Made and Sold by Willm. and Saml. Jones, No. 135, next Furnival's-Inn, Holborn, London* (London, c. 1793), 7.
89. When Cavallo attended an auction by Christie of the surveyor William Roy's books and instruments in December 1790, he thought "they sold very dear." Cavallo to Lind, December 11, 1790, British Library, Add MSS 22897, f. 107r.
90. Cavallo to Lind, February 12, 1793, British Library, Add MSS 22898, f. 29v.

CHAPTER EIGHT

1. Leis, "Sarah Sophia Banks," 13; Elizabeth Eger, *Bluestockings: Women of Reason from Enlightenment to Romanticism* (Basingstoke, UK: Palgrave Macmillan, 2010).
2. See chapter 6.
3. Gascoigne, *Joseph Banks and the English Enlightenment,* 21–24, 70–71; Leis, "Sarah Sophia Banks."
4. Deborah R. Coen, "The Common World: Histories of Science and Domestic Intimacy," *Modern Intellectual History* 11 (2014): 417–38.
5. Michael Charles Boulter, *Darwin's Garden: Down House and The Origin of Species* (London: Constable & Robinson, 2008); Paul White, "Darwin's Home of Science and the Nature of Domesticity," in Opitz, Bergwik, and Van Tiggelen, *Domesticity in the Making of Modern Science,* 61–83.
6. Janet Browne, *Charles Darwin: The Power of Place,* vol. 2 (Princeton: Princeton University Press, 2002), 464. On Darwin's thrift, see also Evelleen Richards, *Darwin and the Making of Sexual Selection* (Chicago: University of Chicago Press, 2017), xxviii, 292–93; Camille de Roquefeuil, *A Voyage round the*

World: Charles Darwin and the Beagle Collections in the University of Cambridge (Cambridge: Cambridge University Press, 2009), 30.

7. David Amigoni and Jeff Wallace, eds., *Charles Darwin's The Origin of Species: New Interdisciplinary Essays* (Manchester: Manchester University Press, 1995), 10.
8. Cooper, "Homes and Households," 224.
9. Sophie Forgan, "The Architecture of Display: Museums, Universities, and Objects in Nineteenth-Century Britain," *History of Science* 32 (1994): 139–62; Graeme Gooday, "Placing or Replacing the Laboratory in the History of Science?" *Isis* 99 (2008): 783–95; Gooday, "Precision Measurement and the Genesis of Physics Teaching Laboratories in Victorian Britain," *British Journal for the History of Science* 23 (1990): 25–51.
10. Gooday, "Placing or Replacing the Laboratory," 787.
11. Frank A. J. L. James and Anthony Peers, "Constructing Space for Science at the Royal Institution of Great Britain," *Physics in Perspective* 9 (2007): 130–85.
12. Morris, *Matter Factory*, 86–96.
13. Ibid., 109–15.
14. Donald L. Opitz, "Domestic Space," in *A Companion to the History of Science*, ed. Bernard Lightman (Oxford: Wiley Blackwell, 2016), 258.
15. Iwan Morus, *When Physics Became King* (Chicago: University of Chicago Press, 2005), 233; Gooday, "Precision Measurement," 27.
16. Caroline Herschel, *Caroline Herschel's Autobiographies*, ed. Michael Hoskin (Cambridge: Science History Publications, 2003), 52.
17. Ibid., 54.
18. Lissa Roberts, Simon Schaffer, and Peter Dear, eds., *The Mindful Hand: Inquiry and Invention from the Late Renaissance to Early Industrialisation* (Amsterdam: Edita; Chicago: University of Chicago Press, 2007); Simon Schaffer, "Enlightened Automata," in *The Sciences in Enlightened Europe*, ed. William Clark, Jan Golinski, and Simon Schaffer (Chicago: University of Chicago Press, 1999), 126–65.
19. Adam Smith, who recommended the division of labor in *Wealth of Nations*, lived with his mother, who provided him "domestic stability" for many years, and relied on his servants to collect papers from correspondents and to transcribe his letters. See Smith's letters in John Rae, *Life of Adam Smith* (London: Macmillan, 1895), 247–48, 307; Ian Simpson Ross, *The Life of Adam Smith*, 2nd ed. (Oxford: Oxford University Press, 2010), xxii; Katrine Marçal, *Who Cooked Adam Smith's Dinner? A Story about Women and Economics* (New York: Pegasus Books, 2016).
20. An identity explored in Ruth Barton, "'Men of Science': Language, Identity and Professionalization in the Mid-Victorian Scientific Community," *History of Science* 41 (2003): 73–119.
21. Humphry Davy, "3 March 1810 lecture," Royal Institution, London, RI MS Pamphlets 1, 37. My thanks to Harriet Lloyd for this reference.

22. Women's agency in these developments and their reaction to the division of the laboratory from the home will not be examined here, but on their reactions to Davy specifically, see Harriet Lloyd, "Rulers of Opinion: Women at the Royal Institution of Great Britain, 1799–1812" (PhD diss., University College London, 2018).
23. Wilkes, *Encyclopaedia Londinensis,* 14: 527.
24. Crookes quoted in Morris, *Matter Factory,* 146.
25. Iwan Morus, "Replacing Victoria's Scientific Culture," *19: Interdisciplinary Studies in the Long Nineteenth Century* 2 (2006): 13.
26. William Nicholson, *The British Encyclopedia; or Dictionary of Arts and Sciences,* 6 vols. (London, 1809), s.v. "Laboratory."
27. Abraham Rees, *The Cyclopædia: Or, Universal Dictionary of Arts, Sciences, and Literature,* 39 vols. (London, 1802–19), s.v. "Laboratory."
28. Ibid.
29. Ibid.
30. Warner, "What Is a Scientific Instrument," 86–89.
31. James Clerk Maxwell, "General Considerations concerning Scientific Apparatus," in *Handbook to the Special Loan Collection of Scientific Apparatus (South Kensington Museum)* (London: Chapman & Hall, 1876), 1–21.
32. Ibid., 2.
33. Ibid., 3.
34. Ibid., 2.
35. Ibid., 3.
36. Christoph Meinel, "Molecules and Croquet Balls," in *Models: The Third Dimension of Science,* ed. Soraya de Chadarevian and Nick Hopwood (Stanford: Stanford University Press, 2004), 242–75.
37. Thomas Romney Robinson, "Inaugural Address," *Proceedings of the Royal Irish Academy* 5 (1850–53): 104.
38. J. W. Queen & Co., *Catalogue of Physical Instruments, Chemical Apparatus, Chemicals and School Apparatus Generally* (Philadelphia, 1881), preface, quoted in Warner, "What Is a Scientific Instrument," 88.
39. M. Norton Wise, ed., *The Values of Precision* (Princeton: Princeton University Press, 1995); Roberts, "Death of the Sensuous Chemist."
40. John Heilbron, "Experimental Philosophy," in *The Oxford Companion to the History of Modern Science,* ed. Heilbron (Oxford: Oxford University Press, 2003), 287.
41. By the 1840s, for example, Pulkovo observatory in Russia was hailed for the lavish spending it entailed. Scientists envied it as the "El Dorado" of astronomy. See Werrett, "Astronomical Capital of the World."
42. Nicholson, *British Encyclopedia,* s.v. "Laboratory."
43. Richard Dennis Hoblyn, *A Manual of Chemistry* (London, 1841), 2.
44. Michael Faraday, *Chemical Manipulation* (London, 1830), 11.
45. Ibid., 11.

46. Catherine M. Jackson, "The 'Wonderful Properties of Glass'": Liebig's Kaliapparat and the Practice of Chemistry in Glass," *Isis* 106 (2015): 43–69.
47. Jeremy Bentham, *Panopticon: or, the Inspection-House. Containing the Idea of a New Principle of Construction Applicable to any Sort of Establishment* (Dublin, 1791).
48. Michel Foucault, *Discipline and Punish: The Birth of the Prison* (New York: Vintage, 1995), 195–230.
49. Jacques-Alain Miller and Richard Miller, "Jeremy Bentham's Panoptic Device," *October* 41 (1987): 8; Simon Werrett, "The Panopticon in the Garden: Samuel Bentham's Inspection House and Noble Theatricality in Eighteenth-Century Russia," *Ab Imperio* 3 (November 2008): 47–70.
50. The essays were collected in Jeremy Bentham, *The Works of Jeremy Bentham, Published under the Superintendence of his Executor, John Bowring*, ed. John Bowring, 11 vols. (Edinburgh, 1838–43), 8: 369–439, and references are to this collected edition.
51. Bentham, *Works*, 8: 373.
52. Ibid., 388.
53. Jeremy Bentham, "A Table of the Springs of Action," *Works*, 1: 214.
54. Jean-Baptiste Say, *A Treatise on Political Economy*, trans. C. R. Prinsep, 2 vols. (London, 1821), 2: 221.
55. James Mill, *Elements of Political Economy* (London, 1821), 1.
56. Neil McKendrick, John Brewer, and J. H. Plumb, eds., *The Birth of a Consumer Society: The Commercialization of Eighteenth-Century England* (Bloomington: Indiana University Press, 1982), 10.
57. Alison Toplis, "A Stolen Garment or a Reasonable Purchase? The Male Consumer and the Illicit Second-Hand Clothing Market in the First Half of the Nineteenth Century," in Stobart and Van Damme, *Modernity and the Second-Hand Trade*, 57.
58. Christophe Bonneuil and Jean-Baptiste Fressoz, *The Shock of the Anthropocene* (London: Verso, 2015), 258–62.
59. Garachon, "Old Repairs of China and Glass," 48–51.
60. Charles Babbage, *On the Economy of Machines and Manufactures*, 4th ed. (London, 1835).
61. Ibid., 6.
62. Ibid.
63. Ibid., 11.
64. Ibid., 393; Werrett, "Household Oeconomy and Chemical Inquiry," 53–54.
65. Anon., "The Foreign Translations of Mr. Babbage's 'Economy of Machines and Manufactures,'" *Mechanics Magazine*, no. 548 (February 8, 1834): 308; on the Prussian experiments, by anatomist and physiologist Karl Friedrich Burdach, see Lynn K. Nyhart, *Biology Takes Form: Animal Morphology and the German Universities, 1800–1900* (Chicago: University of Chicago Press, 1995), 35–64.

66. Babbage, *On the Economy of Machines*, 395. Even the rats in the establishment were recycled: "The fresh carcass of a horse is placed at night in a room, which has a number of openings near the floor. The rats are attracted to it, and the openings then closed. 16,000 rats were killed in one room in four weeks . . . The furriers purchase the rat skins at about 3 s. the hundred."
67. Lorraine Daston and Peter Galison, *Objectivity* (New York: Zone, 2007); Anson Rabinbach, *The Human Motor: Energy, Fatigue, and the Origins of Modernity* (New York: Basic Books, 1990).
68. See chapter 5.
69. van Driel, "Ashes to Ashes."
70. Crosbie Smith and M. Norton Wise, "Work and Waste: Political Economy and Natural Philosophy in Nineteenth-Century Britain," *History of Science* 27 (1989): 263–301, 391–449; 28 (1990): 221–61.
71. Peter Lund Simmonds, *Waste Products and Undeveloped Substances, or, Hints for Enterprise in Neglected Fields* (London, 1862).
72. Anon., "Waste Products and Undeveloped Substances," *Popular Science Review* 2, no. 5 (January 1863): 254.
73. Timothy Cooper, "Peter Lund Simmonds and the Political Ecology of Waste Utilization in Victorian Britain," *Technology and Culture* 52 (2011): 21–44.
74. Simmonds, *Waste Products*, 363.
75. Ibid., 333–39.
76. Simmonds, quoted in Cooper, "Peter Lund Simmonds," 42.
77. Nicholas Goddard, "'A Mine of Wealth'? The Victorians and the Agricultural Value of Sewage," *Journal of Historical Geography* 22 (1996): 274–90; Christopher Hamlin, "William Dibdin and the Idea of Biological Sewage Treatment," *Technology and Culture* 29 (1988): 189–218; Gary Bryan Magee, *Productivity and Performance in the Paper Industry: Labour, Capital and Technology in Britain and America, 1860–1914* (Cambridge: Cambridge University Press, 1997), 104–44.
78. Anon., "Waste Products and Undeveloped Substances," 254.
79. William Crookes, "Chemical Products: The Application of Waste," *Popular Science Review* 2, no. 5 (January 1863): 58.
80. Anthony S. Travis, "Perkin's Mauve: Ancestor of the Organic Chemical Industry," *Technology and Culture* 31 (1990): 51–82; Anthony S. Travis, ed., *150 Years of the Coal-Tar Dye Industry, 1856–2006*, special issue of *History and Technology* 2 (2006).
81. Anon., "Fortunes from Waste Products," *Living Age*, July 21, 1917, 188; Anon., "The Waste Products of Coal," *Scientific American* 27, no. 7 (August 17, 1872): 97; John J. Beer, "Coal Tar Dye Manufacture and the Origins of the Modern Industrial Research Laboratory," *Isis* 49 (1958): 123–31.
82. Anon., "Waste Not!" *Chambers's Journal of Popular Literature, Science and Arts* 312 (December 1869): 807.
83. Lyon Playfair, "The Conquest of Waste," *Current Literature* 14, no. 1 (September 1893): 81.

84. Lyon Playfair, "A Course of Six Lectures on Some of the Chemical Arts, with Reference to their Progress between the Two Great Exhibitions of 1851 and 1862 . . . Lecture II (Thursday, May 15, 1862)," *Chemical News*, June 14, 1862, 327.
85. Martin V. Melosi, "Technology Diffusion and Refuse Disposal: The Case of the British Destructor," in *Technology and the Rise of the Networked City in Europe and America*, ed. Joel A. Tarr and Gabriel Dupuy (Philadelphia: Temple University Press, 1988), 207–26.
86. Emily Hobhouse, "Dust-Women," *Economic Journal* 10 (1900): 411–20.
87. Christopher Hamlin, "The City as a Chemical System? The Chemist as Urban Environmental Professional in France and Britain, 1780–1880," *Journal of Urban History* 33 (2007): 702–28.
88. Materials relating to the "Shoreditch Destructor" may be found in Hackney Archives, London, REPORT M4281 1987; Shoreditch Cuttings vol. 1A S/LD/1/1 1895–1908; Hackney Vestry, 1899, Electricity Supply, H/CC/1/90; see also anon., "House Refuse as Fuel," *Daily News*, no. 15993 (June 30, 1897); William Henry Maxwell, *The Removal and Disposal of Town Refuse* (London, 1898), 324–28.
89. George Weiss, "The Conservation of Waste," *Forum* 57 (February 1917): 242.
90. Busch, "Throwaway Ethic in America."
91. Emily J. North and Rolf U. Halden, "Plastics and Environmental Health: The Road Ahead," *Reviews on Environmental Health* 28 (2013): 1.
92. Samuel Smiles, *Thrift* (London, 1875), 1–2.
93. Ibid., 10.
94. Ibid., 7.
95. Samuel Smiles, *Self-Help* (London, 1859), 76.
96. Ira Remsen, "The Simple Origin of Great Discoveries," in *The Mechanic Arts*, ed. Richard Cockburn Maclaurin (Boston: Hall and Locke, 1911), 370–79.
97. Ibid., 370.
98. See James A. Secord, "Newton in the Nursery: Tom Telescope and the Philosophy of Tops and Balls, 1761–1838," *History of Science* 23 (1985): 127–51; Melanie Keene, *Science in Wonderland: The Scientific Fairy Tales of Victorian Britain* (Oxford: Oxford University Press, 2015); Keene, "Domestic Science: Making Chemistry Your Cup of Tea," *Endeavour* 32 (2008): 16–19.
99. John Smith, *Lessons on Words and Objects, with Easy and Amusing Experiments, for the Parlour or School*, 2nd ed. (London, 1834).
100. Ibid., 140.
101. Charles Foote Gower, *The Scientific Phenomena of Domestic Life* (London, 1847).
102. John A. Bower, *Simple Experiments for Science Teaching* (London: Society for Promoting Christian Knowledge, 1894), v.
103. Ibid., v.
104. IFL Science, "20 Awesome Science Experiments You Can Do Right Now at Home," accessed March 15, 2016, http://www.iflscience.com/chemistry/unfinished-20-fun-science-experiments-you-can-do-home.

105. John Lubbock, *Pre-historic Times: As Illustrated by Ancient Remains, and the Manners and Customs of Modern Savages* (New York, 1872), 356–57.
106. William Emerson, *The Principles of Mechanics. Explaining and Demonstrating the General Laws of Motion, the Laws of Gravity, Motion of Descending Bodies*, 2nd ed. (London, 1758), i.
107. James Greenwood, *Curiosities of Savage Life*, 3rd ed. (London, 1865), 300.
108. Anon., "A Description and Historical Account of Hydraulic and other Machines for Raising Water, Ancient and Modern," *Brother Jonathan* 3 (1842): 75.
109. Anon., "Sure as a Gun (A Dialogue for the Day)," *Punch* 90 (May 22, 1886): 243.
110. Michael Adas, *Machines as the Measure of Men: Science, Technology, and Ideologies of Western Dominance* (Ithaca, NY: Cornell University Press, 1989), 216.
111. Simon Schaffer, "Physics Laboratories and the Victorian Country House," in *Making Space for Science*, ed. Crosbie Smith and Jon Agar (London: Macmillan, 1998), 149–80; Donald Opitz, "'This House Is a Temple of Research': Country House Research for Late Victorian Science," in *Repositioning Victorian Sciences: Shifting Centres in Nineteenth Century Scientific Thinking*, ed. David Clifford et al. (London: Anthem, 2006), 143–53.
112. Gooday, "Placing or Replacing the Laboratory," 790–91.
113. Mrs. J. E. H. Gordon [Alice Gordon], *Decorative Electricity* (London, 1891), 156.
114. Ibid., 158.
115. Schaffer, "Physics Laboratories," 164.
116. Ibid., 167.
117. Rayleigh, quoted in ibid., 166.
118. Donald L. Opitz, "'Not merely wifely devotion': Collaborating in the Construction of Science at Terling Place," in *For Better or for Worse? Collaborative Couples in the Sciences*, ed. Annette Lykknes, Donald L. Opitz, and Brigitte Van Tiggelen (Basel: Springer, 2012), 33–56.
119. Anon., "Scientific News," *English Mechanic and World of Science* 83, no. 2137 (March 9, 1906): 104.
120. Quoted in Alexander Wood, *The Cavendish Laboratory* (Cambridge: Cambridge University Press, 1946), 18; See also Opitz, "'Not merely wifely devotion,'" 35, 47.
121. Schaffer, "Physics Laboratories."
122. Rayleigh, quoted in Opitz, "'Not merely wifely devotion,'" 48.
123. Jeff Hughes, "Plasticine and Valves: Industry, Instrumentation and the Emergence of Nuclear Physics," in *The Invisible Industrialist: Manufacture and the Construction of Scientific Knowledge*, ed. Jean-Paul Gaudillière and Ilana Löwy (Basingstoke, UK: Palgrave, 1998), 59.
124. Crowther quoted in ibid., 58.
125. On the features of Big Science, see Hevly and Galison, *Big Science*.

126. Harry Collins, *Tacit and Explicit Knowledge* (Chicago: University of Chicago Press, 2010).
127. Karin Knorr-Cetina, *The Manufacture of Knowledge: An Essay on the Constructivist and Contextual Nature of Science* (Oxford: Pergamon Press, 1981), 34; part of the quotation is cited by Knorr-Cetina from F. Jacob, "Evolution and Tinkering," *Science* 196 (1977): 1161–66; see also Aharon Kantorovich, *Scientific Discovery: Logic and Tinkering* (Albany: SUNY Press, 1993), 225.

CONCLUSION

1. Remsen, "Simple Origin of Great Discoveries," 379.
2. Material culture in field science is typically a mix of adapted and dedicated apparatus. A recent text advises that to kill insects, one should put them in "kill jars": "Crumpled up newspaper can be placed in the kill jars to prevent insects from damaging each other. The newspaper should be removed when jars are recharged with ethyl acetate . . . Another method of storing unpinned insects is to place them in a cigar box." R. M. Sonoda, "Collection and Protection of Insects and Pathogenic Organisms," in *Handbook for the Collection, Preservation and Characterization of Tropical Forage Germplasm Resources*, ed. G. O. Mott (Cali, Colombia: Centro Internacional de Agricultura Tropical, 1979), 30; on the sciences of the field, see Robert E. Kohler and Jeremy Vetter, "The Field," in *A Companion to the History of Science*, ed. Bernard Lightman (Oxford: Wiley-Blackwell, 2016), 282–95.
3. See, e.g., North and Halden, "Plastics and Environmental Health"; Holly Ahern, "In Today's Labs, Disposable Plastics Play a Supporting, but Essential, Role," *The Scientist* (July 11, 1994), accessed March 29, 2016, http://www.the-scientist.com/?articles.view/articleNo/28522/title/In-Today-s-Labs—Disposable-Plastics-Play-A-Supporting—But-Essential—Role/.
4. GenSpace NYC, "Mars Ice House: A New Habitat Design for the Exploration of Mars," accessed July 7, 2016, https://www.youtube.com/watch?v=G-vklabGBdE.
5. Rob Carlson, "Splice It Yourself," *Wired Magazine*, accessed January 5, 2005, http://www.wired.com/2005/05/splice-it-yourself/.
6. London Biohackspace, "London Biohackspace," accessed April 3, 2016, https://biohackspace.org/.
7. Morgana Matus, "Makerspaces: A Revolution in Sustainable Production," CustomMade.com, accessed April 3, 2016, http://www.custommade.com/blog/makerspaces/.
8. Hasok Chang, *Inventing Temperature: Measurement and Scientific Progress* (New York: Oxford University Press, 2004).
9. Hasok Chang, "How Historical Experiments Can Improve Scientific Knowledge and Science Education: The Cases of Boiling Water and Electrochemistry," *Science and Education* 20 (2011): 317–41.

10. Trevor Pinch has investigated one musical innovation emerging from hacking, the practice of "circuit bending." "'Bring On Sector Two!' The Sounds of Bent and Broken Circuits," *Journal of Sound Studies* 2 (2016): 36–51.
11. For an introduction to various definitions, see Robert Goodland, "The Concept of Environmental Sustainability," *Annual Review of Ecology and Systematics* 26 (1995): 1–24.
12. The benefits of "citizen science" and "participatory research" are already in evidence. See, e.g., Louise Fortmann, ed., *Participatory Research in Conservation and Rural Livelihoods: Doing Science Together* (Oxford: Wiley-Blackwell, 2008).

Bibliography

Location of Manuscripts and Collections Cited

British Library, London
British Museum, London
Hackney Archives, London
Historical Society of Pennsylvania, Philadelphia
Museum of the History of Science, Oxford
National Archives, London
National Maritime Museum, London
Othmer Library, Chemical Heritage Foundation, Philadelphia
Royal Society, London
Science Museum, London
Van Pelt Library, University of Pennsylvania, Philadelphia
Victoria and Albert Museum, London
Wellcome Library, London
Whipple Museum of the History of Science, Cambridge

Periodicals

Daily Courant, London
Daily Journal, London
Daily News, London
Felix Farley's Bristol Journal
General Advertiser, London
London Courant Or the New Advertiser
London Evening Post
Morning Post and Daily Advertiser, London
Pennsylvania Gazette, Philadelphia

Public Advertiser, London
Public Ledger, London
Weekly Journal or Saturday's Post, London

Primary Published Sources

Adams, George. *An Essay on Electricity, explaining the Theory and Practice of that Useful Science*. 3rd ed. London, 1787.

———. *Essays on the Microscope; Containing a Practical Description of the Most Improved Microscopes*. London, 1784.

———. *Lectures on Natural and Experimental Philosophy, considered in its Present State of Improvement*. 5 vols. London, 1794.

Allen, Charles. *The Polite Lady; or, a Course of Female Education: in a Series of Letters, from a Mother to her Daughter*. London, 1788.

Allen, John. *Specimina ichnographica: or, a Brief Narrative of several New Inventions, and Experiments*. London, 1730.

Allestree, Richard. *The Gentleman's Calling*. London, 1660.

Anon. "Account of a Buyer of Bargains." *The American Museum, or, Repository of Ancient and Modern Fugitive Pieces, &c. Prose and Poetical* 1, no. 1 (April 1787): 307–8.

Anon. *An Address to Mothers, under the Following Heads: Maternal Authority. Domestic Attention. Diligence and Activity. Oeconomy. Simplicity*. Oxford, 1784.

Anon. *The Art of Thriving, or the Way to Get and Keep Money*. London, 1674.

Anon. *The Best Way of Using the True Salt Polychrest of Messieurs Seignette of Rochel*. London, 1685.

Anon. *The British Jewel; or Complete Housewife's Best Companion*. London, c. 1785.

Anon. *A Catalogue of all the Large and Valuable Stock in Trade, of Mr. John Urings, Deceased; At his Late House, No. 174, in Fenchurch-Street . . . Consisting of a Great Variety of Mathematical, Philosophical and Optical Instruments . . . Which will be Sold by Auction by T. Aldersey, and S. Martin*. London, 1773.

Anon. *A Catalogue of . . . Astronomical and Mathematical Instruments of the Late Alexander Aubert, Esq. of Highbury, Islington . . . Sold at Auction by Leigh and Sotheby*. London, 1806.

Anon. *A Catalogue of the Large and Valuable Stock of the Well-known and Ingenious Mr. James Short, F.R.S., Late of Surry-Street, in the Strand . . . Which will be Sold by Auction, by Mr. Langford and Son, at their House in the Great Piazza, Covent-Garden, (by Order of the Executors), on Wednesday and Thursday the 5th and 6th of April 1769*. London, 1769.

Anon. *A Catalogue of the Portland Museum, lately the Property of The Duchess Dowager of Portland, Deceased: Which will be Sold by Auction by Mr. Skinner and Co. on Monday the 24th April, 1786*. London, 1786.

Anon. *A Catalogue of the Very Large, Capital, and Genuine Stock in Trade of Mr. Benjamin Martin, Optical, Mathematical, and Philosophical Instrument-Maker, No. 171, facing Serjeant's Inn, Fleet-Street, a Bankrupt, deceased, Which will be Sold by Auction, By Mr. Herring, On Friday, March 8, 1782, and the Four Following Days.* London, 1782.

Anon. *The Countryman's Guide to London. Or, Villainy Detected. Being a Clear Discovery of all the Various Tricks and Frauds that are Daily Practiced in that Great City*. London, 1775.

Anon. "A Description and Historical Account of Hydraulic and other Machines for Raising Water, Ancient and Modern." *Brother Jonathan* 3 (1842): 75.

Anon. *England's Happiness Improved: or, an Infallible Way to get Riches, Encrease Plenty, and Promote Pleasure*. London, 1699.

Anon. "The Foreign Translations of Mr. Babbage's 'Economy of Machines and Manufactures.'" *Mechanics Magazine*, no. 548 (February 8, 1834): 306–14.

Anon. "Fortunes from Waste Products." *The Living Age*, July 21, 1917, 186–89.

Anon. *A Glasse for Housholders wherin thei maye se, bothe howe to rule Theim Selfes [and] Ordre their Housholde verye Godly and Fruytfull*. London, 1542.

Anon. *The Good Hows-holder*. London, 1607.

Anon. "Inventory of Dr. Priestley's Laboratory in 1791." In *The Scientific Correspondence of Joseph Priestley: Ninety-seven Letters addressed to Josiah Wedgwood, Sir Joseph Banks . . . and Others*, ed. Henry Carrington Bolton, 221–33. New York, 1892.

Anon. *A New and Complete Dictionary of the Arts and Sciences*. 2nd ed. 5 vols. London, 1763–64.

Anon. *Observations on Some Papers in that Very Useful Collection: Intitled, Museum Rusticum*. London, 1766.

Anon. [poss. Richard Burridge]. "Oeconomie and Extravagance." *Gentleman's Magazine* 1, no. 11 (November 1731): 489.

Anon. "On Sham Auctions." *The Yearly Chronicle . . . Essays . . . which appeared in the St. James's Chronicle*, no. 96 (October 22, 1761): 331–33.

Anon. "Scientific News." *English Mechanic and World of Science* 83, no. 2137 (March 9, 1906): 104.

Anon. "Sure as a Gun (A Dialogue for the Day)." *Punch* 90 (May 22, 1886): 242–43.

Anon. "Waste Not!" *Chambers's Journal of Popular Literature, Science and Arts* 312 (December 1869): 807–9.

Anon. "Waste Products and Undeveloped Substances." *Popular Science Review* 2, no. 5 (January 1863): 254.

Anon. "The Waste Products of Coal." *Scientific American* 27, no. 7 (August 17, 1872): 97.

Anon. *A Word in Season: or, an Essay to promote Good-Husbandry in Hard and Difficult Times: Being, in part, Advice from a Gentleman, to his Son a Tradesman in London*. London, 1697.

Anthony, Francis. *The Apologie, or Defence of a Verity heretofore published concerning a Medicine called Aurum Potabile*. London, 1616.

Arderon, William, and Henry Baker. "Abstract of a Letter from Mr. William Arderon, F.R.S. to Mr. Henry Baker, F.R.S. on the Giving Magnetism and Polarity to Brass." *Philosophical Transactions* 50 (1757–58): 774–77.

Aristotle. *Nicomachean Ethics*. Trans. and ed. Roger Crisp. Cambridge: Cambridge University Press, 2000.

Ash, John. *The New and Complete Dictionary of the English Language*. 2 vols. London, 1775.

Astell, Mary. *An Essay in Defence of the Female Sex in which are Inserted the Characters of a Pedant, a Squire, a Beau, a Vertuoso, a Poetaster, a City-critick, &c.* London, 1696.

Aubrey, John. *Scientific Lives*. London: Hesperus Press, 2011.

Babbage, Charles. *On the Economy of Machines and Manufactures*. 4th ed. London, 1835.

Bacon, Francis. *The Works of Francis Bacon*. Ed. James Spedding, Robert Leslie Ellis, and Douglas Denon Heath. 14 vols. London, 1858–74.

Bacon, Matthew. *A Treatise on Leases and Terms for Years*. London, 1798.

Baird & Tatlock, Ltd. *Physical Apparatus*. London: Baird & Tatlock, 1912.

Ballard, Martha. "Diary of Martha Ballard." Accessed December 10, 2015. http://dohistory.org/diary/.

Banks, Joseph. *The Endeavour Journal of Joseph Banks: The Australian Journey*. Ed. Paul Brunton. Sydney: Angus & Robertson in association with State Library of New South Wales, 1998.

Barbauld, Mrs. [Anna Letitia]. *Poems*. London, 1773.

Barbon, Nicholas. *A Discourse of Trade*. London, 1690.

Barker, Anne. *The Complete Servant Maid: Or Young Woman's Best Companion. Containing Full, Plain, and Easy Directions for Qualifying Them for Service*. London, c. 1770.

Barlow, William. *Magneticall Aduertisements: or Diuers Pertinent Obseruations, and Approued Experiments, concerning the Natures and Properties of the Load-stone*. London, 1618.

Bateman, Christopher, and John Cooper. *A Catalogue of the Library, Antiquities, &c. of the Late Learned Dr. Woodward . . . Which will begin to be Sold by Auction, at Mr Cooper's in the Great Piazza, Covent Garden, on Monday the 11th Day of November, 1728*. London, 1728.

Battam, Anne. *The Lady's Assistant in the Oeconomy of the Table*. London, 1759.

Baxter, Richard. *A Christian Directory*. London, 1673.

[Beale, John]. "An Experiment to Examine, What Figure, and Celerity of Motion Begetteth, or Encreaseth Light and Flame." *Philosophical Transactions* 1 (1665–66): 226–28.

Beattie, James. "Copy of the last Will and Testament of James Beattie, LL.D. written by his own Hand, and dated 20th July, 1799." In Sir William Forbes, *An Account of the Life and Writings of James Beattie*, 512–13. New York, 1807.

Bentham, Jeremy. *Panopticon: or, the Inspection-House. Containing the Idea of a New Principle of Construction Applicable to any Sort of Establishment*. Dublin, 1791.

———. *The Works of Jeremy Bentham, Published under the Superintendence of his Executor, John Bowring*. Ed. John Bowring. 11 vols. Edinburgh, 1838–43.

Berkenhout, John. *First Lines of the Theory and Practice of Philosophical Chemistry*. London, 1788.

Best, Henry. *Rural Economy in Yorkshire in 1641, being the Farming and Account Books of Henry Best of Elmswell, in the East Riding of the County of York*. Durham, 1857.

Bidstrup, J. *A Catalogue of Optical, Mathematical & Philosophical Instruments, Made and Sold by J. Bidstrup, (no. 36.) St. Martin's Street, Leicester Square*. London, c. 1793.

Birch, Thomas. *The History of the Royal Society of London*. 4 vols. London, 1760.

Black, Joseph. "An Analysis of the Waters of some Hot Springs in Iceland." *Transactions of the Royal Society of Edinburgh* 3 (1794): 95–126.

Boerhaave, Herman. *Elements of Chemistry*. 2 vols. London, 1735.

———. *A New Method of Chemistry; Including the History, Theory, and Practice of the Art*. Trans. Peter Shaw. 2 vols. London, 1741.

Bower, John A. *Simple Experiments for Science Teaching*. London: Society for Promoting Christian Knowledge, 1894.

Boyle, Robert. *The Correspondence of Robert Boyle*. Ed. Michael Hunter, Antonio Clericuzio and Lawrence M. Principe. 6 vols. London: Pickering & Chatto, 2001.

———. "New Pneumatical Experiments about Respiration." *Philosophical Transactions* 5 (1670): 2011–31.

———. "New Pneumatical Experiments about Respiration. Continued." *Philosophical Transactions* 5 (1670): 2035–56.

———. "Some Observations about Shining Flesh, Made by the Honourable Robert Boyle; Febr. 15. 1671/72." *Philosophical Transactions* 7 (1672): 5108–16.

———. *The Works of the Honourable Robert Boyle*. Ed. Thomas Birch. 6 vols. London, 1772. Reprinted, Hildesheim: Georg Olms, 1966.

Bradley, Richard. *The Country Housewife, and Lady's Director, for Every Month of the Year. Both in the Frugal Management of the House, and in the Delights and Profits of the Farm. Published for the Good of the Public*. 6th ed. London, 1732.

Buffon, Georges Louis Leclerc, comte de. *Buffon's Natural History: Containing a Theory of the Earth*. 10 vols. London, 1797.

Carter, Charles. *The London and Country Cook, or, Accomplished Housewife*. 3rd ed. London, 1749.

Cavallo, Tiberius. *A Complete Treatise on Electricity, in Theory and Practice, with Original Experiments*. London, 1777.

———. *A Treatise on Magnetism, in Theory and Practice, with Original Experiments*. London, 1795.

Cavendish, Margaret, Duchess of Newcastle. *Ground of Natural Philosophy Divided into Three Parts*. London, 1668.

Celia Pym, "Home page." Accessed August 17, 2017. http://celiapym.com/.

Chatterton, Thomas. *The Auction: A Poem*. London, 1770.

Chomel, Noel. *Dictionaire Oeconomique: or, the Family Dictionary*. Trans. Richard Bradley. 2 vols. London, 1725.

Clarke, Samuel. *A Collection of Papers, Which passed between the late Learned Mr. Leibnitz, and Dr. Clarke, In the Years 1715 and 1716*. London, 1717.

———. *A Paraphrase of the Four Evangelists*. London, 1738.

Cole, Mary. *The Lady's Complete Guide, or Cookery in all its Branches*. London, 1788.

Colquhoun, Patrick. *A Treatise on the Police of the Metropolis*. 6th ed. London, 1800.

Crookes, William. "Chemical Products: The Application of Waste," *Popular Science Review* 2, no. 5 (January 1863): 58–70.

Defoe, Daniel. *The Complete English Tradesman, in Familiar Letters*. London, 1725.

Delany, Mary. *The Autobiography and Correspondence of Mrs. Delany*. Ed. Sarah Chauncey Woolsey. 2 vols. Boston: Roberts Brothers, 1879.

Delbridge, T. G. "The Cracking of Petroleum." *Journal of the Franklin Institute* 202, no. 5 (November 1926): 569–88.

Derham, William. "Experiments about the Motion of Pendulums in Vacuo." *Philosophical Transactions* 24 (1704–5): 1785–89.

Desaguliers, Jean Theophilus. *A Dissertation Concerning Electricity*. London, 1742.

Digby, Sir Kenelm. *Two Treatises: In the One of which, the Nature of Bodies; In the Other the Nature of Mans Soule, is Looked Into, in Way of Discovery of the Immortality of Reasonable Soules*. London, 1645.

Dossie, Robert. *The Elaboratory Laid Open: or the Secrets of Modern Chemistry and Pharmacy Revealed*. London, 1758.

———. *The Handmaid to the Arts*. 2 vols. London, 1758.

Dreyer, J. L. E. "A Short Account of Sir William Herschel's Life and Work, chiefly from unpublished sources." In *The Scientific Papers of William Herschel*, ed. John Louis Emil Dreyer, 1: xiii–lxiv. Cambridge: Cambridge University Press, 1912.

Ellis, William. *The Country Housewife's Family Companion: or Profitable Directions for Whatever relates to the Management and Good Oeconomy of the Domestick Concerns of a Country Life*. London, 1750.

Emerson, William. *The Principles of Mechanics. Explaining and Demonstrating the General Laws of Motion, the Laws of Gravity, Motion of Descending Bodies*. 2nd ed. London, 1758.

Englefield, Sir H. C. "Description of a New Transit Instrument." *Philosophical Magazine and Journal* 43, no. 189 (January 1814): 1–6.

Estienne, Charles, and Jean Liébault. *Maison rustique; or, The Countrie Farme*. Trans. Richard Surflet. London, 1600.

Evelyn, John. *The Diary and Correspondence of John Evelyn*. Ed. William Bray. 4 vols. London, 1857.

———. *The Letterbooks of John Evelyn*. Ed. Douglas D. C. Chambers and David Galbraith. Toronto: University of Toronto Press, 2014.

———. *Numismata, a Discourse of Medals, Ancient and Modern Together With Some Account of Heads and Effigies of Illustrious, and Famous Persons*. London, 1697.

———. *Silva, or a Discourse on Forest-Trees and the Propagation of Timber in His Majesties Dominions*. 2nd ed. London, 1670.

Faraday, Michael. *Chemical Manipulation*. London, 1830.

Flamsteed, John. *The Correspondence of John Flamsteed, the First Astronomer Royal*. Ed. Eric Gray Forbes, Lesley Murdin, and Frances Wilmoth. 3 vols. Bristol, UK: Institute of Physics Publishing, 1995–2002.

Foote Gower, Charles. *The Scientific Phenomena of Domestic Life*. London, 1847.

Franklin, Benjamin. *Autobiography, Poor Richard, and Later Writings*. Ed. Joseph A. Leo Lemay. New York: Library of America, 1997.

———. *Experiments and Observations on Electricity, Made at Philadelphia in America*. 5th ed. London, 1774.

———. *The Life of Benjamin Franklin Written by Himself*. Ed. John Bigelow. 3 vols. Cambridge: Cambridge University Press, 2011.

———. *The Papers of Benjamin Franklin*. Ed. Leonard W. Labaree. 41 vols. New Haven: Yale University Press, 1959–.

———. "The Way to Wealth." In anon., *Miscellanies in Prose and Verse: Selected from Pope, Swift, Addison, Goldsmith, Sterne, Hume, Smollet, Gay, Shenstone, Prior, Murphy, and Brooke*, 61–72. Leominster, c. 1770.

French, John. *The Art of Distillation, or, A Treatise of the Choicest Spagiricall Preparations performed by Way of Distillation*. London, 1653.

Fulhame, Elizabeth. *An Essay On Combustion with a View to a New Art of Dying and Painting, wherein the Phlogistic and Antiphlogistic Hypotheses are Proved Erroneous*. London, 1794.

Galton, Francis. *The Art of Travel, or, Shifts and Contrivances Available in Wild Countries*. London, 1855.

Gilbert, William. *De magnete*. Trans. P. Fleury Mottelay. New York: Dover, 1958.

Glasse, Hannah. *The Art of Cookery, Made Plain and Easy*. London, 1796.

"Glass-house Clerk." *The Plate-Glass Book, consisting of the following Tables*. London, 1757.

Gordon, Mrs. J. E. H. [Alice Gordon]. *Decorative Electricity*. London, 1891.

Greenwood, James. *Curiosities of Savage Life*. 3rd ed. London, 1865.

Grew, Nehemiah. *The Anatomy of Plants with an Idea of a Philosophical History of Plants*. London, 1682.

———. *Musaeum regalis societatis, or, A Catalogue and Description of the Natural and Artificial Rarities belonging to the Royal Society and preserved at Gresham Colledge*. London, 1685.

Grey, Elizabeth, Countess of Kent. *A Choice Manuall, or Rare and Select Secrets in Physick and Chyrurgery*. London, 1653.

Guild, Reuben Aldridge. *The Life, Times, and Correspondence of James Manning and the Early History of Brown University*. Boston, MA, 1864.

Gunther, R. T., ed. *Early Science in Oxford*. Vol. 10, *The Life and Work of Robert Hooke (Part IV)*. Oxford: for the author, 1935.

Hales, Stephen. *A Description of Ventilators, whereby Great Quantities of Fresh Air May with Ease be Conveyed into Mines, Goals, Hospitals, Work-Houses and Ships in Exchange for their Noxious Air*. London, 1743.

———. *Philosophical Experiments: Containing Useful, and Necessary Instructions for Such as Undertake Long Voyages at Sea*. London, 1739.

———. *Statical Essays: containing Haemastatics; or, an Account of Some Hydraulic and Hydrostatical Experiments Made on the Blood*. London, 1740.

Harris, John. *Lexicon Technicum: Or, An Universal English Dictionary of Arts and Sciences*. 2nd ed. 2 vols. London, 1723.

Hartlib, Samuel. *A Description of the Famous Kingdome of Macaria, Shewing its Excellent Government wherein the Inhabitants Live in Great Prosperity, Health, and Happiness*. London, 1641.

———. *The Reformed Common-Wealth of Bees, Presented in Severall Letters and Observations to Sammuel Hartlib*. London, 1655.

———. *Samuel Hartlib his Legacy: Or an Enlargement of the Discourse of Husbandry used in Brabant and Flandres*. London, 1651.

———. *Samuel Hartlib, his Legacy of Husbandry wherein are Bequeathed to the Common-wealth of England, not onely Braband and Flanders, but also Many More Outlandish and Domestick Experiments and Secrets (of Gabriel Plats and Others)*. 3rd ed. London, 1655.

Hatton, Edward. *An Intire System of Arithmetic: Or Arithmetic in all its Parts*. London, 1721.

Hauksbee, Francis. *Physico-Mechanical Experiments on Various Subjects. Containing an Account of several Surprizing Phenomena touching Light and Electricity*. London, 1709.

Haywood, Eliza Fowler. *A New Present for a Servant-Maid: Containing Rules for her Moral Conduct both with Respect to Herself and her Superiors*. London, 1771.

Head, Richard, and Francis Kirkman. *The English Rogue Described in the Life of Meriton Latroon*. 4 vols. London, 1665–71.

Henley, Walter de. *The Booke of Thrift, containing a Perfite Order, and Right Methode to Profite Lands, and other Things belonging to Husbandry*. London, 1589.

[Henshaw, Thomas]. "Some Observations and Experiments on May-Dew." *Philosophical Transactions* 1 (1665–66): 33–36.

Herschel, Caroline. *Caroline Herschel's Autobiographies*. Ed. Michael Hoskin. Cambridge: Science History Publications, 2003.

Hjelm, Peter Jacob. "Experiments on Molybdaena, with a view to its reduction." *Crell's Chemical Journal*. 3 vols. London, 1791-93. 3: 40–51, 166–88, 220–51.

Hobhouse, Emily. "Dust-Women." *Economic Journal* 10 (1900): 411–20.

Hoblyn, Richard Dennis. *A Manual of Chemistry*. London, 1841.

Hoby, Lady Margaret. *The Private Life of an Elizabethan Lady: The Diary of Lady Margaret Hoby, 1599–1605*. Ed. Joanna Moody. Stroud, UK: Sutton Publishing, 1998.

Hood, John. *Tables of Difference of Latitude and Departure for Navigators, Land Surveyors, &c. with their Application to Plane Trigonometry*. Dublin, 1772.

Hooke, Robert. *The Diary of Robert Hooke, MA, MD, FRS, 1672–1680*. Ed. Henry W. Robinson and Walter Adams. London: Wykeham, 1968.

———. "Dr. Hook's Description of some Instruments for Sounding the great Depths of the Sea, and bringing Accounts of several Kinds from the Bottom of it. Being the Substance of some of his Lectures, in December, 1691." In *Philosophical Experiments and Observations of the late Eminent Dr. Robert Hooke*, ed. William Derham, 225–48. London, 1726.

———. *Micrographia, or, Some Physiological Descriptions of Minute Bodies made by Magnifying Glasses with Observations and Inquiries thereupon*. London, 1665.

IFL Science. "20 Awesome Science Experiments You Can Do Right Now at Home." Accessed March 15, 2016. http://www.iflscience.com/chemistry/unfinished-20-fun-science-experiments-you-can-do-home.

Jenner, Charles. *Town Eclogues*. London, 1772.

Johnson, Samuel. *A Dictionary of the English Language: In which the Words are Deduced from their Originals, and Illustrated in their Different Significations*. 2 vols. London, 1755–56.

Jones, William, and Samuel Jones. *A Catalogue of Optical, Mathematical, and Philosophical Instruments, Made and Sold by Willm. and Saml. Jones, No. 135, next Furnival's-Inn, Holborn, London*. London, c. 1793.

Jurin, James. "An Account of Some Experiments Relating to the Specifick Gravity of Human Blood." *Philosophical Transactions* 30 (1717–19): 1000–1014.

Kellett, Edward. *Tricoenivm Christi in nocte proditionis suae, The Threefold Svpper of Christ in the Night that He vvas Betrayed*. London, 1641.

L. C. *Fundamenta Chymica; or, A Sure Guide into the High and Rare Mysteries of Alchemy*. London, 1658.

Leibniz, G. W. *New Essays on Human Understanding*. Trans. and ed. Peter Remnant and Jonathan Francis Bennett. Cambridge: Cambridge University Press, 1996.

London Biohackspace. "London Biohackspace." Accessed April 3, 2016. https://biohackspace.org/.

Longfield, John, and Neville Maskelyne. "Extracts of Three Letters from John Longfield, M.D. at Corke in Ireland, to the Astronomer Royal, Containing Some Astronomical Observations; Together with the Longitude of Corke, Deduced from the Said Observations, by the Astronomer Royal." *Philosophical Transactions* 69 (1779): 163–81.

Löwenstern, Hermann Ludwig von. *The First Russian Voyage around the World: The Journal of Hermann Ludwig von Löwenstern, 1803–1806*. Ed. Victoria Joan Moessner. Fairbanks: University of Alaska Press, 2003.

Lubbock, John. *Pre-historic Times: As Illustrated by Ancient Remains, and the Manners and Customs of Modern Savages*. New York, 1872.

Lucas, Richard. *The Duty of Servants*. London, 1710.

Luckombe, Philip, and William Caslon. *A Concise History of the Origin and Progress of Printing: With Practical Instructions to the Trade in General*. London, 1770.

Mackenzie, Sir George. *The Moral History of Frugality, with its opposite Vices, Covetousness, Niggardliness, Prodigality, and Luxury*. London, 1711. Printed in Sir George Mackenzie, *Essays Upon Several Moral Subjects*, 286–309. London, 1713.

Mallet, David. *The Life of Francis Bacon, Lord High Chancellor of England*. London, 1740.

Mandeville, Bernard. *The Fable of the Bees: or, Private Vices, Publick Benefits*. London, 1732.

Markham, Gervase. *Cheape and Good Husbandry for the VVell-ordering of all Beasts, and Fowles*. London, 1614.

———. *Markhams Farwell to Husbandry or, The Inriching of all Sorts of Barren and Sterill Grounds in our Kingdome*. London, 1620.

Marshall, William. *Experiments and Observations Concerning Agriculture and the Weather*. London, 1779.

Mascall, Leonard. *A Profitable Booke, declaring Diuers Approoued Remedies, to take out Spots and Staines in Silkes, Veluets, Linnen and Woollen Clothes*. London, 1583.

Maxwell, James Clerk. "General Considerations concerning Scientific Apparatus." In *Handbook to the Special Loan Collection of Scientific Apparatus (South Kensington Museum)*, 1–21. London: Chapman & Hall, 1876.

Maxwell, William Henry. *The Removal and Disposal of Town Refuse*. London, 1898.

Mill, James. *Elements of Political Economy*. London, 1821.

Monro, Alexander. "An Essay on the Method of Preparing and Preserving the Parts of Animal Bodies for Anatomical Uses." *The Works of Alexander Monro*, 11–24. Edinburgh, 1781.

Montague, Peregrine. *The Family Pocket-book: or, Fountain of True and Useful Knowledge*. London, 1760.

Morrice, Alexander. *A Treatise on Brewing*. 2nd ed. London, 1802.

Moxon, Joseph. *Mechanick Exercises, or, The Doctrine of Handy-works*. 2nd ed. London, 1693.

Neale, John. *Directions for Gentlemen, who have Electrical Machines, how to Proceed in Making their Experiments*. London, 1747.

Newton, Isaac. "An Extract of Another Letter of the Same to the Publisher, Dated March 30. 1672. by Way of Answer to Some Objection, Made by an Ingenious French Philosopher to the New Reflecting Telescope." *Philosophical Transactions* 7 (1672): 4034–35.

———. *Opticks; Or, A Treatise of the Reflections, Refractions, Inflections and Colours of Light*. London, 1721.

———. "Original Letters of Sir Isaac Newton." *Monthly Magazine* 49 (1820): 234–36.

Nichols, John, ed. *Illustrations of the Literary History of the Eighteenth Century*. Vol. 1. London, 1817.

———. *Literary Anecdotes of the Eighteenth Century*. 6 vols. London, 1812.

Nicholson, William. *The British Encyclopedia; or Dictionary of Arts and Sciences*. 6 vols. London, 1809.

———. *An Introduction to Natural Philosophy*. 2 vols. London, 1807.

North, Dudley [4th Baron North]. *Observations and Advices Oeconomical*. London, 1669.

Oldenburg, Henry. *The Correspondence of Henry Oldenburg*. Ed. A. R. Hall and M. B. Hall. 13 vols. Madison: University of Wisconsin Press; London: Mansel; London: Taylor & Francis, 1965–86.

Pajot des Charmes, C. *The Art of Bleaching Piece-Goods, Cottons, and Threads of Every Description*. London, 1799.

Paterson, Samuel. *A Catalogue of the Entire and Valuable Collection of Minerals, Ores, Fossils, Earths, Petrefactions, Gems, and other Stones, made in Great Britain and Ireland by Mr. David Main, Deceased*. London, 1766.

———. *A Catalogue of the Genuine Household Furniture, Philosophical and Mathemati-*

cal Instruments, Various Improvements in Mechanics . . . Plate, Watches, Books . . . of Henry Mill . . . Which Will be Sold by Auction, by Samuel Paterson . . . on Thursday, April the 18th, 1771. London, 1771.

———. *Museum Falconarianum. A Catalogue of the Entire and Capital Museum of Anatomical Preparations, . . . Instruments; Medicaments, . . . And Other Effects; of the Late Mr. Magnus Falconar, Surgeon*. London, 1778.

Peacham, Henry. *The Worth of a Penny; or, A Caution to Keep Money. With the Causes of the Scarcity and Misery of the Want Thereof*. London, 1703.

Philiatros [Alethea Talbot]. *Natura exenterata: or Nature unbowelled by the most exquisite anatomizers of her*. London, 1655.

Plattes, Gabriel. *The Profitable Intelligencer, Communicating his Knowledge for the Generall Good of the Common-wealth and All Posterity*. London, 1644.

Playfair, Lyon. "The Conquest of Waste." *Current Literature* 14, no. 1 (September 1893): 79–81.

———. "A Course of Six Lectures on Some of the Chemical Arts, with Reference to their Progress between the Two Great Exhibitions of 1851 and 1862 . . . Lecture II (Thursday, May 15, 1862)." *Chemical News*, June 14, 1862, 327–32.

Powell, Hilary. *Urban Alchemy*. London: by the author, 2015.

Price, Richard. The *Correspondence of Richard Price*, vol. 3: *February 1786 to February 1791*. Ed. W. Bernard Leach. Durham: Duke University Press, 1994.

Priestley, Joseph. *Experiments and Observations on Different Kinds of Air*. 3 vols. Birmingham, 1790; Kraus Reprint, 1970.

———. *The History and Present State of Electricity*. 2nd ed. London, 1769.

Ralph, James. *The Touch-Stone; or, Historical, Critical, Political, Philosophical, and Theological Essays on the Reigning Diversions of the Town*. London, 1728.

Ravenscroft, Edward. *The Anatomist, or, The Sham-Doctor*. London, 1697.

Ray, John. *Miscellaneous Discourses concerning the Dissolution and Changes of the World*. London, 1692.

Rees, Abraham. *The Cyclopædia: Or, Universal Dictionary of Arts, Sciences, and Literature*. 39 vols. London, 1802–19.

Remsen, Ira. "The Simple Origin of Great Discoveries." In *The Mechanic Arts*, ed. Richard Cockburn Maclaurin, 370–79. Boston: Hall and Locke, 1911.

Robinson, Thomas Romney. "Inaugural Address." *Proceedings of the Royal Irish Academy* 5 (1850–53): 101–12.

Say, Jean-Baptiste. *A Treatise on Political Economy*. Trans. C. R. Prinsep. 2 vols. London, 1821.

Shackleford, Ann. *The Modern Art of Cookery Improved; Or, Elegant, Cheap, and Easy Methods, of Preparing Most of the Dishes Now in Vogue*. London, 1767.

Shadwell, Thomas. *The Virtuoso*. Ed. Marjorie Hope Nicholson and David Stuart Rodes. Lincoln: University of Nebraska Press, 1966.

Shakespeare, William. *Richard II*. Ed. Anthony B. Dawson and Paul Yachnin. Oxford: Oxford University Press, 2011.

Shaw, Peter. *Chemical Lectures, Publickly Read at London in the Years 1731, and 1732, and at Scarborough, in 1733*. 2nd ed. London, 1755.

Shaw, Peter, and Francis Hauksbee. *An Essay for Introducing a Portable Laboratory: By Means whereof all the Chemical Operations are Commodiously Perform'd*. London, 1731.

Sheridan, Thomas. *A Complete Dictionary of the English Language*. 2nd ed. London, 1789.

Simmonds, Peter Lund. *Waste Products and Undeveloped Substances, or, Hints for Enterprise in Neglected Fields*. London, 1862.

Skeete, Thomas. *Experiments and Observations on Quilled and Red Peruvian Bark*. London, 1786.

Smiles, Samuel. *Lives of Boulton and Watt*. London, 1865.

———. *Self-Help*. London, 1859.

———. *Thrift*. London, 1875.

Smith, Adam. *An Inquiry into the Nature and Causes of the Wealth of Nations*. 3 vols. Dublin, 1776.

Smith, John. *Lessons on Words and Objects, with Easy and Amusing Experiments, for the Parlour or School*. 2nd ed. London, 1834.

Smith, Robert. *A Compleat System of Opticks*. 2 vols. London, 1738.

Spilsbury, Francis. *Free Thoughts on Quacks and their Medicines, occasioned by the Death of Dr. Goldsmith and Mr. Scawen*. London, 1776.

Sprat, Thomas. *The History of the Royal Society of London for the Improving of Natural Knowledge*. London, 1667.

Sri Mahaalakshmi Scientific Company. "Biotechnology Equipments." Accessed June 30, 2016. http://www.mahaascience.com/biotechnology-equipments.html.

Stallard, J. H. "On the Economical Construction of Workmen's Dwellings and Especially in Reference to Improving the Health and Habits of the Class." *Journal of the Society of Arts* 19 (1871): 461–70.

The Statutes at Large From the Tenth Year of the Reign of King George the Third, Inclusive. 11 vols. London, 1774.

Steuart, Sir James. *An Inquiry into the Principles of Political Economy. Being an Essay on the Science of Domestic Policy in Free Nations*. 2 vols. London, 1767.

Swift, Jonathan. *Travels into Several Remote Nations of the World . . . by Lemuel Gulliver*. Dublin, 1752.

Symmer, Robert. "New Experiments and Observations concerning Electricity." *Philosophical Transactions* 51 (1759–60): 340–93.

Talling, Paul. "Derelict London." Accessed December 15, 2015. http://www.derelictlondon.com/.

Tomlins, Thomas Edlyne, ed. *The Law-Dictionary: Explaining the Rise, Progress, and Present State of the English Law, in Theory and Practice*. 2 vols. London, 1797.

Tusser, Thomas. *Fiue Hundreth Points of Good Husbandry vnited to as many of Good Huswiferie first deuised*. London, 1573.

———. *A Hundreth Good Pointes of Husbandry lately Maried vnto a Hundreth Good Poynts of Huswifery*. London, 1570.

Urquhart, Thomas. *The Trissotetras: or, a Most Exquisite Table for Resolving all Manner of Triangles*. 2nd ed. London, 1645.

US Environmental Protection Agency / US Department of Energy. "Laboratories for the 21st Century: An Introduction to Low-Energy Design." PDF, 2008. Accessed May 20, 2016. www.i2sl.org/documents/toolkit/lowenergy_508.pdf.

Warren, Thomas. *A True and Exact Particular and Inventory of All and Singular the Estate and Effects Whatsoever, of Thomas Warren*. London, 1732.

Webster, John. *Academiarum examen, or, The Examination of Academies wherein is Discussed and Examined the Matter, Method and Customes of Academick and Scholastick Learning*. London, 1654.

Weiss, George. "The Conservation of Waste." *Forum* 57 (February 1917): 241–51.

White, John. *Arts Treasury: or, A Profitable and Pleasing Invitation to the Lovers of Ingenuity Contained in Many Extraordinary Experiments, Rareties, and Curious Inventions*. London, 1688.

Whytt, Robert. "Of the various Strength of different Lime-waters." *Essays and Observations, Physical and Literary: Read Before a Society in Edinburgh and Published by Them* 1 (1754): 372–85.

Wilkes, John. *Encyclopaedia Londinensis, or, Universal Dictionary of Arts, Sciences, and Literature*. 24 vols. London, 1801–28.

Wilson, George. *A Compleat Course of Chymistry*. London, 1703.

Woodhouse, James. "Dr Woodhouse's Economical Apparatus." Appendix to James Parkinson, *The Chemical Pocket-Book, or Memoranda Chemica, arranged in a Compendium of Chemistry*, 201–3. Philadelphia, 1802.

Woolley, Hannah. *The Compleat Servant-Maid; or, Young Maiden's Tutor*. London, 1677.

———. *The Cook's Guide: or, Rare Receipts for Cookery Published and Set Forth Particularly for Ladies and Gentlwomen*. London, 1664.

———. *The Queen-like Closet, or Rich Cabinet stored with All Manner of Rare Receipts for Preserving, Candying and Cookery*. London, 1675.

Xenophon. *Memorabilia; Oeconomicus; Symposium; Apology*. Trans. E. C. Marchant, O. J. Todd, Jeffrey Henderson. Cambridge: Harvard University Press, 2013.

———. *Xenophons Treatise of House Holde*. London, 1532.

Young, Arthur. *Rural Oeconomy, or Essays on the Practical Parts of Husbandry: Designed to Explain Several Methods of Conducting Different Farms*. London, 1770.

Secondary Published Sources

Ackermann, Robert John. *Data, Instruments, and Theory: A Dialectical Approach to Understanding Science*. Princeton: Princeton University Press, 1985.

Adas, Michael. *Machines as the Measure of Men: Science, Technology, and Ideologies of Western Dominance*. Ithaca, NY: Cornell University Press, 1989.

Addyman, Tom. "Materia Chemica: Excavation of the Early Chemistry Stores at Old College, University of Edinburgh." Typescript, 2012.

Agar, Jon. *Science in the Twentieth Century and Beyond*. Cambridge: Polity Press, 2012.

Ahern, Holly. "In Today's Labs, Disposable Plastics Play a Supporting, but Essential, Role." *The Scientist* (July 11, 1994). Accessed March 29, 2016. http://www.the-scientist.com/?articles.view/articleNo/28522/title/In-Today-s-Labs—Disposable-Plastics-Play-A-Supporting—But-Essential—Role/.

Albala, Ken. *Food in Early Modern Europe*. Westport, CT; London: Greenwood Press, 2003.

Algazi, Gadi. "Scholars in Households: Refiguring the Learned Habitus, 1480–1550." *Science in Context* 16 (2003): 9–42.

Amigoni, David, and Jeff Wallace, eds. *Charles Darwin's The Origin of Species: New Interdisciplinary Essays*. Manchester: Manchester University Press, 1995.

Appadurai, Arjun, ed. *The Social Life of Things: Commodities in Cultural Perspective*. Cambridge: Cambridge University Press, 1986.

Appleby, John H. "Erasmus King: Eighteenth-Century Experimental Philosopher." *Annals of Science* 47 (1990): 375–92.

———. "Humphrey Jackson, F.R.S., 1717–1801: A Pioneering Chemist." *Notes and Records of the Royal Society of London* 40 (1986): 147–68.

Ashworth, William J. *Customs and Excise: Trade, Production, and Consumption in England, 1640–1845*. Oxford: Oxford University Press, 2003.

Atkinson, A. D. "William Derham, F.R.S. (1657–1735)." *Annals of Science* 8 (1952): 368–92.

Aughterson, Kate. " 'Strange Things So Probably Told': Gender, Sexual Difference and Knowledge in Bacon's *New Atlantis*." In *Francis Bacon's New Atlantis: New Interdisciplinary Essays*, ed. Glynn White, 156–79. Manchester: Manchester University Press, 2002.

Avramov, Iordan, Michael Hunter, Hideyuki Yoshimoto. "Boyle's Books: The Evidence of His Citations." *Robert Boyle Project Occasional Papers* 4 (2010): i–xxviii, 1–35.

Ayres, James. *Building the Georgian City*. New Haven: Yale University Press, 1998.

Baker, Alexi. " 'Precision,' 'Perfection,' and the Reality of British Scientific Instruments on the Move during the 18th Century." *Material Culture Review* 74–75 (Spring 2012): 14–29.

———. "Symbiosis and Style: The Production, Sale and Purchase of Instruments in the Luxury Markets of Eighteenth-Century London." In *How Scientific Instruments Have Changed Hands*, ed. Sara J. Schechner, Alison D. Morrison-Low, and Paolo Brenni, 1–20. Leiden: Brill, 2016.

Baker, David. "Optical Connections." *Optician*, February 10, 2012, 29–30.

Barton, Ruth. " 'Men of Science': Language, Identity and Professionalization in the Mid-Victorian Scientific Community." *History of Science* 41 (2003): 73–119.

Bayly, C. A. *Imperial Meridian: The British Empire and the World, 1780–1830*. London: Routledge, 2016.

Bedini, Silvio A. "Christina of Sweden and the Sciences." In *Making Instruments Count: Essays on Historical Scientific Instruments presented to Gerard L'Estrange Turner*, ed. Robert Anderson, James A. Bennett, W. F. Ryan, 99–117. Aldershot, UK: Variorum, 1993.

———. *Thinkers and Tinkers: Early American Men of Science*. New York: Scribner, 1975.

Beer, John J. "Coal Tar Dye Manufacture and the Origins of the Modern Industrial Research Laboratory." *Isis* 49 (1958): 123–31.

Ben-Amos, Ilana Krausman. *The Culture of Giving: Informal Support and Gift-Exchange in Early Modern England*. Cambridge: Cambridge University Press, 2008.

Bennett, Jim. "The Art of Polishing: Practice and Prose in Eighteenth-Century Telescope Making." In *From Earth-Bound to Satellite: Telescopes, Skills and Networks*, ed. Alison D. Morrison-Low, Sven Dupré, Stephen Johnston, and Giorgio Strano, 103–21. Leiden: Brill, 2011.

———. "Hooke's Instruments for Astronomy and Navigation." In *Robert Hooke: New Studies*, ed. Michael Hunter and Simon Schaffer, 21–32. Woodbridge, UK: Boydell Press, 1989.

———. "Shopping for Instruments in Paris and London." In *Merchants and Marvels: Commerce, Science, and Art in Early Modern Europe*, ed. Pamela H. Smith and Paul Findlen, 370–98. New York: Routledge, 2002.

Bennett, J. A., S. A. Johnston, and A. V. Simcock. *Solomon's House in Oxford: New Finds from the First Museum*. Oxford: Museum of the History of Science, 2000.

Bennett, Jim, and Scott Mandelbrote. *The Garden, the Ark, the Tower, the Temple: Biblical Metaphors of Knowledge in Early Modern Europe*. Oxford: Museum of the History of Science, 1998.

Berg, Maxine. "The British Product Revolution of the Eighteenth Century." In *Reconceptualizing the Industrial Revolution*, ed. Jeff Horn, Leonard N. Rosenband, and Merritt Roe Smith, 47–64. Cambridge: MIT Press, 2010.

———. "From Imitation to Invention: Creating Commodities in Eighteenth-Century Britain." *Economic History Review* 55 (2002): 1–30.

———. "In Pursuit of Luxury: Global History and British Consumer Goods in the Eighteenth Century." *Past and Present* 182 (2004): 85–142.

———. *Luxury and Pleasure in Eighteenth-Century Britain*. Oxford: Oxford University Press, 2005.

Bergman, Robert E., and Thomas V. Moore. *Managing Interactive Video/Multimedia Projects*. Englewood Cliffs, NJ: Educational Technology Publications, 1990.

Bergwik, Staffan. "An Assemblage of Science and Home: The Gendered Lifestyle of Svante Arrhenius and Early Twentieth-Century Physical Chemistry." *Isis* 105 (2014): 265–91.

Berry Drago, Elisabeth. "Thomas Wijck's Painted Alchemists at the Intersection of Art, Science, and Practice." PhD diss., University of Delaware, 2016.

Bertucci, Paola. "Medical and Animal Electricity in the Work of Tiberius Cavallo, 1780–1795." In *Luigi Galvani International Workshop Proceedings*, ed. Marco Bresadola and Giuliano Pancaldi, 147–66. Bologna: Universita di Bologna

Dipartimento di Filosofia Centro Internazionale per la Storia delle Universita e della Scienza, 1999.

Biagioli, Mario. *Galileo, Courtier: The Practice of Science in the Age of Absolutism.* Chicago: University of Chicago Press, 1993.

———. *Galileo's Instruments of Credit: Telescopes, Images, Secrecy.* Chicago: University of Chicago Press, 2006.

Bleichmar, Daniela. *Visible Empire: Botanical Expeditions and Visual Culture in the Hispanic Enlightenment.* Chicago: University of Chicago Press, 2012.

Bolt, Marvin, and Michael Korey. "The World's Oldest Surviving Telescopes." In *The Origins of the Telescope*, ed. Albert van Helden, Sven Dupré, Rob van Gent, and Huib J. Zuidervaart, 231–56. Amsterdam: KNAW, 2010.

Bonneuil, Christophe, and Jean-Baptiste Fressoz. *The Shock of the Anthropocene.* London: Verso, 2015.

Borsay, Peter. *The English Urban Renaissance: Culture and Society in the Provincial Town, 1660–1770.* Oxford: Clarendon Press, 1989.

Bossche, Willy Van den. *Antique Glass Bottles: Their History and Evolution (1500–1850).* Woodbridge, UK: Antique Collectors' Club, 2001.

Boulter, Michael Charles. *Darwin's Garden: Down House and The Origin of Species.* London: Constable & Robinson, 2008.

Brooker, Will. *Star Wars.* London: BFI, Palgrave Macmillan, 2009.

Brown, Frank E. "Continuity and Change in the Urban House: Developments in Domestic Space Organisation in Seventeenth-Century London." *Comparative Studies in Society and History* 28 (1986): 558–90.

Browne, Janet. *Charles Darwin: The Power of Place.* Vol. 2. Princeton: Princeton University Press, 2002.

Brownsey, P. J. "The Banks and Solander Collections—A Benchmark for Understanding the New Zealand Flora." *Journal of the Royal Society of New Zealand* 42 (2012): 131–37.

Busch, Jane Celia. "The Throwaway Ethic in America." PhD diss., University of Pennsylvania, 1983.

Buxton, Anthony. *Domestic Culture in Early Modern England.* Woodbridge, UK: Boydell Press, 2015.

Calder, Lendol. "Spending and Saving." In *The Oxford Handbook of the History of Consumption*, ed. Frank Trentmann, 348–75. Oxford: Oxford University Press, 2012.

Cambers, Andrew. *Godly Reading: Print, Manuscript and Puritanism in England, 1580–1720.* Cambridge: Cambridge University Press, 2011.

Canadian Broadcasting Company. "How to Think about Science: Episode 4—Ian Hacking & Andrew Pickering." Accessed February 25, 2009. http://www.cbc.ca/ideas/features/science/index.html#episode2.

Carlson, Rob. "Splice It Yourself." *Wired Magazine.* Accessed January 5, 2005. http://www.wired.com/2005/05/splice-it-yourself/.

Carpenter, Audrey T. *John Theophilus Desaguliers: A Natural Philosopher, Engineer and Freemason in Newtonian England.* London: Bloomsbury, 2011.

Chalmers-Hunt, J. M., ed. *Natural Historical Auctions, 1700–1972: A Register of Sales in the British Isles*. London: Sotheby Parke Bernet, 1976.

Chang, Hasok. "How Historical Experiments Can Improve Scientific Knowledge and Science Education: The Cases of Boiling Water and Electrochemistry." *Science and Education* 20 (2011): 317–41.

———. *Inventing Temperature: Measurement and Scientific Progress*. New York: Oxford University Press, 2004.

Chaplin, Simon. "Anatomy and the 'Museum Oeconomy': William and John Hunter as Collectors." In *William Hunter's World: The Art and Science of Eighteenth-Century Collecting*, ed. E. Geoffrey Hancock, Nick Pearce, and Mungo Campbell, 29–44. Farnham, UK: Ashgate, 2015.

Chapman, Allan. *England's Leonardo: Robert Hooke and the Seventeenth-Century Scientific Revolution*. Bristol: Institute of Physics Publishing, 2005.

———. "From Alchemy to Airpumps: The Foundations of Oxford Chemistry." In *Chemistry at Oxford: A History from 1600 to* 2005, ed. Robert J. P. Williams, Allan Chapman, and John S. Rowlinson, 17–51. Cambridge: Royal Society of Chemistry, 2009.

Cipolla, Carlo M. *Before the Industrial Revolution: European Society and Economy, 1000–1700*. New York: Norton, 1976.

Clifton, Gloria. *Directory of British Scientific Instrument Makers, 1550–1851*. London: Zwemmer, National Maritime Museum, 1995.

Cockayne, Emily. *Hubbub: Filth, Noise and Stench in England, 1600–1770*. New Haven: Yale University Press, 2007.

Cockerell, T. D. A. "Dru Drury, an Eighteenth Century Entomologist." *Scientific Monthly* 14 (1922): 67–82.

Coen, Deborah R. "The Common World: Histories of Science and Domestic Intimacy." *Modern Intellectual History* 11 (2014): 417–38.

———. "A Lens of Many Facets: Science through a Family's Eyes." *Isis* 97 (2006): 395–419.

———. *Vienna in the Age of Uncertainty: Science, Liberalism, and Private Life*. Chicago: University of Chicago Press, 2007.

Cohen, I. B., ed. *Puritanism and the Rise of Modern Science: The Merton Thesis*. New Brunswick, NJ: Rutgers University Press, 1990.

Colaco, Jill. "The Window Scenes in 'Romeo and Juliet' and Folk Songs of the Night Visit." *Studies in Philology* 83 (1986): 138–57.

Collins, Harry. *Tacit and Explicit Knowledge*. Chicago: University of Chicago Press, 2010.

Cook, Harold J. "Time's Bodies: Crafting the Preparation and Preservation of Naturalia." In *Merchants and Marvels: Commerce, Science and Art in Early Modern Europe*, ed. Pamela H. Smith and Paula Findlen, 223–47. London: Routledge, 2001.

Cooper, Alix. "Homes and Households." In *The Cambridge History of Science*, vol. 3: *Early Modern Science*, ed. Katharine Park and Lorraine Daston, 224–37. Cambridge: Cambridge University Press, 2006.

———. "'The Possibilities of the Land': The Inventory of 'Natural Riches' in the Early Modern German Territories." *History of Political Economy* 35, special issue: *Oeconomies in the Age of Newton*, ed. Margaret Schabas and Neil De Marchi (2003): 129–53.

Cooper, Jeremy. *Under the Hammer: The Auctions and Auctioneers of London*. London: Constable, 1979.

Cooper, Michael. *A More Beautiful City: Robert Hooke and the Rebuilding of London after the Great Fire*. Stroud, UK: Sutton Publishing, 2003.

Cooper, Timothy. "Peter Lund Simmonds and the Political Ecology of Waste Utilization in Victorian Britain." *Technology and Culture* 52 (2011): 21–44.

———. "Rags, Bones and Recycling Bins." *History Today* 56 (2006): 17–18.

Cowan, Brian. *The Social Life of Coffee: The Emergence of the British Coffee House*. New Haven: Yale University Press, 2005.

Crosland, Maurice. "Priestley Memorial Lecture: A Practical Perspective on Joseph Priestley as a Pneumatic Chemist." *British Journal for the History of Science* 16 (1983): 223–38.

Crossley, D. W. "The English Glassmaker and His Search for Materials in the 16th and 17th Centuries." In *The Prehistory and History of Glassmaking Technology*, ed. Patrick McCray, 167–79. Westerville, OH: American Ceramic Society, 1998.

Dance, S. P. "The Authorship of the Portland Catalogue (1786)." *Journal of the Society for the Bibliography of Natural History* 4 (1962): 30–34.

Dannehl, Karin. "'To Families Furnishing Kitchens': Domestic Utensils and Their Use in the Eighteenth-Century Home." In *Buying for the Home: Shopping for the Domestic from the Seventeenth Century to the Present*, ed. David Hussey and Margaret Ponsonby, 27–46. Aldershot, UK: Ashgate, 2008.

Daston, Lorraine, ed. *Biographies of Scientific Objects*. Chicago: University of Chicago Press, 2000.

———, ed. *Things That Talk: Object Lessons from Art and Science*. New York: Zone Books, 2004.

Daston, Lorraine, and Peter Galison. *Objectivity*. New York: Zone, 2007.

Davis, James Calvin, and Charles Mathewes. "Saving Grace and Moral Striving: Thrift in Puritan Theology." In *Thrift and Thriving in America: Capitalism and Moral Order from the Puritans to the Present*, ed. Joshua Yates and James Davison Hunter, 88–116. Oxford: Oxford University Press, 2011.

Daybell, James. *The Material Letter in Early Modern England: Manuscript Letters and the Culture and Practices of Letter-Writing, 1512–1635*. Basingstoke, UK: Palgrave Macmillan, 2012.

de Clercq, Peter. "Private Instrument Collections Sold at Auction in London in the Late 18th Century." *Bulletin of the Scientific Instrument Society* 95 (2007): 28–36.

———. "Private Instrument Collections Sold at Auction in London in the Late 18th Century." *Bulletin of the Scientific Instrument Society* 100 (2009): 27–35.

———. "Second-Hand Instruments in a W. & S. Jones Catalogue of 1795." *Bulletin of the Scientific Instrument Society* 85 (2005): 38–39.

———. "Transcription of the Ludlam sale catalogue, in an untitled PDF transcription of fourteen sale catalogues in the University of Oxford Museum of the History of Science." Accessed October 21, 2015. http://www.mhs.ox.ac.uk/collections/library/ephemera/auction-catalogues/auction03/.

Delbourgo, James. *A Most Amazing Scene of Wonders: Electricity and Enlightenment in Early America.* Cambridge: Harvard University Press, 2006.

Delbourgo, James, Kapil Raj, Lissa Roberts, and Simon Schaffer, eds. *The Brokered World: Go-Betweens and Global Intelligence, 1770–1820.* Sagamore Beach, MA: Science History Publications, 2009.

de Vries, Jan. *The Industrious Revolution: Consumer Behavior and the Household Economy, 1650 to the Present.* Cambridge: Cambridge University Press, 2008.

Dillon, Brian. *Ruin Lust.* London: Harry N. Abrams, 2014.

DiMeo, Michelle. "Katherine Jones, Lady Ranelagh (1615–91): Science and Medicine in a Seventeenth-Century Englishwoman's Writing." PhD diss., University of Warwick, 2009.

———. "Such a Sister became such a Brother. Lady Ranelagh's Influence on Robert Boyle." *Intellectual History Review* 25 (2015): 21–36.

DiMeo, Michelle, and Sara Pennell, eds. *Reading and Writing Recipe Books, 1550–1800.* Manchester: Manchester University Press, 2013.

DiMeo, Michelle, and Joanna Warren. "The Countess of Kent's Powder: A Seventeenth-Century 'Cure-all.'" Blog entry for *The Recipes Project.* Accessed September 22, 2015. http://recipes.hypotheses.org/tag/cure-all.

Di Palma, Vittoria. *Wasteland: A History.* New Haven: Yale University Press, 2014.

Dobbs, Betty Jo Teeter. *The Foundations of Newton's Alchemy, or, 'The Hunting of the Greene Lyon.'* Cambridge: Cambridge University Press, 1975.

———. "Newton's Alchemy and His Theory of Matter." *Isis* 73 (1982): 512–28.

Douglas, Mary. *Purity and Danger: An Analysis of Concepts of Pollution and Taboo.* London: Routledge Classics, 2002.

Drayton, Richard. *Nature's Government: Science, Imperial Britain, and the "Improvement" of the World.* New Haven: Yale University Press, 2000.

Dungworth, D. "Vauxhall, London: The Scientific Examination of Glass and Glassworking Materials from the Late Seventeenth Century Glasshouse, Research Department Report 83/2006 (English Heritage, 2006)." Accessed December 10, 2015. http://research.historicengland.org.uk/Report.aspx?i=14503&ru=%2FResults.aspx%3Fp%3D619.

Dungworth, D., and C. Mortimer. "Examination of Glassworking Materials from Cheese Lane, Bristol. Centre for Archaeology Report 6/2005 (English Heritage, 2005)." Accessed December 10, 2015. services.english-heritage.org.uk/ResearchReportsPdfs/006-2005.pdf.

Dunn, Richard. "Touching and Cleaning: The Routine Work of an East London Instrument Supplier." *Bulletin of the Scientific Instrument Society* 89 (2006): 21–26.

Dunn, Richard, and Rebekah Higgitt. *Ships, Clocks and Stars: The Quest for Longitude*. Glasgow: HarperCollins, 2014.

Dupré, Sven, and Christoph Herbert Lüthy, eds. *Silent Messengers: The Circulation of Material Objects of Knowledge in the Early Modern Low Countries*. Berlin: LIT Verlag, 2011.

Eamon, William. *Science and the Secrets of Nature*. Princeton: Princeton University Press, 1994.

Edgerton, David. *The Shock of the Old: Technology and Global History since 1900*. Oxford: Oxford University Press, 2007.

Edwards, Clive. *Turning Houses into Homes: A History of the Retailing and Consumption of Domestic Furnishings*. Aldershot, UK: Ashgate, 2005.

Edwards, Clive, and Margaret Ponsonby. "Desirable Commodity or Practical Necessity? The Sale and Consumption of Second-Hand Furniture, 1750–1900." In *Buying for the Home: Shopping for the Domestic from the Seventeenth Century to the Present*, ed. David Hussey and Margaret Ponsonby, 117–38. Aldershot, UK: Ashgate, 2008.

Eger, Elizabeth. *Bluestockings: Women of Reason from Enlightenment to Romanticism*. Basingstoke, UK: Palgrave Macmillan, 2010.

Ehrman, Edwina. "Dressing Well in Old Age: The Clothing Accounts of Martha Dodson, 1746–1765." *Costume* 40 (2006): 28–37.

Evans, David. "Thrifty, Green or Frugal: Reflections on Sustainable Consumption in a Changing Economic Climate." *Geoforum* 42 (2011): 550–57.

Fara, Patricia. *Pandora's Breeches: Women, Science and Power in the Enlightenment*. London: Pimlico, 2004.

Feather, John. "The Merchants of Culture: Bookselling in Early Industrial England." *Studies on Voltaire and the Eighteenth Century* 217 (1983): 11–21.

Feenburg, Andrew. *Transforming Technology: A Critical Theory Revisited*. Oxford: Oxford University Press, 2002.

Feingold, Mordechai. *Before Newton: The Life and Times of Isaac Barrow*. Cambridge: Cambridge University Press, 1990.

Fennetaux, Ariane. "Sentimental Economics: Textile Recycling in Eighteenth-Century Britain." In *The Afterlife of Used Things*, ed. Ariane Fennetaux, Sophie Vasset, and Amélie Junqua, 122–41. New York: Routledge, 2014.

Fennetaux, Ariane, Sophie Vasset, and Amélie Junqua, eds. *The Afterlife of Used Things: Recycling in the Long Eighteenth Century*. New York: Routledge, 2014.

Field, Rachel. *Irons in the Fire: A History of Cooking Equipment*. Marlborough, UK: Crowood Press, 1984.

Findlen, Paula. "Sites of Anatomy, Botany, and Natural History." In *The Cambridge History of Science*, vol. 3: *Early Modern Science*, ed. Katherine Park and Lorraine Daston, 272–89. Cambridge: Cambridge University Press, 2005.

Flather, Amanda. *Gender and Space in Early Modern England*. Woodbridge, UK: Boydell Press, 2007.

Fontaine, Laurence, ed. *Alternative Exchanges: Second-Hand Circulations from the Sixteenth Century to Today*. Oxford: Berghahn Books, 2008.

Foote, Helen S. "Silver in the Service of Medicine." *Bulletin of the Medical Library Association* 32 (1944): 369–75.

Forgan, Sophie. "The Architecture of Display: Museums, Universities, and Objects in Nineteenth Century Britain." *History of Science* 32 (1994): 139–62.

Fortmann, Louise, ed. *Participatory Research in Conservation and Rural Livelihoods: Doing Science Together*. Oxford: Wiley-Blackwell, 2008.

Foucault, Michel. *Discipline and Punish: The Birth of the Prison*. New York: Vintage, 1995.

———. *The History of Sexuality*, vol. 2: *The Uses of Pleasure*. New York: Vintage, 1990.

———. *The Order of Things: An Archaeology of the Human Sciences*. London: Routledge, 1991.

Fowkes Tobin, Beth. *The Duchess's Shells: Natural History Collecting in the Age of Cook's Voyages*. New Haven: Yale University Press, 2014.

Galison, Peter, and Bruce Hevly, eds. *Big Science: The Growth of Large-Scale Research*. Stanford: Stanford University Press, 1992.

Garachon, Isabelle. "Old Repairs of China and Glass." *Rijksmuseum Bulletin* 58 (2010): 35–54.

Gascoigne, John. *Joseph Banks and the English Enlightenment: Useful Knowledge and Polite Culture*. Cambridge: Cambridge University Press, 1994.

Gaudillière, Jean Paul, and Ilana Löwy, eds. *The Invisible Industrialist: Manufactures and the Production of Scientific Knowledge*. New York: Palgrave Macmillan, 1998.

Gaukroger, Stephen. *Francis Bacon and the Transformation of Early-Modern Philosophy*. Cambridge: Cambridge University Press, 2001.

Geismar, Haidy. "Alternative Market Values? Interventions into Auctions in Aotearoa/New Zealand." *Contemporary Pacific* 20 (2008): 291–327.

GenSpace NYC. "Mars Ice House: A New Habitat Design for the Exploration of Mars." Accessed July 7, 2016. https://www.youtube.com/watch?v=G-vklabGBdE.

Gerhold, Dorian. *Carriers and Coachmasters: Trade and Travel before the Turnpikes*. Chichester, UK: Phillimore, 2005.

———. *Road Transport before the Railways: Russell's London Flying Waggons*. Cambridge: Cambridge University Press, 1993.

Gerritsen, Anne, and Giorgio Riello, eds. *Writing Material Culture History*. London: Bloomsbury, 2014.

Gingerich, Owen. "Astronomical Paper Instruments with Moving Parts." In *Making Instruments Count: Essays on Historical Scientific Instruments presented to Gerard L'Estrange Turner*, ed. Robert Anderson, James A. Bennett, W. F. Ryan, 63–74. Aldershot, UK: Variorum, 1993.

———. "A Tusi Couple from Schöner's *De Revolutionibus*?" *Journal for the History of Astronomy* 15 (1984): 128–33.

Goddard, Nicholas. '"A Mine of Wealth'? The Victorians and the Agricultural Value of Sewage." *Journal of Historical Geography* 22 (1996): 274–90.

Golinski, Jan. *British Weather and the Climate of Enlightenment*. Chicago: University of Chicago Press, 2007.

———. "A Noble Spectacle: Phosphorus and the Public Cultures of Science in the Early Royal Society." *Isis* 80 (1989): 11–39.

———. *Science as Public Culture: Chemistry and Enlightenment in Britain, 1760–1820*. Cambridge: Cambridge University Press, 1992.

Gooday, Graeme. "Placing or Replacing the Laboratory in the History of Science?" *Isis* 99 (2008): 783–95.

———. "Precision Measurement and the Genesis of Physics Teaching Laboratories in Victorian Britain." *British Journal for the History of Science* 23 (1990): 25–51.

Goodland, Robert. "The Concept of Environmental Sustainability." *Annual Review of Ecology and Systematics* 26 (1995): 1–24.

Gordon, Alexander. "John Horsley (1685–1732)." *Oxford Dictionary of National Biography*. 60 vols. 9: 1276–77. Oxford: Oxford University Press, 2013.

Graham, Stephen, and Nigel Thrift. "Out of Order: Understanding Repair and Maintenance." *Theory, Culture and Society* 24 (2007): 1–25.

Gray, Annie. "'A Practical Art': An Archaeological Perspective on the Use of Recipe Books." In *Reading and Writing Recipe Books, 1550–1800,* ed. Michelle DiMeo and Sara Pennell, 47–67. Manchester: Manchester University Press, 2013.

Greengrass, Mark, Michael Leslie, and Timothy Raylor, eds. *Samuel Hartlib and Universal Reformation: Studies in Intellectual Communication*. Cambridge: Cambridge University Press, 1994.

Gregson, N., A. Metcalfe, and L. Crewe. "Practices of Object Maintenance and Repair: How Consumers Attend to Consumer Objects within the Home." *Journal of Consumer Culture* 9 (2009): 248–72.

Gröger, B. Lisa. "Of Men and Machines: Co-operation among French Family Farmers." *Ethnology* 20 (1981): 163–76.

Guerrini, Anita. *Experimenting with Humans and Animals: From Galen to Animal Rights*. Baltimore: Johns Hopkins University Press, 2003.

———. "The Ghastly Kitchen." *History of Science* 54 (2016): 71–97.

Hackmann, W. D. "The Relationship between Concept and Instrument Design in Eighteenth-Century Experimental Science." *Annals of Science* 36 (1979): 205–24.

Hall, M. B. *Promoting Experimental Learning: Experiment and the Royal Society, 1660–1727*. Cambridge: Cambridge University Press, 1991.

Hamlin, Christopher. "The City as a Chemical System? The Chemist as Urban Environmental Professional in France and Britain, 1780–1880." *Journal of Urban History* 33 (2007): 702–28.

———. "William Dibdin and the Idea of Biological Sewage Treatment." *Technology and Culture* 29 (1988): 189–218.

Hankins, Thomas, and Robert Silverman. *Instruments and the Imagination*. Princeton: Princeton University Press, 1995.

Hannaway, Owen. "Laboratory Design and the Aim of Science: Andreas Libavius versus Tycho." *Isis* 77 (1986): 584–610.

Harkness, Deborah E. *The Jewell House: Elizabethan London and the Scientific Revolution*. New Haven: Yale University Press, 2007.

———. "Managing an Experimental Household: The Dees of Mortlake and the Practice of Natural Philosophy." *Isis* 88 (1997): 247–62.

Harris, Frances. "Living in the Neighbourhood of Science: Mary Evelyn, Margaret Cavendish and the Greshamites." In *Women, Science, and Medicine, 1500–1700: Mothers and Sisters of the Royal Society*, ed. Lynette Hunter and Sarah Hutton, 198–217. Stroud, UK: Sutton Publishing, 1997.

Harrison, Molly. *The Kitchen in History*. Reading, UK: Osprey, 1972.

Harvey, Karen. *History and Material Culture: A Student's Guide to Approaching Alternative Sources*. London: Routledge, 2009.

———. *The Little Republic: Masculinity and Domestic Authority in Eighteenth-Century Britain*. Oxford: Oxford University Press, 2012.

———. "Oeconomy and the Eighteenth-Century 'House': A Cultural History of Social Practice." In *Domestic Practice in the Past: Historical Sources and Methods*, ed. Alison Blunt, special issue of the journal *Home Cultures* 11 (2014): 375–90.

Heal, Felicity. *The Power of Gifts: Gift Exchange in Early Modern England*. Oxford: Oxford University Press, 2015.

Heilbron, John L. *Electricity in the 17th and 18th Centuries: A Study of Early Modern Physics*. Berkeley: University of California Press, 1979.

———. "Experimental Philosophy." In *The Oxford Companion to the History of Modern Science*, ed. J. L. Heilbron, 286–88. Oxford: Oxford University Press, 2003.

———. "Robert Symmer and the Two Electricities." *Isis* 67 (1976): 7–20.

Herrmann, Frank. "Christie, James (1730–1803)." *Oxford Dictionary of National Biography*. Oxford: Oxford University Press, 2004. Accessed October 20, 2015. http://www.oxforddnb.com/view/article/5362.

Hickman, Clare. "The Garden as a Laboratory: The Role of Domestic Gardens as Places of Scientific Exploration." *Post-Medieval Archaeology* 48 (2014): 229–47.

Hill, Bridget. *Servants: English Domestics in the Eighteenth Century*. Oxford: Clarendon Press, 1996.

Hindle, Steve. *On the Parish? The Micro-politics of Poor Relief in Rural England, c.1550–1750*. Oxford: Oxford University Press, 2004.

Hitchcock, Tim. *Down and Out in Eighteenth-Century London*. London: Hambledon and London, 2004.

Hont, Istvan. "The Early Enlightenment Debate on Commerce and Luxury." In *The Cambridge History of Eighteenth-Century Political Thought*, ed. Mark Goldie and Robert Wokler, 379–418. Cambridge: Cambridge University Press, 2006.

Hoskin, Michael A. *The Herschel Partnership: As Viewed By Caroline*. Cambridge: Science History Publications, 2003.

Hoskins, W. G. "The Rebuilding of Rural England, 1570–1640." *Past and Present* 4 (1953): 44–59.

Hounsell, Peter. *London's Rubbish: Two Centuries of Dirt, Dust and Disease in the Metropolis*. Stroud, UK: Amberley, 2013.

Howse, Derek. *Greenwich Observatory: The Royal Observatory at Greenwich and Herstmonceux, 1675–1975*, vol. 3: *The Buildings and Instruments*. London: Taylor & Francis, 1975.

Hughes, Jeff. "Plasticine and Valves: Industry, Instrumentation and the Emergence of Nuclear Physics." In *The Invisible Industrialist: Manufacture and the Construction of Scientific Knowledge*, ed. Jean-Paul Gaudillière and Ilana Löwy, 58–101. Basingstoke, UK: Palgrave, 1998.

Hunt, Margaret. *The Middling Sort: Commerce, Gender and the Family in England, 1680–1780*. Berkeley: University of California Press, 1996.

Hunter, Lynette. "Sisters of the Royal Society: The Circle of Katherine Jones, Lady Ranelagh." In *Women, Science, and Medicine, 1500–1700: Mothers and Sisters of the Royal Society*, ed. Lynette Hunter and Sarah Hutton, 178–97. Stroud, UK: Sutton Publishing, 1997.

———. "Women and Domestic Medicine: Lady Experimenters, 1570–1620." In *Women, Science, and Medicine, 1500–1700: Mothers and Sisters of the Royal Society*, ed. Lynette Hunter and Sarah Hutton, 89–107. Stroud, UK: Sutton Publishing, 1997.

Hunter, Michael. *Boyle: Between God and Science*. New Haven: Yale University Press, 2009.

———. "Hooke's Possessions at His Death: A Hitherto Unknown Inventory." In *Robert Hooke: New Studies*, ed. Michael Hunter and Simon Schaffer, 287–94. Woodbridge, UK: Boydell Press, 1989.

———, ed. *Robert Boyle: By Himself and His Friends: With a Fragment of William Wotton's 'Lost Life of Boyle.'* London: William Pickering, 1994.

———. *Science and the Shape of Orthodoxy: Intellectual Change in Late Seventeenth-Century Britain*. Woodbridge, UK: Boydell Press, 1995.

Hunter, Michael, and Alison Wiggins, eds. "The Workdiaries of Robert Boyle." Accessed June 30, 2016. http://www.livesandletters.ac.uk/wd/index.html.

Iliffe, Rob. "Material Doubts: Hooke, Artisan Culture and the Exchange of Information in 1670s London." *British Journal for the History of Science* 28 (1995): 285–318.

Iliffe, Rob, and Frances Willmoth. "Astronomy and the Domestic Sphere: Margaret Flamsteed and Caroline Herschel as Assistant-Astronomers." In *Women, Science, and Medicine, 1500–1700: Mothers and Sisters of the Royal Society*, ed. Lynette Hunter and Sarah Hutton, 235–65. Stroud, UK: Sutton Publishing, 1997.

Insley, Jane. "James Watt's Cookbook Chemistry." *Notes and Records of the Royal Society* 65 (2011): 301–8.

Inwood, Stephen. *The Man Who Knew Too Much: The Strange and Inventive Life of Robert Hooke, 1635–1703*. London: Macmillan, 2002.

Jackson, Catherine M. "The 'Wonderful Properties of Glass': Liebig's Kaliapparat and the Practice of Chemistry in Glass." *Isis* 106 (2015): 43–69.

James, Frank A. J. L., and Anthony Peers. "Constructing Space for Science at the Royal Institution of Great Britain." *Physics in Perspective* 9 (2007): 130–85.

Jardine, Lisa, and Alan Steward. *Hostage to Fortune: The Troubled Life of Francis Bacon*. London: Victor Gollancz, 1998.

Johnson, Matthew. *English Houses, 1300–1800: Vernacular Architecture, Social Life*. London: Routledge, 2014.

———. "Rethinking the Great Rebuilding." *Oxford Journal of Archaeology* 12 (1993): 117–25.

Jones, Matthew. *Reckoning with Matter: Calculating Machines, Innovation, and Thinking about Thinking from Pascal to Babbage*. Chicago: University of Chicago Press, 2016.

Jungnickel, Christa, and Russell McCormmach. *Cavendish*. Philadelphia: American Philosophical Society, 1996.

Junqua, Amélie. "Unstable Shades of Grey: Cloth and Paper in Addison's Periodicals." In *The Afterlife of Used Things*, ed. Ariane Fennetaux, Sophie Vasset, and Amélie Junqua, 184–98. New York: Routledge, 2014.

Kantorovich, Aharon. *Scientific Discovery: Logic and Tinkering*. Albany: SUNY Press, 1993.

Katz, Dana E. "'Clamber not you up to the Casements': On Ghetto Views and Viewing." *Jewish History* 24 (2010): 127–53.

Kay, E. Alison. "The Reverend John Lightfoot, Daniel Solander, and the Portland Catalogue." *Nautilus* 79 (1965): 10–19.

Keene, Melanie. "Domestic Science: Making Chemistry Your Cup of Tea." *Endeavour* 32 (2008): 16–19.

———. *Science in Wonderland: The Scientific Fairy Tales of Victorian Britain*. Oxford: Oxford University Press, 2015.

Klein, Ursula, and Emma C. Spary, eds. *Materials and Expertise in Early Modern Science*. Chicago: University of Chicago Press, 2010.

Knapp, Peggy A. "Thrift." In *Art and Context in Late Medieval English Narrative: Essays in Honor of Robert Worth Frank, Jr.*, ed. Robert Edwards, 193–205. Oxford: Boydell & Brewer, 1994.

Knorr-Cetina, Karin. *The Manufacture of Knowledge: An Essay on the Constructivist and Contextual Nature of Science*. Oxford: Pergamon Press, 1981.

———. "Objectual Practice." In *The Practice Turn in Contemporary Theory*, ed. Theodore R. Schatzki, Karin Knorr-Cetina, and Eike von Savigny, 175–88. London: Routledge, 2001.

Koerner, Lisbet. "Linnaeus' Floral Transplants." *Representations* 47 (1994): 144–69.

———. *Linnaeus: Nature and Nation*. Cambridge: Harvard University Press, 1999.

Kohler, Robert E., and Jeremy Vetter. "The Field." In *A Companion to the History of Science*, ed. Bernard Lightman, 282–95. Oxford: Wiley-Blackwell, 2016.

Kohn, Alan J. *Conus of the Southeastern United States and Caribbean*. Princeton: Princeton University Press, 2014.

Korda, Natasha. *Shakespeare's Domestic Economies: Gender and Property in Early Modern England*. Philadelphia: University of Pennsylvania Press, 2012.

Krajewski, Markus. *Paper Machines: About Cards & Catalogs, 1548–1929*. Cambridge: MIT Press, 2011.

Kubrin, David. "Newton and the Cyclical Cosmos: Providence and the Mechanical Philosophy." *Journal of the History of Ideas* 27 (1968): 325–46.

Kuhn, Thomas. *The Structure of Scientific Revolutions*. 4th ed. Chicago: University of Chicago Press, 2012.

Kwass, Michael. Review of *The Industrious Revolution: Consumer Behavior and the Household Economy, 1650 to the Present*, by Jan de Vries. *American Historical Review* 114, no. 3 (2009): 705–8.

Lacey, Robert. *Sotheby's: Bidding for Class*. London: Warner, 1998.

Lambert, Miles. "'Cast-off Wearing Apparell': The Consumption and Distribution of Second-Hand Clothing in Northern England during the Long Eighteenth Century." *Textile History* 35 (2004): 1–26, 38–48.

Langford, John. *A Century of Birmingham Life*. 2 vols. Birmingham, 1868.

Latham, Siena Louise. "'Lady Alcumy': Elizabethan Gentlewomen and the Practice of Chymistry." PhD diss., Victoria University, Wellington, 2010.

Lawler, John. *Book Auctions in England in the Seventeenth Century (1676–1700): With a Chronological List of the Book Auctions of the Period*. London: Elliot Stock, 1898.

Learmont, Brian. *A History of the Auction*. Iver, UK: Barnard and Learmont, 1985.

Lefebvre E., and J. G. de Bruijn, eds. *Martinus Van Marum: Life and Work*. 6 vols. Haarlem and Leyden: Noordhoff International and H. D. Tjeenk Willink for Hollandsche Maatschappij der Wetenschappen, 1969–76.

Leis, Arlene C. "Sarah Sophia Banks: Femininity, Sociability and the Practice of Collecting in Late Georgian England." PhD diss., University of York, 2013.

Lemay, J. A. Leo. *The Life of Benjamin Franklin*, vol. 3: *Soldier, Scientist, and Politician, 1748–1757*. Philadelphia: University of Pennsylvania Press, 2009.

Lemire, Beverly. *The Business of Everyday Life: Gender, Practice and Social Politics in England, c.1600–1900*. Manchester: Manchester University Press, 2005.

———. "Consumerism in Pre-industrial and Early Industrial England: The Trade in Second-Hand Clothes." *Journal of British Studies* 27 (1988): 1–24.

———. "Plebeian Commercial Circuits and Everyday Material Exchange in England, c. 1600–1900." In *Buyers & Sellers: Retail Circuits and Practices in Mediaeval and Early Modern Europe*, ed. Bruno Blondé, Peter Stabel, Jon Stobart, and Ilja Van Damme, 245–66. Turnhout: Brepols, 2006.

Leong, Elaine. "Collecting Knowledge for the Family: Recipes, Gender and Practical Knowledge in the Early Modern English Household." *Centaurus* 55 (2013): 81–103.

———. "Making Medicines in the Early Modern Household." *Bulletin of the History of Medicine* 82 (2008): 145–68.

———. *Recipes and Everyday Knowledge: Medicine, Science, and the Household in Early Modern England*. Chicago: University of Chicago Press, 2018.

Leshem, Dotan. "Retrospectives: What Did the Ancient Greeks Mean by *Oikonomia?*" *Journal of Economic Perspectives* 30 (2016): 225–38.

Levine, Joseph M. *Dr. Woodward's Shield: History, Science, and Satire in Augustan England*. Ithaca: Cornell University Press, 1977.

Li, Ming-Hsun. *The Great Recoinage of 1696 to 1699*. London: Weidenfeld and Nicolson, 1963.

Livingstone, David N. *Putting Science in Its Place: Geographies of Scientific Knowledge*. Chicago: University of Chicago Press, 2003.

Lloyd, Harriet. "Rulers of Opinion: Women at the Royal Institution of Great Britain, 1799–1812." PhD diss., University College London, 2018.

Louw, H. J. "The Origins of the Sash-Window." *Architectural History* 26 (1983): 49–72, 144–50.

Love, Harold. *Scribal Publishing in Seventeenth-Century England*. Oxford: Clarendon Press, 1993.

Loveman, Kate. "Books and Sociability: The Case of Samuel Pepys's Library." *Review of English Studies* 61 (2010): 214–33.

Lykknes, Annette, Donald L. Opitz, and Brigitte Van Tiggelen, eds. *For Better or for Worse? Collaborative Couples in the Sciences*. Basel: Springer, 2012.

MacArthur, Rosie, and Jon Stobart. "Going for a Song? Country House Sales in Georgian England." In *Modernity and the Second-Hand Trade: European Consumption Cultures and Practices, 1700–1900*, ed. Jon Stobart and Ilja Van Damme, 175–95. Basingstoke, UK: Palgrave Macmillan, 2010.

Magee, Gary Bryan. *Productivity and Performance in the Paper Industry: Labour, Capital and Technology in Britain and America, 1860–1914*. Cambridge: Cambridge University Press, 1997.

Malmendier, Ulrike, and Young Han Lee. "The Bidder's Curse." *American Economic Review* 101, no. 2 (2011): 749–87.

Mandelbrote, Giles. "The Organization of Book Auctions in Late Seventeenth-Century London." In *Under the Hammer: Book Auctions since the Seventeenth Century*, ed. Robin Myers, Michael Harris, and Giles Mandelbrote, 15–36. New Castle, DE: Oak Knoll Press; London: British Library, 2001.

Marçal, Katrine. *Who Cooked Adam Smith's Dinner? A Story about Women and Economics*. New York: Pegasus Books, 2016.

Margócsy, Dániel. *Commercial Visions: Science, Trade, and Visual Culture in the Dutch Golden Age*. Chicago: University of Chicago Press, 2015.

Märker, Anna. "From Cookery to Chemistry: The Development of Preservation Techniques for Anatomical Research." Unpublished typescript, 2009.

Marples, Alice. "Collecting and Correspondence in the Papers of Sir Hans Sloane (1660–1753)." PhD diss., Kings College London, 2016.

Marr, Alexander. "Learned Benefaction: Science, Civility and Donations of Books and Instruments to the Bodleian Library before 1605." In *Documenting the Early Modern Book World: Inventories and Catalogues in Manuscript and Print*, ed. Malcolm Walsby and Natasha Constantinidou, 27–50. Leiden: Brill, 2013.

Martin, A. R. *Project Daedalus: The Final Report on the BIS Starship Study*. London: British Interplanetary Society, 1978.

Martin, Joanna. *Wives and Daughters: Women and Children in the Georgian Country House*. London: Hambledon and London, 2004.

Matus, Morgana. "Makerspaces: A Revolution in Sustainable Production." *CustomMade.com*. Accessed April 3, 2016. http://www.custommade.com/blog/makerspaces/.

Maunder, E. Walter. *The Royal Observatory Greenwich: A Glance at Its History and Work*. Cambridge: Cambridge University Press, 1900.

McConnell, Anita. *Barometers*. 2nd ed. Risborough, UK: Shire, 2003.

———. "Instruments and Instrument-Makers, 1700–1850." In *The Oxford Handbook of the History of Physics*, ed. Jed Z. Buchwald and Robert Fox, 326–57. Oxford: Oxford University Press, 2013.

McKendrick, Neil, John Brewer, and J. H. Plumb. *The Birth of a Consumer Society: The Commercialization of Eighteenth-Century England*. Bloomington: Indiana University Press, 1982.

McKeon, Michael. *The Secret History of Domesticity: Public, Private, and the Division of Knowledge*. Baltimore: Johns Hopkins University Press, 2009.

McKie, Douglas. "Priestley's Laboratory and Library and Other of His Effects." *Notes and Records of the Royal Society* 12 (1956): 114–36.

McShane, Angela, and Nigel Jeffries. "I say 'shard,' you say 'sherd': Contrasting and Complementary Approaches to a Piece of Early Modern 'Venice Glass.'" In *The Routledge Handbook of Material Culture in Early Modern Europe*, ed. Catherine Richardson, Tara Hamling, and David Gaimster, 452–65. London: Routledge, 2016.

Meinel, Christoph. "Molecules and Croquet Balls." In *Models: The Third Dimension of Science*, ed. Soraya de Chadarevian and Nick Hopwood, 242–75. Stanford: Stanford University Press, 2004.

Melosi, Martin V. "Technology Diffusion and Refuse Disposal: The Case of the British Destructor." In *Technology and the Rise of the Networked City in Europe and America*, ed. Joel A. Tarr and Gabriel Dupuy, 207–26. Philadelphia: Temple University Press, 1988.

Merchant, Carolyn. *The Death of Nature: Women, Ecology and the Scientific Revolution*. San Francisco: HarperCollins, 1980.

———. "The Scientific Revolution and the Death of Nature." *Isis* 97 (2006): 513–33.

Merton, Robert K., and Elinor Barber. *The Travels and Adventures of Serendipity: A Study in Sociological Semantics and the Sociology of Science*. Princeton: Princeton University Press, 2004.

Michaelis, Ronald F. "English Pewter Porringers." *Pewter Collectors Club of America Bulletin* 7 (February 1976): 115–21, and 7 (August 1976): 155–62.

Milburn, John R. *Adams of Fleet Street: Instrument-Makers to King George III*. Aldershot, UK: Ashgate, 2000.

Miller, Jacques-Alain, and Richard Miller. "Jeremy Bentham's Panoptic Device." *October* 41 (1987): 3–29.

Mirabella, M. Bella, ed. *Ornamentalism: The Art of Renaissance Accessories*. Ann Arbor: University of Michigan Press, 2011.

Mitchell, Ian. "'Old Books—New Bound'? Selling Second-Hand Books in England, c. 1680–1850." In *Modernity and the Second-Hand Trade: European Consumption Cultures and Practices, 1700–1900*, ed. Jon Stobart and Ilja Van Damme, 139–57. Basingstoke, UK: Palgrave Macmillan, 2010.

———. *Tradition and Innovation in English Retailing, 1700 to 1850: Narratives of Consumption*. Farnham, UK: Ashgate, 2014.

Morar, Florin-Stefan. "Reinventing Machines: The Transmission History of the Leibniz Calculator." *British Journal for the History of Science* 48 (2015): 123–46.

Morris, Peter. *The Matter Factory: A History of the Chemical Laboratory*. London: Reaktion, 2015.

Morrison-Low, A. D. "Early Navigational Instruments in Scotland." In *Making Instruments Count: Essays on Historical Scientific Instruments presented to Gerard L'Estrange Turner*, ed. Robert Anderson, James A. Bennett, and W. F. Ryan, 218–31. Aldershot, UK: Variorum, 1993.

———. *Making Scientific Instruments in the Industrial Revolution*. Aldershot, UK: Ashgate, 2007.

Morrison-Low, Alison D., Sven Dupré, Stephen Johnston, Giorgio Strano, eds. *From Earth-Bound to Satellite: Telescopes, Skills and Networks*. Leiden: Brill, 2011.

Morton, Alan Q., and Jane A. Wess. *Public and Private Science: The King George III Collection*. Oxford: Oxford University Press/Science Museum, 1993.

Morus, Iwan. "Replacing Victoria's Scientific Culture." *19: Interdisciplinary Studies in the Long Nineteenth Century* 2 (2006): 1–19.

———. *When Physics Became King*. Chicago: University of Chicago Press, 2005.

Mörzer Bruyns, W. F. J., and Richard Dunn. *Sextants at Greenwich: A Catalogue of the Mariner's Quadrants, Mariner's Astrolabes, Cross-staffs, Backstaffs, Octants, Sextants, Quintants, Reflecting Circles and Artificial Horizons in the National Maritime Museum, Greenwich*. Oxford: Oxford University Press, 2009.

Mukherjee, Ayesha. *Penury into Plenty: Dearth and the Making of Knowledge in Early Modern England*. London: Routledge, 2015.

Muldrew, Craig. *The Economy of Obligation: The Culture of Credit and Social Relations in Early Modern England*. London: Palgrave Macmillan, 1998.

Müller-Wille, Staffan. "Nature as a Marketplace: The Political Economy of Linnaean Botany." *History of Political Economy* 35, special issue: *Oeconomies in the Age of Newton*, ed. Margaret Schabas and Neil De Marchi (2003): 154–72.

Multhauf, Robert P. *A Catalogue of Instruments and Models in the Possession of the American Philosophical Society*. Philadelphia: American Philosophical Society, 1961.

Neeson, J. M. *Commoners: Common Right, Enclosure and Social Change in England, 1700–1820*. Cambridge: Cambridge University Press, 1993.

Newcomb, Sally. *The World in a Crucible: Laboratory Practice and Geological Theory at the Beginning of Geology*. Boulder, CO: Geological Society of America, 2009.

Noblett, William. "Samuel Paterson and the London Auction Market for Second-Hand Books, 1755–1802." *Papers of the Bibliographical Society of America* 108 (2014): 139–90.

North, Emily J., and Rolf U. Halden. "Plastics and Environmental Health: The Road Ahead." *Reviews on Environmental Health* 28 (2013): 1–8.

Nyhart, Lynn K. *Biology Takes Form: Animal Morphology and the German Universities, 1800–1900*. Chicago: University of Chicago Press, 1995.

O'Brien, Patrick, Trevor Griffiths, and Philip Hunt. "Political Components of the Industrial Revolution: Parliament and the English Cotton Textile Industry, 1660–1774." *Economic History Review* 44 (1991): 395–423.

Ogborn, Miles. *Spaces of Modernity: London's Geographies, 1680–1780*. New York: Guilford, 1998.

Oldenziel, Ruth, and Helmuth Trischler, eds. *Cycling and Recycling: Histories of Sustainable Practices*. New York: Berghahn Books, 2016.

Ophir, Adi. "The Place of Knowledge Recreated: The Library of Michel de Montaigne." *Science in Context* 4 (1991): 163–89.

Opitz, Donald L. "Domestic Space." In *A Companion to the History of Science*, ed. Bernard Lightman, 252–67. Oxford: Wiley Blackwell, 2016.

———. "'Not merely wifely devotion': Collaborating in the Construction of Science at Terling Place." In *For Better or for Worse? Collaborative Couples in the Sciences*, ed. Annette Lykknes, Donald L. Opitz, and Brigitte Van Tiggelen, 33–56. Basel: Springer, 2012.

———. "'This House Is a Temple of Research': Country House Research for Late Victorian Science." In *Repositioning Victorian Sciences: Shifting Centres in Nineteenth Century Scientific Thinking*, ed. David Clifford et al., 143–53. London: Anthem, 2006.

Opitz, Donald L., Staffan Bergwik, and Brigitte Van Tiggelen. "Introduction: Domesticity and the Historiography of Science." In *Domesticity in the Making of Modern Science*, ed. Opitz, Bergwik, and Van Tiggelen, 1–18. Basingstoke: Palgrave Macmillan, 2016.

Oxford English Dictionary. "Oxford English Dictionary Online." Accessed June 30, 2016. http://www.oed.com.libproxy.ucl.ac.uk/.

Packard, Vance. *The Waste-Makers*. New York: D. Mackay, 1960.

Papadopoulos, G. "The Admirable Effects of Panaceas: Ideas between Antiquity and Early Modern Times." *Wurzburg Medizinhistorische Mitteilungen* 30 (2011): 228–45.

Parker, Karen. "A Use for Broken Pipe Stems." *Society for Clay Pipe Research Newsletter* 17 (1988): 12–13.

Parthasarathi, Prasannan. *Why Europe Grew Rich and Asia Did Not: Global Economic Divergence, 1600–1850*. Cambridge: Cambridge University Press, 2011.

Paskins, Matthew. "One of These Things Is Just Like the Others: Substitution as a Motivator in Eighteenth-Century Chemistry." In *Theory Choice in the History of Chemical Practices*, ed. Emma Tobin and Chiara Ambrosio, 55–70. Dordrecht: Springer, 2016.

———. "Sentimental Industry: The Society of Arts and the Encouragement of Public Useful Knowledge, 1754–1848." PhD diss., University College London, 2014.

Pearson, Robin. "The Impact of Fire and Fire Insurance on Eighteenth-Century English Towns and Populations." In *Investing in the Early Modern Built Environment: Europeans, Asians, Settlers and Indigenous Societies*, ed. Carole Shammas, 67–93. Leiden: Brill, 2012.

Pennell, Sara. "'All but the Kitchen Sink': Household Sales and the Circulation of Second-Hand Goods in Early Modern England." In *Modernity and the Second-Hand Trade: European Consumption Cultures and Practices, 1700–1900*, ed. Jon

Stobart and Ilja Van Damme, 37–56. Basingstoke, UK: Palgrave Macmillan, 2010.

———. *The Birth of the English Kitchen, 1600–1850*. London: Bloomsbury, 2016.

———. "'For a Crack and a Flaw Despis'd": Thinking about Ceramic Semi-durability and the 'Everyday' in Early Modern England." In *Everyday Things: Medieval and Early Modern Material Culture*, ed. Tara Hamling and Catherine Richardson, 27–40. Farnham, UK: Ashgate, 2010.

———. "Material Culture in Seventeenth-Century 'Britain': The Matter of Domestic Consumption." In *The Oxford Handbook of the History of Consumption*, ed. Frank Trentmann, 64–84. Oxford: Oxford University Press, 2012.

———. "Perfecting Practice? Women, Manuscript Recipes and Knowledge in Early Modern England." In *Early Modern Women's Manuscript Writing*, ed. Victoria E. Burke and Jonathan Gibson, 237–58. Aldershot, UK: Ashgate, 2004.

———. "'Pots and Pans History': The Material Culture of the Kitchen in Early Modern England." *Journal of Design History* 11 (1998): 201–16.

Pennell, Sara, and Elaine Leong. "Recipe Collections and the Currency of Medical Knowledge in the Early Modern 'Medical Marketplace.'" In *Medicine and the Market in England and Its Colonies, c.1450–c.1850*, ed. Mark Jenner and Patrick Wallis, 133–52. Basingstoke: Palgrave Macmillan, 2007.

Pinch, Trevor. "'Bring On Sector Two!' The Sounds of Bent and Broken Circuits." *Journal of Sound Studies* 2 (2016): 36–51.

Pinch, Trevor, and Nelly Oudshoorn, eds. *How Users Matter: The Co-construction of Users and Technology*. Cambridge: MIT Press, 2003.

Pollock, Linda A. *With Faith and Physic: The Life of a Tudor Gentlewoman*. New York: St. Martin's Press, 1993.

Ponsonby, Margaret. *Stories from Home: English Domestic Interiors, 1750–1850*. Aldershot, UK: Ashgate, 2007.

Poole, Will, Felicity Henderson, and Yelda Nasifoglu, eds. "Robert Hooke's Books." Accessed March 20, 2016. http://www.hookesbooks.com/.

Potofsky, Allan. "Recycling the City: Paris 1760–1800." In *The Afterlife of Used Things*, ed. Ariane Fennetaux, Sophie Vasset, and Amélie Junqua, 71–88. New York: Routledge, 2014.

Power, M. J. "East London Housing in the Seventeenth Century." In *Crisis and Order in English Towns, 1500–1700*, ed. Peter Clark and Paul Slack, 237–62. London: Routledge, 2007.

Principe, Lawrence M. "Apparatus and Reproducibility in Alchemy." In *Instruments and Experimentation in the History of Chemistry*, ed. Frederic L. Holmes and Trevor H. Levere, 55–74. Cambridge: MIT Press, 2000.

———. *The Aspiring Adept: Robert Boyle and His Alchemical Quest*. Princeton: Princeton University Press, 1998.

Quarrell, W. H., and Margaret Ware, eds. *London in 1710: From the Travels of Zacharias Conrad von Uffenbach*. London, 1934.

Rabinbach, Anson. *The Human Motor: Energy, Fatigue, and the Origins of Modernity*. New York: Basic Books, 1990.

Rae, John. *Life of Adam Smith*. London: Macmillan, 1895.

Raj, Kapil. *Relocating Modern Science: Circulation and the Constitution of Scientific Knowledge, South Asia & Europe, 17th–18th Century*. London: Palgrave Macmillan, 2007.

Ramsey, Rachel. "The Literary History of the Sash Window." *Eighteenth-Century Fiction* 22 (2009–10): 171–94.

Rankin, Alisha. "Empirics, Physicians, and Wonder Drugs in Early Modern Germany: The Case of the Panacea Amwaldina." *Early Science and Medicine* 14 (2009): 680–710.

———. "Medicine for the Uncommon Woman: Remedy Exchange among Noblewomen in Early Modern Germany." PhD diss., Harvard University, 2005.

———. *Panaceia's Daughters: Noblewomen as Healers in Early Modern Germany*. Chicago: University of Chicago Press, 2013.

———. "Women in Science and Medicine, 1400–1800." In *The Ashgate Research Companion to Women and Gender in Early Modern Europe*, ed. Allyson M. Poska, Jane Couchman, and Katherine A. McIver, 407–22. London: Routledge,2016.

Raven, James. *London Booksellers and American Customers: Transatlantic Literary Community and the Charleston Library Society, 1748–1811*. Columbia: University of South Carolina Press, 2002.

Reeves, Nicky. "Constructing an Instrument: Nevil Maskelyne and the Zenith Sector, 1760–1774." PhD diss., University of Cambridge, 2008.

Rheinberger, Hans-Jörg. *Towards a History of Epistemic Things: Synthesizing Proteins in the Test Tube*. Stanford: Stanford University Press, 1997.

Richards, Evelleen. *Darwin and the Making of Sexual Selection*. Chicago: University of Chicago Press, 2017.

Richardson, R. C. *Household Servants in Early Modern England*. Manchester: Manchester University Press, 2010.

Roberts, Lissa, ed. *Centres and Cycles of Accumulation in and around the Netherlands during the Early Modern Period*. Berlin: LIT Verlag, 2011.

———. "The Death of the Sensuous Chemist: The 'New' Chemistry and the Transformation of Sensuous Technology." *Studies in History and Philosophy of Science* 26 (1995): 503–29.

———. "Practicing Oeconomy during the Second Half of the Long Eighteenth Century: An Introduction." *History and Technology* 30 (2014): 133–48.

Roberts, Lissa, Simon Schaffer, and Peter Dear, eds. *The Mindful Hand: Inquiry and Invention from the Late Renaissance to Early Industrialisation*. Amsterdam: Edita; Chicago: University of Chicago Press, 2007.

Roberts, Lissa, and Simon Werrett. "Introduction: A More Intimate Acquaintance." In *Compound Histories: Materials, Production, Governance, 1760*–1840, ed. Roberts and Werrett, 1–32. Leiden: Brill, 2017.

Roberts, William. *Memorials of Christie's: A Record of Art Sales from 1766 to 1896*. 2 vols. London: George Bell and Sons, 1897.

Robinson, Eric, and Douglas McKie, eds. *Partners in Science: Letters of James Watt and Joseph Black*. London: Constable, 1970.

Rogers, Joyce. *The Second Best Bed: Shakespeare's Will in a New Light.* Westport, CT: Greenwood Press, 1993.

Roos, Anna-Marie. "A Speculum of Chymical Practice: Isaac Newton, Martin Lister (1639 –1712), and the Making of Telescopic Mirrors." *Notes and Records of the Royal Society* 64 (2010): 105–20.

———. *Web of Nature: Martin Lister (1639–1712), the First Arachnologist.* Leiden: Brill, 2011.

Roquefeuil, Camille de. *A Voyage round the World: Charles Darwin and the Beagle Collections in the University of Cambridge.* Cambridge: Cambridge University Press, 2009.

Rosenband, Leonard N. *Papermaking in Eighteenth-Century France: Management, Labor, and Revolution at the Montgolfier Mill, 1761–1805.* Baltimore: Johns Hopkins University Press, 2000.

Ross, Ian Simpson. *The Life of Adam Smith.* 2nd ed. Oxford: Oxford University Press, 2010.

Rostenberg, Leona. *The Library of Robert Hooke: The Scientific Book Trade of Restoration England.* Santa Monica, CA: Modoc Press, 1989.

Russell, Andrew, and Lee Vinsel. "Hail the Maintainers." *Aeon*, April 7, 2016. Accessed December 16, 2017. https://aeon.co/essays/innovation-is-overvalued-maintenance-often-matters-more.

Sanderson, Elizabeth C. "Nearly New: The Second-Hand Clothing Trade in Eighteenth-Century Edinburgh." *Costume* 31 (1997): 38–48.

Schabas, Margaret. *The Natural Origins of Economics.* Chicago: University of Chicago Press, 2005.

Schabas, Margaret, and Neil De Marchi. Introduction. *History of Political Economy* 35, special issue: *Oeconomies in the Age of Newton*, ed. Schabas and De Marchi (2003): 1–13.

Schaffer, Simon. "Easily Cracked: Scientific Instruments in States of Disrepair." *Isis* 102 (2011): 706–17.

———. "Enlightened Automata." In *The Sciences in Enlightened Europe*, ed. William Clark, Jan Golinski, and Simon Schaffer, 126–65. Chicago: University of Chicago Press, 1999.

———. "Glass Works: Newton's Prisms and the Uses of Experiment." In *The Uses of Experiment: Studies in the Natural Sciences*, ed. David Gooding, Trevor Pinch, and Simon Schaffer, 67–104. Cambridge: Cambridge University Press, 1989.

———. "Natural Philosophy and Public Spectacle in the Eighteenth Century." *History of Science* 21 (1983): 1–43.

———. "Physics Laboratories and the Victorian Country House." In *Making Space for Science*, ed. Crosbie Smith and Jon Agar, 149–80. London: Macmillan, 1998.

Schechner, Sara J. "Instrumentation." In *A Companion to the History of American Science*, ed. Georgina M. Montgomery and Mark A. Largent, 408–19. Oxford: Wiley, 2016.

Schibeci, Lynn Frances. "The London Auction Market and the Commodification of English Taste, 1766–1823." PhD diss., Northwestern University, 1999.

Schiebinger, Londa. *The Mind Has No Sex? Women in the Origins of Modern Science*. Cambridge: Harvard University Press, 1989.

Schofield, Robert. *The Enlightenment of Joseph Priestley: A Study of His Life and Work from 1733 to 1773*. University Park: Pennsylvania State University Press, 1997.

Schumacher, E. F. *Small Is Beautiful: A Study of Economics As If People Mattered*. New York: Vintage, 1993.

Secord, James A. "Knowledge in Transit." *Isis* 95 (2004): 654–72.

———. "Newton in the Nursery: Tom Telescope and the Philosophy of Tops and Balls, 1761–1838." *History of Science* 23 (1985): 127–51.

Serrano, Elena. "Chemistry in the City: The Scientific Role of Female Societies in Late Eighteenth-Century Madrid." *Ambix* 60 (2013): 139–59.

———. "Making *Oeconomic* People: The Spanish *Magazine of Agriculture and Arts for Parish Rectors* (1797–1808)." *History and Technology* 30 (2014): 149–76.

Shammas, Carole. "The Domestic Environment in Early Modern England and America." *Journal of Social History* 14 (1980): 1–24

Shapin, Steven. "The Accidental Scientist." *American Scientist* 92 (2004): 374–76.

———. "The House of Experiment in 17th-Century England." *Isis* 79 (1988): 373–404.

———. "The Invisible Technician." *American Scientist* 77 (1989): 554–63.

———. " 'The Mind is its own Place': Science and Solitude in Seventeenth-Century England." *Science in Context* 4 (1991): 191–218.

———. "Pump and Circumstance: Robert Boyle's Literary Technology." *Social Studies of Science* 14 (1984): 481–520.

———. *A Social History of Truth: Civility and Science in Seventeenth-Century England*. Chicago: University of Chicago Press, 1994.

Shapin, Steven, and Simon Schaffer. *Leviathan and the Air-Pump: Hobbes, Boyle and the Experimental Life*. Princeton: Princeton University Press, 1985.

Sheeran, George. *Medieval Yorkshire Towns: Buildings, People, and Spaces*. Edinburgh: Edinburgh University Press, 1998.

Sheppard, F. H. W., ed. *Survey of London*, vol. 29: *The Parish of St. James Westminster, Part One, South of Piccadilly*. London: Athlone Press, 1960.

Sibum, H. Otto. "Reworking the Mechanical Equivalent of Heat: Instruments of Precision and Gestures of Accuracy in Early Victorian England." *Studies in History and Philosophy of Science* 26 (1995): 73–106.

Simmons, Anna. "Sites, Production and Networks: Wholesale Pharmaceutical Manufacturing in London, c. 1760–c. 1840." In *Compound Histories: Materials, Production, Governance, 1760–1840*, ed. Lissa Roberts and Simon Werrett, 289–311. Leiden: Brill, 2017.

Sivasundaram, Sujit, ed. "Global Histories of Science." Special section of *Isis* 101 (2010): 95–158.

Slade, Giles. *Made to Break: Technology and Obsolescence in America*. Cambridge: Harvard University Press, 2006.

Sloboda, Stacey. "Displaying Materials: Porcelain and Natural History in the Duchess of Portland's Museum." *Eighteenth-Century Studies* 43 (2010): 455–72.

Smith, Crosbie, and M. Norton Wise. "Work and Waste: Political Economy and Natural Philosophy in Nineteenth-Century Britain." *History of Science* 27 (1989): 263–301, 391–449; 28 (1990): 221–61.

Smith, Pamela H. *The Body of the Artisan: Art and Experience in the Scientific Revolution*. Chicago: University of Chicago Press, 2004.

———. "The Making and Knowing Project. " http://www.makingandknowing.org/.

Sofia, Zoë. "Container Technologies." *Hypatia* 15 (2000): 181–201.

Sonoda, R. M. "Collection and Protection of Insects and Pathogenic Organisms." In *Handbook for the Collection, Preservation and Characterization of Tropical Forage Germplasm Resources*, ed. G. O. Mott, 27–32. Cali, Colombia: Centro Internacional de Agricultura Tropical, 1979.

Sorrenson, Richard. *Perfect Mechanics: Instrument Makers at the Royal Society of London in the Eighteenth Century*. Boston, MA: Docent Press, 2013.

———. "The Ship as a Scientific Instrument in the Eighteenth Century." *Osiris* 11 (1996): 221–36.

Spary, Emma C. *Feeding France: New Sciences of Food, 1760–1815*. Cambridge: Cambridge University Press, 2014.

———. "'Peaches Which the Patriarchs Lacked': Natural History, Natural Resources, and the Natural Economy in France." *History of Political Economy* 35, special issue: *Oeconomies in the Age of Newton*, ed. Margaret Schabas and Neil De Marchi (2003): 14–41.

———. "Political, Natural, and Bodily Economies." In *Cultures of Natural History*, ed. Nicholas Jardine, James A. Secord, and Emma C. Spary, 178–96. Cambridge: Cambridge University Press, 1996.

Spiller, Elizabeth, ed. *Seventeenth-Century English Recipe Books: Cooking, Physic and Chirurgery in the Works of Elizabeth Talbot Grey and Aletheia Talbot Howard*. Aldershot, UK: Ashgate, 2008.

Stapelbroek, Koen, and Jani Marjanen, eds. *The Rise of Economic Societies in the Eighteenth Century: Patriotic Reform in Europe and North America*. Basingstoke, UK: Palgrave, 2012.

Stewart, Alan. "The Early Modern Closet Discovered." *Representations* 50 (1995): 76–100.

Stewart, Larry. "Other Centres of Calculation, or, Where the Royal Society Didn't Count: Commerce, Coffee-Houses and Natural Philosophy in Early Modern London." *British Journal for the History of Science* 32 (1999): 133–53.

———. *The Rise of Public Science: Rhetoric, Technology, and Natural Philosophy in Newtonian Britain, 1660–1750*. Cambridge: Cambridge University Press, 1992.

Stine, Jennifer K. "Opening Closets: The Discovery of Household Medicine in Early Modern England." PhD diss., Stanford University, 1996.

Stobart, Jon. "Clothes, Cabinets and Carriages: Second-Hand Dealing in Eighteenth-Century England." In *Buyers & Sellers: Retail Circuits and Practices in Mediaeval and Early Modern Europe*, ed. Bruno Blondé, Peter Stabel, Jon Stobart, and Ilja Van Damme, 225–44. Turnhout: Brepols, 2006.

Stobart, Jon, and Ilja Van Damme, eds. *Modernity and the Second-Hand Trade: European Consumption Cultures and Practices, 1700–1900*. Basingstoke, UK: Palgrave Macmillan, 2010.

Stoesser, Alison. "Robert Hooke's Montagu House." In *Robert Hooke: Tercentennial Studies*, ed. Michael Cooper and Michael Hunter, 165–80. Aldershot, UK: Ashgate, 2006.

Strasser, Susan. *Waste and Want: A Social History of Trash*. New York: Metropolitan Books, 1999.

Styles, John. "Product Innovation in Early Modern London." *Past & Present* 168 (2000): 124–69.

Summit, Jennifer. *Memory's Library: Medieval Books in Early Modern England*. Chicago: University of Chicago Press, 2008.

Tarlow, Sarah. *The Archaeology of Improvement in Britain, 1750–1850*. Cambridge: Cambridge University Press, 2007.

Terrall, Mary. *Catching Nature in the Act: Réaumur and the Practice of Natural History in the Eighteenth Century*. Chicago: University of Chicago Press, 2013.

Tharp, Lars. *Hogarth's China: Hogarth's Paintings and Eighteenth-Century Ceramics*. London: Merrell, 1997.

Thomas, Keith. "Cleanliness and Godliness in Early Modern England." In *Religion, Culture and Society in Early Modern Britain: Essays in Honour of Patrick Collinson*, ed. Anthony Fletcher and Peter Roberts, 56–83. Cambridge: Cambridge University Press, 1994.

Thornton, Dora. *The Scholar in His Study: Ownership and Experience in Renaissance Italy*. New Haven: Yale University Press, 1998.

Thornton, Jonathan. "A Brief History and Review of the Early Practice and Materials of Gap-Filling in the West." *Journal of the American Institute of Conservation* 37 (1998): 3–22.

Thornton, Peter. *Authentic Decor: The Domestic Interior, 1620–1920*. London: Weidenfeld and Nicolson, 1984.

———. *Seventeenth-Century Interior Decoration in England, France and Holland*. New Haven: Yale University Press, 1978.

Tinniswood, Adrian. *His Invention So Fertile: A Life of Christopher Wren*. Oxford: Oxford University Press, 2001.

Toplis, Alison. "A Stolen Garment or a Reasonable Purchase? The Male Consumer and the Illicit Second-Hand Clothing Market in the First Half of the Nineteenth Century." In *Modernity and the Second-Hand Trade: European Consumption Cultures and Practices, 1700–1900*, ed. Jon Stobart and Ilja Van Damme, 57–72. Basingstoke, UK: Palgrave Macmillan, 2010.

Torrens, Hugh S. "The Geological Work of Gregory Watt, His Travels with William Maclure in Italy 1801–02 and Watt's 'Proto-Geological' Map of Italy of 1804." In *The Origins of Geology in Italy*, ed. Gian Battista Vai and W. Glen E. Caldwell. *Geological Society of America Special Papers*, 179–97. Boulder, CO: Geological Society of America, 2006.

Touwaide, Alain. "Quid pro Quo: Revisiting the Practice of Substitution in Ancient Pharmacy." In *Herbs and Healers from the Ancient Mediterranean through the Medieval West: Essays in Honor of John M. Riddle*, ed. Anne Van Arsdall and Timothy Graham, 19–61. Ashgate, UK: Aldershot, 2011.

Travis, Anthony S., ed. *150 Years of the Coal-Tar Dye Industry, 1856–2006*. Special issue of *History and Technology* 2 (2006).

———. "Perkin's Mauve: Ancestor of the Organic Chemical Industry." *Technology and Culture* 31 (1990): 51–82.

Turner, Gerard L'Estrange. "The Auction Sales of the Earl of Bute's Instruments, 1793." *Annals of Science* 23 (1967): 213–42.

Tutton, Michael, and Elizabeth Hirst, eds. *Windows: History, Repair and Conservation*. London: Routledge, 2015.

Van Damme, Ilja. "Second-Hand Trade and Respectability: Mediating Consumer Trust in Old Textiles and Used Clothing (Low Countries, Seventeenth–Eighteenth Centuries)." In *Selling Textiles in the Long Eighteenth Century: Comparative Perspectives from Western Europe*, ed. Jon Stobart and Bruno Blondé, 193–209. London: Palgrave Macmillan, 2014.

van Driel, Joppe. "Ashes to Ashes: The Stewardship of Waste and Oeconomic Cycles of Agricultural and Industrial Improvement, 1750–1800." *History and Technology* 30 (2014): 177–206.

———. "The Filthy and the Fat: Oeconomy, Chemistry and Resource Management in the Age of Revolutions." PhD diss., University of Twente, 2016.

Vickery, Amanda. *Behind Closed Doors: At Home in Georgian England*. New Haven: Yale University Press, 2009.

———. *The Gentleman's Daughter: Women's Lives in Georgian England*. New Haven: Yale University Press, 1998.

Viney, William. *Waste: A Philosophy of Things*. London: Bloomsbury, 2014.

Wall, Cynthia. "The English Auction: Narratives of Dismantlings." *Eighteenth-Century Studies* 31 (1997): 1–25.

Wall, Wendy. *Recipes for Thought: Knowledge and Taste in the Early Modern Kitchen*. Philadelphia: University of Pennsylvania Press, 2015.

Walters, Alice N. "Conversation Pieces: Science and Politeness in Eighteenth-Century England." *History of Science* 35 (1997): 121–54.

Warner, Deborah Jean. "What Is a Scientific Instrument, When Did It Become One, and Why?" *British Journal for the History of Science* 23 (1990): 83–93.

Watson, Rowan. "Some Non-textual Uses of Books." In *A Companion to the History of the Book*, ed. Simon Eliot and Jonathan Rose, 480–92. Oxford: Wylie-Blackwell, 2007.

Wear, Andrew. *Knowledge and Practice in English Medicine, 1550–1680*. Cambridge: Cambridge University Press, 2000.

Webster, Mary. *Johan Zoffany, 1733–1810*. New Haven: Yale University Press, 2011.

Wennerlind, Carl. *Casualties of Credit: The English Financial Revolution, 1620–1720*. Cambridge: Harvard University Press, 2011.

Werrett, Simon. "The Astronomical Capital of the World: Pulkovo Observatory in the Russia of Tsar Nicholas I." In *The Heavens on Earth: Observatories and Astronomy in Nineteenth-Century Science and Culture*, ed. Charlotte Bigg, Otto Sibum, and David Aubin, 33–57. Durham, NC: Duke University Press, 2010.

———. *Fireworks: Pyrotechnic Arts and Sciences in European History*. Chicago: University of Chicago Press, 2010.

———. "From the Grand Whim to the Gasworks: Philosophical Fireworks in Georgian England." In *The Mindful Hand: Inquiry and Invention from the Late Renaissance to Early Industrialisation*, ed. Lissa Roberts, Simon Schaffer, and Peter Dear, 325–48. Amsterdam: Edita; Chicago: University of Chicago Press, 2007.

———. "Household Oeconomy and Chemical Inquiry." In *Compound Histories: Materials, Production, Governance, 1760–1840*, ed. Lissa Roberts and Simon Werrett, 35–56. Leiden: Brill, 2017.

———. "Matter and Facts: Material Culture in the History of Science." In *Material Evidence: Learning from Archaeological Practice*, ed. Robert Chapman and Alison Wylie, 339–52. New York: Routledge, 2014.

———. "The Panopticon in the Garden: Samuel Bentham's Inspection House and Noble Theatricality in Eighteenth-Century Russia." *Ab Imperio* 3 (November 2008): 47–70.

———. "Recycling in Early Modern Science." *British Journal for the History of Science* 46 (2013): 627–46.

———. "Russian Responses to the Voyages of Captain Cook." In *Captain Cook: Explorations and Reassessments*, ed. Glyn Williams, 179–200. New York: Boydell & Brewer, 2004.

Westfall, Richard S. *Never at Rest: A Biography of Isaac Newton*. Cambridge: Cambridge University Press, 1980.

Wheatland, David Pingree. *The Apparatus of Science at Harvard, 1765–1800*. Cambridge: Harvard University Press, 1968.

White, Paul. "Darwin's Home of Science and the Nature of Domesticity." In *Domesticity in the Making of Modern Science*, ed. Donald L. Opitz, Staffan Bergwik, and Brigitte Van Tiggelen, 61–83. Basingstoke, UK: Palgrave Macmillan, 2016.

Whittle, Jane, and Elizabeth Griffiths. *Consumption and Gender in the Seventeenth-Century Household: The World of Alice Le Strange*. Oxford: Oxford University Press, 2012.

Wiesner, Merry E. *Women and Gender in Early Modern Europe*. 2nd ed. Cambridge: Cambridge University Press, 2000.

Wigston Smith, Chloe. *Women, Work, and Clothes in the Eighteenth-Century Novel*. Cambridge: Cambridge University Press, 2013.

Wilson, C. Anne, ed. *Waste Not, Want Not: Food Preservation in Britain from Early Times to the Present Day*. Edinburgh: Edinburgh University Press, 1991.

Wilson, Gillian. "The Kedleston Fountain: Its Development from a Seventeenth-Century Vase." *J. Paul Getty Museum Journal* 11 (1983): 1–12.

Wise, M. Norton, ed. *The Values of Precision*. Princeton: Princeton University Press, 1995.

Wood, Alexander. *The Cavendish Laboratory*. Cambridge: Cambridge University Press, 1946.

Woodward, Donald. "Straw, Bracken and the Wicklow Whale: The Exploitation of Natural Resources in England since 1500." *Past & Present* 159 (1998): 43–76.

———. "Swords into Ploughshares: Recycling in Pre-industrial England." *Economic History Review* 38 (1985): 175–91.

Woolf, D. R. *Reading History in Early Modern England*. Cambridge: Cambridge University Press, 2000.

Worsley, Lucy. *If Walls Could Talk: An Intimate History of the Home*. London: Faber and Faber, 2011.

Worster, Donald. *Nature's Economy: A History of Ecological Ideas*. 2nd ed. Cambridge: Cambridge University Press, 1994.

Yale, Elizabeth. *Sociable Knowledge: Natural History and the Nation in Early Modern Britain*. Philadelphia: University of Pennsylvania Press, 2016.

———. "With Slips and Scraps: How Early Modern Naturalists Invented the Archive." *Book History* 12 (2009): 1–36.

Yarwood, Doreen. *The British Kitchen: Housewifery since Roman Times*. London: Batsford, 1981.

Yates, Joshua, and James Davison Hunter, eds. *Thrift and Thriving in America: Capitalism and Moral Order from the Puritans to the Present*. Oxford: Oxford University Press, 2011.

Zuidervaart, H. J. "The 'Invisible Technician' Made Visible: Telescope Making in the Seventeenth- and Early Eighteenth-Century Dutch Republic." In *From Earth-Bound to Satellite: Telescopes, Skills and Networks*, ed. Alison D. Morrison-Low, Sven Dupré, Stephen Johnston, and Giorgio Strano, 41–102. Leiden: Brill, 2012.

Index